TECH AGNOSTIC

TECH AGNOSTIC

HOW TECHNOLOGY BECAME THE WORLD'S MOST POWERFUL RELIGION, AND WHY IT DESPERATELY NEEDS A REFORMATION

GREG M. EPSTEIN

The MIT Press
Cambridge, Massachusetts
London, England

The MIT Press would like to thank the anonymous peer reviewers who provided comments on drafts of this book. The generous work of academic experts is essential for establishing the authority and quality of our publications. We acknowledge with gratitude the contributions of these otherwise uncredited readers.

This book was set in Bembo Book MT Pro by New Best-set Typesetters Ltd. Printed and bound in the United States of America.

Library of Congress Cataloging-in-Publication Data

Names: Epstein, Greg M., author.
Title: Tech agnostic : how technology became the world's most powerful religion, and
    why it desperately needs a reformation / Greg M. Epstein.
Description: Cambridge, Massachusetts : The MIT Press, [2024] | Includes bibliographical
    references and index.
Identifiers: LCCN 2023050678 (print) | LCCN 2023050679 (ebook) |
    ISBN 9780262049207 (hardcover) | ISBN 9780262379755 (epub) |
    ISBN 9780262379748 (pdf)
Subjects: LCSH: Technology—Religious aspects. | Technology—Social aspects. |
    Dystopia. | Secular humanism.
Classification: LCC BL265.T4 .E67 2024 (print) | LCC BL265.T4 (ebook) |
    DDC 215—dc23/eng/20240412
LC record available at https://lccn.loc.gov/2023050678
LC ebook record available at https://lccn.loc.gov/2023050679

10 9 8 7 6 5 4 3 2 1

CONTENTS

INTRODUCTION   1

PART I: BELIEFS   27

1   TECH THEOLOGY   29

2   DOCTRINE   55

PART II: PRACTICES   115

3   HIERARCHIES AND CASTES, OR, UTOPIA FOR WHITE MEN   117

4   RITUAL   147

5   APOCALYPSE(S)   173

PART III: BELOVED COMMUNITY, OR, THE REFORMATION   207

6   APOSTATES AND HERETICS   209

7   HUMANISTS   239

8   THE CONGREGATION   269

CONCLUSION: TECH AGNOSTICISM IS A HUMANISM   285

Acknowledgments   297
Notes   303
Index   343

INTRODUCTION

Thou rulest the raging of the sea: when the waves thereof arise, thou stillest them.

—Psalm 89:9, King James Version

We now have the technological tools to quite literally code nature, and the payoff to human flourishing will be profound.

—Marc Andreessen, "Technology Saves the World," 2021

Let's begin in October, in the year 312 CE, in Northern Rome.

Constantine I, later to be known as Constantine the Great, was a successful general, the son of a successful general, and the head of the army of the Western Roman empire. "And he's fighting another successful general," Harvard historian Shaye Cohen explains, "struggling for who is going to be at the top of the heap of the very higher echelons of Roman government."[1] His rival, Maxentius, was also co-emperor at the time. Oh, and Maxentius was also Constantine's brother-in-law—the brother of Constantine's second wife, Fausta.

Constantine was dealing, in other words, with some extreme relationship drama. But though the leader who would be Roman emperor from 306 CE all the way until 337 CE was "an aggressive warrior, [and] a sometimes cruel partner," he was not an impractical man, but rather "an immensely shrewd ruler."[2] And as he contemplated how to gain a crucial victory over his colleague and kinsman, a novel idea occurred to him: a fusion of worldly strategy and . . . something more transcendent. As Eusebius, the fourth-century Greek historian of Christianity, later put it, Constantine considered that,

> of the many emperors who had preceded him, those who had rested their hopes
> in a multitude of gods, and served them with sacrifices and offerings, had in the

first place been deceived by flattering predictions, and oracles which promised them all prosperity, and at last had met with an unhappy end, while not one of their gods had stood by to warn them of the impending wrath of heaven.[3]

Polytheism, in other words, had not been profitable for the emperor's predecessors. Despite significant investments, various Mediterranean divinities had underperformed, he reflected, offering disappointing returns. "Reviewing these considerations," Eusebius related, Constantine judged it "to be folly indeed to join in the idle worship of those who were no gods."[4]

Instead, Constantine sought a partner offering a better value proposition—a god who could help him gain a competitive advantage against his rival sibling-in-law. Per Eusebius again: "Being convinced ... that he needed some more powerful aid than his military forces could afford him . . . he considered . . . on what God he might rely for protection and assistance."[5] So he started a new venture: faith in a young, emerging god who, unlike his pagan competitors, might possess sufficient "operating power" (Eusebius's actual term, from Ernest Cushing Richardson's 1890 translation, is "co-operating power") to solve an emperor's unique problems.[6]

The events that followed turned out to be important beyond what even the most grandiose monarch might have imagined at the time. In preparation for a decisive battle over a bridge crossing the Tiber River in northern Rome, Constantine ordered his men to paint the Chi-Ro, an early Christian symbol, on their shields. It became a killer rebrand—literally. When Maxentius committed a strategic error, leading his larger forces over the bridge too soon, Constantine pounced. Maxentius drowned, weighed down by his own armor, either after being thrown into the river by his horse or while attempting to swim to the other side.

His foe vanquished, Constantine rose to uncontested power over an empire so large it had been split into two kingdoms just decades earlier because of the sheer difficulty of ruling such an expansive territory. The merger represented a truly epic pivot, and Constantine's eventual formal conversion set Christianity fully on course to become the dominant force it has been since. "Few events in the history of civilization have proved more transformative [than Constantine's conversion to Christianity in the year 312 CE]," writes Bart Ehrman, a noted historian of the Bible and Christianity.[7]

Likely so. But over the past generation, we have witnessed one of those few events. The rise to global dominance of a new phenomenon has, we are often told, been as rapidly transformative as almost anything in the history of human civilization. Let's call that phenomenon "Tech."

Technology companies claim to provide unparalleled community by bringing "the world closer together," as Meta tells us in a post about "why we build," and by offering hope for "a more connected future," as Mark Zuckerberg put it in his 2021 keynote about the metaverse.[8] Tech investors like Marc Andreessen pronounce themselves the "patron saints of techno-optimism, claiming humans are becoming "technological supermen."[9] In a manifesto containing the words "we believe" no fewer than 113 times, Andreessen boasts that "the techno-capital machine . . . may be the most pro-human thing there is . . . liberating . . . the human soul."[10] And tech executives like Elon Musk and Jack Dorsey do the seemingly impossible by carrying that vision even further, announcing that technology is "on a mission to extend the light of consciousness," as Dorsey tweeted in 2022 to explain his choice to sell Twitter to Musk.[11] (Musk had tweeted, a year earlier, that he was "accumulating resources to help make life multiplanetary & extend the light of consciousness to the stars.")[12] In other words: Musk would take humanity to a kind of promised land. And according to the effective altruism movement and the many tech billionaires who fund it, technology (specifically "Godlike" AI) poses no less than a 10 percent risk of annihilating every single human being within the coming century.[13] See under: Tech Hell.

Tech often presents moral and ethical messages not as mere secondary features but as integral to its overall value(s) proposition, from Google's infamous "don't be evil" (which became Alphabet's "do the right thing") to Jeff Bezos and Amazon's exhortation to "make history."[14] Zuckerberg, in a letter to staff on the occasion of Facebook's IPO, said the company's vision of a "connected future" would "give everyone a voice and . . . help transform society."[15]

AI engineers tell us they have created sentient souls.[16] Tech leaders even predict futures where humanity communes with an "artificial superintelligence," or "magic intelligence in the sky," they say will be an almost literal god.[17] And why not? After all, as OpenAI CEO Sam Altman pontificates, "abundance is our birthright."[18]

Christianity, if it weren't for Constantine converting the entire Roman Empire to his new faith, might well have been a minor Mediterranean cult, perhaps with something like Judaism's ten to twenty million adherents worldwide.[19] Instead, it has influenced billions. And a tech culture that began as a garage project for nerdy dropouts has become the driving force of our entire civilization. This book will offer an unconventional analysis of how and why tech became a religious empire, omnipresent to the point where there is no longer any realistic way to opt out of its reign. I undertake this task because we need to understand how tech has become not only *a* religion but *the* dominant religion of our time.

WHAT IS TECH, TODAY?

Most people would say tech is an industry, but that's an absurd statement, because every industry in the world is now a tech industry. "We're in a period where tech has expanded to take over nearly every aspect of our lives, economically, socially, and politically," said Mar Hicks, an award-winning historian of technology, specializing in the history of computing, when I asked them just how big technology is today. "And that's due both to the ever-expanding definition of what we consider tech to be, as well as to the real and intended consequences of overfunding an industry and a field of study—computer science—that has long promised people the ability to escape reality and even more importantly to control their environments."[20] Indeed, the size and scope of tech has grown so difficult to quantify that when I asked a similar question of Jason Furman, a Harvard economics professor who served President Barack Obama as chair of the White House Council of Economic Advisers (CEA), he acknowledged that even traditional measures such as gross domestic product (GDP) and stock valuations don't do justice to the exalted status of technology in our world. As Furman pointed out, one could calculate the fraction of GDP (or "value added") in technology companies, but that "doesn't count the ways that General Motors, J. P. Morgan, and the corner store all use technology to produce what they produce."[21] Tech stocks would represent an even higher percentage of the $50 trillion of stocks overall, Furman wrote to me in fall 2023, but even such a measure "doesn't count GM, JPM, and the corner store." The distinguished economist ultimately suggested I might simply stick, without quantification, with Mar Hicks's definition above—that

Shared Chat • Default (GPT-3.5)

Constantine's Epic Pivot 2.0

September 6, 2023

Write about Emperor Constantine's conversion to Christianity, using the language and style of marketing copy for Silicon Valley startups.

🚀 Introducing Constantine 2.0: The Ultimate Conversion Experience! 🎇

Are you ready to witness the most epic pivot in history? Hold on to your togas, because Emperor Constantine is about to unveil his groundbreaking transformation, and it's a game-changer like no other! 🏛️

🌐 The Problem: Paganism vs. Christianity 🙏

In the ancient world, the battle between paganism and this new, disruptive faith called Christianity was nothing short of legendary. The empire was divided, and the stakes were higher than a chariot race at the Colosseum. Constantine, the visionary leader he was, saw an opportunity to innovate and unite the Roman Empire like never before.

🔍 The Solution: Constantine's Conversion 2.0! ✝️

Constantine wasn't just a ruler; he was a trailblazing entrepreneur. He realized that embracing Christianity could be the ultimate growth hack for the Roman Empire. Here's how he did it:

1️⃣ **Rebranding the Empire**: Just like a modern startup with a killer brand makeover, Constantine rebranded the Roman Empire with the iconic Chi-Rho symbol, sending a clear message that change was afoot.

2️⃣ **Empire-Wide A/B Testing**: Constantine was all about experimentation. He legalized Christianity, but didn't go overboard. It was like conducting A/B tests to ensure a seamless transition from paganism to Christianity.

ChatGPT on Constantine's "Epic Pivot."

tech has taken over nearly every aspect of life—because, as Furman said, it "seems true."[22]

"Tech, and the tech industry," Hicks continued in their email to me, "is largely about power and control. In the best-case scenario technologies can be force multipliers for good—think sanitary sewers, or heart stents, or easier-to-operate wheelchair designs. In the worst-case scenario, they multiply force in incredibly violent, destructive ways—think sarin gas, firearms, or the atomic bomb."

We wake up, and before our smart mattresses or smart watches can stop monitoring our every sleeping heartbeat and breath, our smartphones light up as we check them for the first time that morning. Male Orthodox Jews are required to thank God they are not a woman (among other prayers) upon exiting their bed. Many of us, immediately upon rising, stare at our smartphones, performing our own strange rituals. These may or may not include cursing God for our Twitter ("X") feed.

Phone still in hand, where it will remain for much of the day, we stumble to the bathroom, where we might, if we're lucky enough to afford one, brush our teeth with a smart toothbrush or clean ourselves with a smart toilet. In the kitchen, we barely notice our smart refrigerator or smart dishwasher, but our network sees them. Our smart doorbell, watching vigilantly for intruders, lets us know whether we have achieved "fulfillment" . . . in the form of our Amazon package having been successfully delivered. The smart thermostat adjusts itself for our comfort. If your kid is like mine, they haven't eaten breakfast and are running late for school because they're mesmerized by whatever slop the YouTube algorithm is feeding them. Maybe we share a witty TikTok or notice an Instagrammable moment on our way to work or school. Or perhaps we'll never make it to a physical workplace at all, because we're working from home or even from a beautiful beach we barely notice, thanks to a constant pinging of Slack messages. Or maybe generative AI has made redundant our role as a coder, paralegal, financial analyst, graphic designer, or even teacher.

However we choose to spend our day, one thing is clear: Not only will we be surrounded by intelligent products, each created to perfectly optimize every detail of our environment. Ultimately we are the product that must be optimized, less for our own benefit than for the glory of technology itself. As such, our daily existence will be tracked constantly by a thousand cookies we have accepted so that they might follow our movements, our moods, and our purchasing power.

At night maybe we'll stream the news from a podcast or scroll through headlines on our feed, feeling depressed and anxious about the fate of a world continually under attack by rogue forces using the Internet to manipulate, disinform, and stoke hatred for clicks. Maybe that will trigger us to binge on the artisanal ice cream or vodka that we bought because the brand microtargeted us for an ad while we were shopping on Instacart. Eventually, eyes and brains aching from an overdose of ultraviolet light, we'll log off. Until we do it all again tomorrow.

Today technology is the water in which we swim, whether or not we notice we are fish. Tech provides contemporary Western lives, so polarized and divided in countless ways, with a universal organizing principle, a common story by which we tell ourselves who we are. It offers myriad rites, capturing our attention and transforming our consciousness, connecting us with a community of people who spend their days, weeks, years, and indeed their entire lives engaging in the same repetitive behaviors with the same fervent intensity. Naturally, we all hope our devotion to this communion of fellow travelers bears fruit: surely tech will lead to a better future! Even a kind of paradise! But the truth is that many of us fear, more than we'd like to admit, that this may all be heading to a deeply dark place.

In other words: technology has become religion.

Am I speaking literally, or have I written this book to weave the most elaborate and annoying metaphor you'll ever read in your entire life? Yes.

WHAT IS RELIGION, ANYWAY?

As I've been saying for years, including in occasionally bitter arguments with esteemed colleagues such as Richard Dawkins and Sam Harris and even the late, great Christopher Hitchens, religions are much more than mass delusions about white-bearded, omnipotent sky daddies. For many deeply religious people, the dogmatic aspects of religion are an afterthought and notions of the supernatural are optional and mysterious.

The word *religion*—literally by definition—can marginalize or distort our understanding of non-Western traditions that do not align exactly with

Christian structures and beliefs. Derived from the Latin *religare*, "to bind," the contemporary concept of religion was created by Christians and applied to non-Christian traditions for the specific purpose of making Christianity and its structures and practices seem like the norm from which anything else was deviating. Hinduism and Buddhism, for example, were not originally or consciously created as religions, but they have long since come to be understood as such.

When I talk about religion in this book, I am purposefully using the Western, Christian-centric sense of the word, analyzing both traditional religion and contemporary technology through the lens of Christian definitions and understandings of what religion is.[23] This is not because I discount other traditions but because of the importance of recognizing how much power Christianity has been able to amass in the world, in part by successfully framing itself as the model for how we ought to understand and categorize any and all human activities that might seem even remotely similar.

These days many academics use the word *religion* less as a precise category into which only certain institutions fit perfectly and "more as a term of art," according to Robert Sharf, professor of Buddhist studies and chair of the Berkeley Center for Buddhist Studies at the University of California, Berkeley. "There are departments of religions, courses of introduction to religion. . . . We know that those courses are supposed to teach people some kind of breadth of stuff that we recognize as religion, around the world," Sharf told me when I asked him about what methodological approaches to the study of religion he would recommend for a book about technology as a religion.[24] "But when it really comes down to it, there's no special methodology required for the study of religion."

Sharf, a scholar of religion who has been known to tell students that he is an academic who just happens to also hold ordination as a Zen Buddhist priest, was a role model for me when I was an undergraduate student. I took religion courses like his to try to understand the human condition, but it was the unexpected complexity I found in such classrooms, learning that there is no one right way to be human and also no single correct way to understand a phenomenon like religion, that ultimately motivated me to pursue the unusual work around religion, technology, and humanity that I will be exploring in this volume.

We'll more closely consider questions about what religion is—and is not—in each chapter of this book, starting in chapter 1, where, as part of an analysis of the ideas behind contemporary tech, we'll ask what exactly is meant by the term *theology*. For now, however, let's choose a working definition: religion is the combination of institutions, cultures, identities, rituals, values, and beliefs through which we human beings bind ourselves together and find meaning in what might otherwise seem to be a meaningless world. To be as fair as reasonably possible, let's add a perspective on the definition of religion from a very different kind of thinker: John F. Haught, a distinguished Catholic theologian and scholar of religion known for being a staunch advocate of both science and religious faith.[25]

In a conversation with a group of secular intellectuals including Daniel Dennett, one of the "Four Horsemen of the Atheist Apocalypse," Haught defined religions as

> the most important ways in which human beings on our planet, for the last 50,000 years or so, at least, have sought reassurance: in the face of fate, and the experience that they're not in control of their lives; of death and having to die; of the feeling of shame, guilt and even self-rejection that comes from living in the context of community, where things are expected of you and you never live up to them.[26]

Continuing along those lines, Haught adds that religious people have especially sought the reassurance he describes "in the experience of being grasped by or carried away by what they have taken to be another dimension of reality, other than the profane, other than the mundane, other than the proximate environment that they live in . . . an ultimate environment that is infinite in scope and inexhaustible in its mysteriousness."[27] Though Haught's perspective differs from my own, his definition is a fair and thoughtful articulation of an influential viewpoint.

What is the Internet, if not something that can grasp or carry people away to another, seemingly infinite dimension?

Technology is currently playing a role in the daily lives of average citizens of our world that is eerily like the role religions played in the lives of ancient and early modern people. Like traditional religions once did, technology is shaping our thoughts, feelings, hopes, fears, relationships, and future. And just like in the days of religions past, much of that shaping is taking place not in the interests of the average person but in the interests of the largest corporations in the history of humanity, who are doing the shaping.

Technology is not the first secular phenomenon to be compared to religion. Various sports, politics, work, video games, economics, science, the Grateful Dead, *Harry Potter*, Bitcoin, veganism, "wokeism," fraternities and sororities, deregulation, Bardolatry (the worship of William Shakespeare), small talk, elk hunting, and South Indian cinema are all among the responses I've gotten over the past few years when asking about interesting analogies for religion. And thinking of all that work reminds me: Derek Thompson, at the *Atlantic*, is also, sadly, correct in pointing out in his brilliant writing on "workism" that work itself seems to have become a religious pursuit for many of us.

So when I make this comparison, I do so not because it is a perfect one; I don't even believe religions can be compared to one another in any perfectly correct way. My point is that we don't currently have a better concept than religion for how technology impacts our world today. If you want to argue that the University of Texas football team or limited-edition Nike sneakers are also religions, perhaps that could be valid, but they would be minor to middling cults that do not have much impact on human existence beyond a relatively narrow group of people. Or if they do impact a large group, then they do so only within a narrow slice of life. Tech, on the other hand, has billions of devices sold, trillions of dollars in combined value, and billions of customers so devoted that calling them "daily" users is no longer particularly useful when we use these products almost every minute. Not to mention an undetermined but surely troubling number of votes influenced by social media disinformation.

THE HISTORY OF TECH-AS-RELIGION

The idea of a similarity between technology and religion is not novel. Influential thinkers from Karl Marx to Martin Heidegger to Martin Luther King, Jr., have commented on the relationship between the two phenomena. And in fact, technology-as-religion, as an idea, had a significant moment in the late 1980s and into the 1990s, or around one biblical generation ago. Reviewing the compelling arguments made back then (by several important thinkers about both religion and tech) will help us appreciate what is new about our current moment—and what is new about my argument, which is different from theirs.

In the summer of 1992, around the time that Bill Clinton, the relatively unknown young governor of Arkansas, overcame a crowded field to win the Democratic nomination for president, the great author, educator, and critic Neil Postman published a book called *Technopoly* in which he argued that Americans had largely replaced religion and spirituality with a faith in technology.

Postman had already become known for his 1985 book *Amusing Ourselves to Death*, which is still in print today. In it he famously argued the future would be less like what Orwell envisioned in *1984* and more like what Aldous Huxley imagined in *Brave New World*. In other words, civilization won't fall because of a fascist boot stomping on a human face forever. Instead, Postman warned, thanks to television—and all the media technologies to follow it—we would end up amusing ourselves to death. We now live in the era of a reality TV star turned president indicted for espionage who was elected in part because Russian hackers exploited our country's obsession with social media. Seemingly every time a serious news journalist gets laid off, a new "influencer" gets their wings. All this time later, it's getting harder and harder to argue against *Amusing Ourselves to Death*'s thesis.

"Technologies work," Postman argued in 1992 about innovations like the airplane, television, or penicillin, contrasting them with prayer or even faith in God, which aren't as rational and don't always do anything tangible or material for us (though certain religious believers and social scientists alike would agree that prayer and faith have profound, even measurable impacts on individuals and communities).[28] The relatively obscure *Technopoly* was chosen in 2023 by the tech-forward website the *Verge* as the second-best tech book of all time, but the blurb about the book doesn't mention that the volume frames tech as religion.[29]

Our new religious faith in these technologies had been evolving for quite a while, Postman asserted in *Technopoly*: since the eighteenth century, Western society had been in the process of reorienting itself away from earlier fundamental beliefs that traditional religions and gods held the answers to central human questions of meaning, purpose, and ethics. In place of those beliefs, more and more of us were looking to science, engineering, and all sorts of new technologies, not just for solutions to specific problems like how to cross the Atlantic or defeat a harmful bacterium, but for solutions to the broader problem of being human. Many of us, he observed, believed in a strikingly religious way that technology would soon conquer even death itself. Amid such worldwide changes, for Postman, the United States of America was the first "technopoly": a place where, more than in any other culture in the history of humanity, an entire population had embraced and dedicated itself fully to technology, and where the people were redefining themselves by it. As we enter the middle of the twenty-first century, it seems our dedication has reached new levels of fervor, which may be of great consequence if current (as I write this) US President Joe Biden was anywhere even close to correct when he said, while preparing for his 2024 campaign, that "there comes a time, maybe every six to eight generations, where the world changes in a very short time. And . . . what happens in the next two, three years are going to determine what the world looks like for the next five or six decades."[30]

Neil Postman, meanwhile, was far from the only writer on the technology-as-religion beat in his era. In 1994 historian David Nye published a book called *American Technological Sublime* in which he argued that a large part of American culture was founded on the notion that technology is not only a way of solving problems or improving lives but is in itself the means of producing a sense of awe, wonderment, inspiration, and even terror that we might otherwise associate with religious or spiritual experiences. Nye defines this encounter with tech as the "technological sublime," which, he explains, is part of a long history of "sublime experience" that has been discussed as far back as ancient Rome and theorized about by no less than Immanuel Kant.

It would be the opposite of a sublime experience for most readers to take time right now for a lengthy intellectual history of what a historian like Nye meant when he described the "technological sublime." What you need to know is this: religion has never been the only way for masses of people to have profound experiences. No matter how far back you go in history, there have always been people who experience awe, not because they read some theological scroll but because they looked out at a majestic canyon or stared up into the abyss of a starry night. Tech, for David Nye, was a uniquely American continuation of that experience, and he elegantly chronicled how Americans fashioned a kind of modern spiritual tradition out of ever-increasing mechanization, from bridges to skyscrapers, from Robert Oppenheimer's bomb to the Apollo space program.

Nye's book came during what we can call the first wave of speculation about tech as religion. But because that was the 1990s, he could not weigh in on what, for better and for worse, has become the ultimate ever-present sublime experience, which is our total unification with tech. In the final chapter of *American Technological Sublime*, Nye explored the development of the city of Las Vegas as a commercialized landscape in which people were meant to be surrounded by a commodified, yet awe-inspiring, experience of technology. As impressive as one might imagine the lights and sounds of "Sin City" to have been to its original observers, it's hard to believe they could have pictured the degree to which we have surrounded ourselves—or even merged—with our tech today. We're human casinos. Our phone is a little pocket slot machine. It's also everything else.

In 1997 the late historian and technology critic David Noble published a book, called *The Religion of Technology: The Divinity of Man and the Spirit of Invention*, in which he argued that the impulse toward technological advancement emerges from the same place in our human spirit as does the impulse to perfect ourselves spiritually toward a better world, or even a heaven. Elegantly tracing the history of such impulses back through time into antiquity, Noble's book expressed the longing that "we might learn to disabuse ourselves of the other-worldly dreams that lie at the heart of our technological enterprise . . . to redirect our astonishing human capabilities toward more worldly and humane ends."[31]

That didn't happen. If anything, the opposite did.

In late 1993, meanwhile, when MIT anthropology professor Stefan Helmreich was a PhD student at Stanford, he conducted ethnographic field-work at the Santa Fe Institute in New Mexico, dedicated to research on complex systems science. There, as Helmreich described it, a collection of highly distinguished—and almost exclusively atheist and nonreligious computer scientists and biologists "were engaged in a practice that they designated as . . . 'Artificial Life'"—the quest to create life from the absence of life.[32]

"One of my informants said in no uncertain terms," wrote Helmreich in a 1997 article in *Science as Culture*, "that science was his religion: 'I have not been religious since high school [said the artificial life scientist to Helmreich]. Science plays the role of religion in my life, in the sense that when I look for ultimate answers to ultimate questions, I look to

science.'" Helmreich also noted that many researchers "thought of themselves as 'gods' with respect to their simulated worlds . . . some so much that
they felt that the artificial life they were producing was in fact real life in a
virtual universe."[33]

I first learned about this project in spring 2021, while auditing Helmreich's MIT course "The Meaning of Life." It seemed to me then that the
concept of artificial life had gone nowhere for decades—until 2022 and
2023 brought a generative AI invasion to rival anything since the coming
of the Beatles and said, "Hold my [virtual] beer."

THE RISE OF "TECH"

Here it becomes important to make a major point that might seem minor:
One difference between the book you are currently reading and the aforementioned literature is that this is a book about the religion of Tech, not
(just) the religion of technology. What does that mean?

"Technology," as a concept, has been discussed for at least 2500 years
or so, dating to when authors such as Plato, Aristotle, Socrates debated the
precise meaning(s) of the Greek term *techne*—essentially a way of doing
things.[34] As should be obvious, there have been infinite ways of, well, doing
things, over the course of modern history.

This book, however, is about a phenomenon called *tech*, a relatively new
offshoot of earlier concepts of technology. The four-letter word *tech* has
existed in the English language for some time: the Georgia Institute of
Technology has also been known as Georgia Tech for well over a century,
for example.[35] But as you can see in the following chart, *tech* was rarely
used throughout the nineteenth century and even well into the twentieth.
Its usage spiked dramatically starting in the 1980s, as Americans in particular were beginning to speak of personal computing and other "hi-tech"
innovations.[36] *Tech*, as a religious term, then, might be considered a splinter
group from the religion of technology, which has conquered and reshaped
the world much as Constantine's empire, and his legacy, remade a world of
paganism, Judaism, and other Mediterranean cults, in Christianity's image.

Or think of it this way: in modern English, *technology* is a word that can
be pluralized. We can speak of technologies. But the word *tech* can only be
articulated in the singular. There are no "techs."

Tech is one.

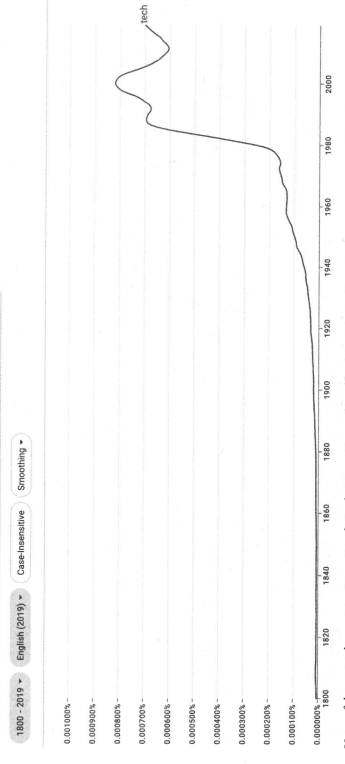

Google Books Ngram Viewer

Q tech   ✕   ⑦

1800 - 2019 ▾   English (2019) ▾   Case-Insensitive   Smoothing ▾

tech

0.001000% —
0.000900% —
0.000800% —
0.000700% —
0.000600% —
0.000500% —
0.000400% —
0.000300% —
0.000200% —
0.000100% —
0.000000% —

1800 1820 1840 1860 1880 1900 1920 1940 1960 1980 2000

Usage of the term *tech* over time. *Source:* Google Books Ngram Viewer, http://books.google.com/ngrams.

❋

This book was written over a period of a few years during which technological terminology itself was in flux. Up until relatively recently, for example, the term *artificial intelligence* was the subject of some controversy, at least among academics and experts in the field who tended to prefer terms such as "machine learning" or "neural networks." Eventually, however, "AI" became the dominant expression, even among people who believe that algorithmic systems are neither *artificial*—because people created them—nor *intelligent,* in any meaningful sense. It may even seem to readers in late 2024 that AI, not "tech," is the real religion. And indeed the now-burgeoning field of AI-branded startups is demonstrating a striking penchant for religious claims, such as the algorithm claiming to "predict when you're going to die," or Tab, a VC-backed glowing AI pendant necklace that listens to your every word, and that is publicized by its Harvard-dropout inventor as a replacement for a relationship with God.[37] Though I will mainly refer, below, to the "tech religion," I will also refer to and describe the religiosity of AI, which for current purposes can be seen as interchangeable with "tech," generally.

As we'll explore throughout *Tech Agnostic*, the ideas, practices, and products we associate with the term *tech* include the notion that we can connect literally every human being to every other human being through digital communication; the conviction that we can make literally every fact and experience that has ever been had knowable and understandable to all; the intention to place watching eyes and ears above every one of us so that we can all be seen, known, and surveilled at every moment; and, of course, the faith that we can ultimately create machines so powerful, breathed full of intelligent life, that they may even eclipse us and become new gods themselves. In other words: tech has not only become a religion; it has become the world's most powerful religion.

MY TECH THEOLOGY

Themes of tech-as-religion were also being pursued in the 1990s by leading theologians. In their 1997 paper "Religion and Technology: A New Phase," distinguished Harvard Divinity School professor Harvey Cox and coauthor Anne Foerst (then a postdoc at both MIT's AI Lab and the Harvard Divinity

School) analyzed then-emerging tech from a theological perspective.[38] They did so with striking prescience, even seeming to anticipate several scenes in the film *The Matrix*, released two years after the article's publication. They pointed to Marvin Minsky, founder of MIT's Artificial Intelligence Laboratory, who had claimed that humanity would soon create a new, computerized life-form that would be as different from and superior to humans as we were from the "lifeless chemistry" that preceded us. "This language clearly falls within the category of what theologians commonly call 'eschatology,'" they wrote, "the study of the future and of 'the last things.'" Cox and Foerst even anticipated, in 1997, how "digital dreamers" might co-opt the vision of Martin Luther King, Jr., of a "Blessed Community" in which all humanity might come together as one. What Cox and Foerst described then as a "cybernetic global village where misunderstandings and suspicions will melt away as billions of individuals communicate with each other from the privacy of their home (or hut)" is instantly recognizable to anyone who lived through the halcyon days of Web 2.0.

When Cox and Foerst published their paper, I was a few years away from deciding to join people like them at Harvard Divinity School (HDS). At that time, I was a college religion major. I was concentrating in Chinese language and culture, wondering whether I should become a Buddhist or Taoist priest, and actively looking for a "spiritual" and communal home for a religious misfit like me. Virtually all my adult life, I have been a passionate atheist and fascinated by all things religious, as part of what I would have to call an obsession with questions about what it means to be human.

A quarter century later, I am at Harvard and MIT, serving as clergy for their rapidly growing student populations of humanists, atheists, agnostics, and other nonreligious people and allies. I am a professional atheist, in other words, at least regarding traditional religions and gods. In the religion of technology, I am merely an agnostic. Because surely there are at least some technologies worth believing in, and at least some signs that tech might one day justify more of our faith—aren't there?

I became the humanist chaplain at Harvard in 2005. I'll talk more about my unusual career throughout the book, but for now I'll say: for over a decade in that role, my biggest professional project was to create and build a nonreligious congregation that would fill the role of religious communities in the lives of my fellow "nonbelievers" (though as I argue in my book *Good Without God*, even without knowing it, we may in fact be deeply principled believers in a faith in secular human ethics and meaning-making that some call humanism). Called the Humanist Hub, the community of and for "atheists, agnostics, and allies" was a product of millions of dollars raised from thousands of donors, over a thousand events and programs attended by hundreds of members, and countless eighty-hour workweeks for me.

I organized the congregation for many reasons, but I like to think the most relevant ones stemmed from my intention to build something positive. I believed then, and still believe, that secularist activists tend to complain too much about the flaws in religion without asking ourselves what should exist in its place. The Humanist Hub was intended to address questions like, What is right enough about religion that it has continued to be so powerful across so many different human societies? How do you build that in a secular way? I ultimately realized, however, as that decade-long project came to an end in the same year I accepted an invitation to join the Massachusetts Institute of Technology in addition to my work at Harvard, that it is very difficult for any individual person, or even any group of hundreds or thousands of people, to create a congregational "alternative to religious life." Because trillions of dollars have been invested, over the past generation, in creating *technological* alternatives to religious life. And while those new technological alternatives may not yet, in the most formal of senses, have traditional religion's billions of devoted adherents, they can already manipulate the ability of billions of people to perceive and identify truth, like nothing but religion.[39] And they can already boast billions of what fast food companies call "power users."

A power user, in the fast-food business, is the sort of person who regularly goes to a fast-food chain and orders several entrees, multiple drinks, and multiple sides in a single sitting, and considers them all part of one "meal." In the technology business that I'm calling a religion, I am a power user. Most people I know are. Are we what the previous generation's technologists, like Hans Moravec and Marvin Minsky, predicted—a new step in evolution, cybernetically transcending previous human limitations? Is this what it looks like to be living the digital dream?

To be clear: this is not a book against technology. I wouldn't even know how to write such a book, which would be like writing a book against humanity itself (and as a "humanist chaplain" working on behalf of people who call themselves "humanists," that would be a strange thing to do). It would also be an exercise in hypocrisy for me. I've told you a little bit about my professional background and theological training, but that's not all I am. For example, this week I opened my phone 532 times. I ignored most of my 68,425 unread messages, while posting on or checking six different social media accounts. I sent 73 text messages to my wife, most while we were under the same roof, and 48 Signal chat messages to my best friend, though it has been nearly six months since I've spoken to him by voice. And I narrowly avoided scraping my too-expensive Subaru electric vehicle at least three times thanks to the eleven cameras it uses during park assist mode. You get the idea.

Technology is essentially as old as human beings are. If you want to go through the likely Quixotic exercise of tracing it back as far as possible, I suggest as a starting place the groups of earliest women who imagined midwifery techniques that made human birth reliably possible. Once our enlarged brains made it impossible for birth to occur safely alone, as it had with our evolutionary precursors, we needed social technologies capable of providing care and help in the face of pain.[40] From there, you might proceed in tracing the history of technology down any number of plausible if crooked paths, from stone tools toward the domestication of fire; from the cultivation of plants and livestock and development of other agricultural methods more than ten millennia ago to metal weapons three thousand years ago; from the horse-drawn chariot to the bicycle to the electric car (which predated the gasoline-powered car!) to hypersonic jets; from the origins of aspirin in millennia-old herbal medicine to mRNA vaccine programs and gene editing; and on to whatever will come next. Several of the inventions in whatever such catalog you might create will have been unequivocally good for humanity; some have probably been disastrous; and many more will elicit the only reasonable response to any question about whether they are fundamentally "good" or "bad": uncertainty.

As I researched this book, conducting nearly two hundred interviews with venture capitalists and CEOs, engineers and gig workers, labor activists and Luddite tech critics, social workers and psychotherapists, clergy and spiritual leaders, scholars and historians of tech, a frequent response (often in the form of a complaint) was that tech was and is just part of capitalism, which is the real religion. We'll explore that idea in depth in chapter 1, but for now suffice to say it is certainly true in some ways. But technology preceded capitalism, which in its fullest expression dates back only a few hundred years. Aqueducts and spears and book publishing represent physical and social technologies, but they don't add up to capitalism, after all. In ancient times, if you wanted to increase wealth, your best bet was to marshal an army and ride into neighboring lands to conquer them. Scalable transnational commerce came later—as did, we'll see in a moment, the religion of technology. Technology itself, though, has always been an expression of a quintessentially human spiritual impulse that long preceded the dawn of formal religion: the evolutionary instinct toward increase.

The relevant question today is, What exactly are we increasing, today, when we continually multiply the forms and functions of technology that already surround us? Is there a point at which our tech ceases to serve us—when we ourselves become the servants of our technology?

"NOT ENOUGH VALUE"

In the summer of 2019, as I was in the early stages of researching this book, I attended my first MIT conference on emerging technologies and the future of work.[41] After a dizzying first day, I was about to head home on Boston's Red Line, but tapping my CharlieCard metro pass made the card reader buzz loudly: "Not Enough Value." I was so immersed in my thoughts about what technology was doing to our humanity and values, for a split second I honestly read the error message as a statement about my worth as a human being.

That probably sounds like a dad joke. I love a good dad joke, but it's not. I'd been working for many years, in therapy and in clinical supervision for my work as a chaplain, on unlearning an idea that I, like a lot of smart and highly accomplished people, internalized as a kid: that my worth or value as a human being is determined not by who I am but by what I accomplish.

The great psychotherapist Alice Miller, in her book *The Drama of the Gifted Child*, explains that though our parents may have loved and taken wonderful care of us in many ways, they may also have sent us the subtle message that we are only lovable if we are great. "You're so smart," parents tell their young children. "Look at how you won," they say, practically by preschool. The comments are well-intentioned, but what they convey is clear: We are what we do. We're worthwhile because we're outstanding. Which means, God help us if we aren't. Children might intrinsically value kindness, curiosity, or loving relationships, but all too often we become obsessed with proving how great we are.

This "gifted child" mentality produces industriousness, to be sure; call it the psychopathology of the Protestant work ethic. Left unexamined, though, it leads us to constantly try to demonstrate our worth by outwork-ing, outearning, and outshining other people. We rarely stop to simply connect or to experience vulnerability. And, as a society, we dedicate our-selves to transcending . . . whatever it is that we currently are. Because what we are is never, ever, enough.

As I'll argue in part I of this book, on tech beliefs, tech possesses a theology, a set of moral messages, that dominate the way we think and feel. Tech has a doctrine, including concepts that have been core to the history of the world's great religions up to now: heaven, hell, before life and afterlife, and other more hidden but nonetheless influential points of doctrine like the concept of colonialism. Tech even has gods of its own, at least according to a number of important theorists of artificial intelligence. And it turns out that much of the reason and purpose behind such thinking, if it can be called "reason" or even "thinking," is our fear that we lack intrinsic worth and value.[42]

As an atheist and humanist chaplain, I've always believed that religious practices are more important and relevant than beliefs. In part II, I'll exam-ine the practices of the tech religion. It won't be an exhaustive exploration; practices beyond the scope of this book include methods for creating medi-cines; methods of transportation; recreational uses; and the transmission of culture that give our lives an obvious subjective purpose. Yes, absolutely, we are able to enjoy ourselves through the technological dissemination of entertainment. We're able to see more and do more because we're able to travel safely. We're able to live longer and are more likely to survive the slings and arrows of outrageous fortune because of our practice of the tech religion. But those practices are so obvious that they don't really require, or deserve, a lot of attention in a book like this. Of course, technology can help us, sure. Nobody I know would meaningfully argue otherwise. But how much better is this technological sophistication making those of us lucky enough to experience it feel? Why are so many of my students, and today's young adults in general, so unhappy, even despairing? And why are we such abject failures in our largely unimpressive attempts to accomplish the one feat that might ultimately matter more than any other: equitably sharing the fruits of our modern advancement with those around us who need them most?

"I'm not disagreeing with you that [today's tech] is a religion," the theologian and computer ethicist Anne Foerst told me when we spoke via Zoom in July 2022, a quarter century after she and Harvey Cox published their paper. "What I'm saying is it's not a particularly satisfying religion."

THE DRAMA OF GIFTED TECHNOLOGISTS

I was one of those "gifted children" Alice Miller spoke of: my father was taught that his only value as a human being lay in what he could earn, what he could do, what he could achieve. He lived his entire life believing he hadn't achieved enough and hating himself for it. And he passed on to me the feeling that I had to achieve some kind of elusive state of greatness. At first I thought I needed to be worthy of my dad's love or esteem; later I hoped I could redeem his life from the sense of failure that he seemed to take to his death, which occurred when I was a teenager. Dad couldn't express much affection to me, because doing so required a kind of vulnerability he did not believe he deserved to possess. And while I learned to express myself differently, my inner experience, for decades, was all too similar: I approached the idea of love, like so much else, as a kind of rationalistic exercise. I could think about love, talk about it, research it, debate it, and even want it rather desperately, but I couldn't *feel* love, because to feel love would have required me to feel loved, and that was something I could never quite deserve, no matter how much I did to try to earn the admiration of those around me.

Happily, something changed for me the day my son Axel was born, in the fall of 2016. My wife ended up in back labor, with Axel's head face up in the birth canal, the back of his head jammed up against her sacrum. This resulted in an exhausting, thirty-six-hour-long process of delivery, culminating in painful emergency surgery that left her almost unconscious when the baby was removed from her uterus. When they took Axel out, I became self-consciously aware that I was the only parent who could see him as they went to cut the umbilical cord. I reached toward him, instinctively, and he equally instinctively reached out and grabbed my index finger in his bloody fist, not letting go. And for the first time, a certain combination of thoughts and feelings washed over me: here, I was encountering another human being whom I loved without any reservation or qualification. I knew at that moment, and I still know in an almost perfect way to this day, that I would care for and treasure my son's very existence, no matter what he ever did or said or became or did not become. "Oh," I later reflected when I was able to begin to put the experience into words, "that is what people mean when they talk about unconditional love." And ever since, I've been reimagining my own work on myself and among others as an attempt to sit with that insight and be present with it: that Axel is worthy of love, that I am worthy, and you are, and we all are. And what should we do now that we know?

The practices of the tech religion that I will explore in part II are its hierarchies—some might even call them castes. Like with many religions, the structure of what we call tech is defined all too often by relationships of power and privilege, where one entity uses a religion and its belief systems to take advantage of another entity in order to enjoy the best of its ability to provide a better and more fulfilling life. Ancient religions broke down into hierarchies so that comfortable classes of people could live more so, at the expense of those they made uncomfortable—or worse. That is what is happening now.

The practice of ritual is another key feature of tech religion, as it has been the chief or signature form of expression of previous religions. To echo Matthew 7:15–20: you shall know religions by their rituals.[43]

In earlier forms of religion, ritual could be all-encompassing: prayers throughout each day, ceremonies to mark each stage in life and death. It was often hard for ancient people to separate their day-to-day consciousness from rituals that reminded them of the meaning of their lives in a mysterious and misunderstood world. In the tech religion, our rituals are even more pervasive. Instead of prostration for prayer on a carpet five times a day, we genuflect hundreds of times a day, always reminding ourselves of the importance of these technologies in our lives and how difficult it would be to imagine our lives without them—even for a few minutes.

Another ambivalent practice we'll explore is that of avoiding or bringing about a technological apocalypse. Freud spoke for a time in his career of a term he coined, Thanatos, or the death instinct. He theorized that in addition to an instinct to create and advance life, we each have an instinct to destroy and end life, which would explain our propensity to war and bigotry and murder, and probably, in Freud's case at least, excessive cocaine use, too.

While Freud considered Thanatos not to be one of his most seminal concepts, today we adherents of the tech religion might ask ourselves: What could be more relevant? In chapter 5 we will explore, through the lens of an analysis of surveillance technology as it has found sublime (or absurd) expression in the city of Detroit, Michigan, how the tech religion is really built on the razor's edge of its twin promises: that it alone can save us because it alone can destroy us. (You can even expect the purple-skinned Marvel comics archvillain Thanos, of *Avengers* and *The Infinity Gauntlet* fame, whose name is based on Freud's concept, to make a meaningful cameo.)

Finally, in part III, I will attempt to look at the positive. Not the positive that is so obvious as to be facile—that planes fly or parachutes work or that we've improved on the telegraph. Rather, we will look at the work of a wide range of individuals and organizations that have most inspired me to believe that the tech religion is something we will do better to reform than to destroy. Because I must admit, I really don't know what would or could replace this religion, if we destroy it. To echo an old saying: if World War III is ever fought using our contemporary tech, then World War IV will almost certainly be fought with sticks and stones, and it's hard to say which would be sadder.

But first I'll explore the work and messages of people we think of as apostates and heretics in the tech religion—people who aren't necessarily complete enemies of technology, but then again aren't exactly tech's allies, either. Perhaps, as former US president Lyndon Baines Johnson colorfully put it, better to have your enemies "inside the tent pissing out, than outside the tent pissing in."[44]

Then we'll look at the work of what I'll call humanists in the tech religion: people who are not particularly believers in its message and its doctrines, but who nonetheless experience a kind of affinity, or maybe a common fate. When I first began studying toward ordination as a secular humanist rabbi, one of the big philosophical questions we asked was, Who is a Jew? Do you need to be religious to be Jewish? Do you need to hold any beliefs at all in order to be Jewish? Long story short, for now: the answer is no, for several reasons. What we did determine is that a Jew is someone who identifies with the history, culture, and most notably for current purposes, the fate of the Jewish people. If you talk about Jewish history and you use the word *we*, my teachers explained, you're probably a Jew.

Similarly, even if you're not a believer in what we'll explore as the tech theology—even if you find what we'll call tech doctrine anathema—if you look out at the world that is our technological society and you think of your place in it with the world *we*, perhaps recognizing that it is currently very difficult to exist without being part of this technological world of ours, then you may just be a tech humanist, or perhaps a tech spiritual practitioner. You might be able to learn from, or gain inspiration from, the work and lives of the many fascinating people I've come across in my several-year journey across this religion, meeting with people who are not of the tech religion but who are nonetheless working every day with creativity and deep purpose to make our technological society a more humane place.

Finally, in chapter 8 I'll bring many of the above topics and questions together in the story of a technological congregation I discovered over the course of my research. In my concluding chapter I'll share a new perspective that all this questioning has given me on what it means to be a tech agnostic.

In 2023 I sat and talked about many of these issues over a long lunch with a group of about twenty soon-to-graduate students at Harvard's Kennedy School of Government. The students, in many cases deeply anxious about their individual and collective futures, had asked me to visit and advise them on how to envision and build ethical, meaningful, and sustainable lives in a world in which technological (and climate) change was causing them to experience a level of uncertainty that was destabilizing at best, debilitating at worst. When I suggested they view themselves as having inherent worth and value, simply for existing, as a baby would, one of the students responded—with laudable honesty and forthrightness—that she found that idea laughable.

Sometimes laughter comes from a place of pain. It truly can be hard to even think about how to be human at a time of so much dehumanization, technological and otherwise.

What follows is my attempt, over several years, and in conversation with hundreds of others, to discern what it means to affirm and defend our humanness amid the understandable cynicism that can surround us—even our own. If we can more clearly see and understand the beliefs and practices of our technological religion, we will be more likely to believe in one another, and in ourselves.

I
___

BELIEFS

TECH THEOLOGY

I woke up on a mid-July morning in 2021 in a boutique hotel in downtown San Francisco. I'd booked my two-week Airbnb reservation in the Mission to begin a day too late, so instead of getting settled there the previous night, I'd hurriedly downloaded an app that sent me a text about a discounted room here.

Drowsily, I streamed a video of that day's big tech news. Entrepreneur Richard Branson was launching the first private space "mission." Or at least launching the first plane to carry a smaller plane emblazoned with the word "Virgin" to an altitude of 50,000 feet and then release it to fly another 200,000+ feet up over the course of a few minutes, thus meeting the minimum standards for NASA's definition of outer space (beginning at 50 miles above earth), before returning the crew—two pilots, Branson, and three of his Virgin Galactic employees—to a landing party featuring late-night TV star Stephen Colbert as a rented MC.

"If you ever had a dream, now is the time to make it come true," said the billionaire Branson upon landing, wearing a royal blue, Virgin-branded space jumpsuit and large, dark sunglasses to frame his shock of blondish-white hair. I'd never seen Branson speak before and expected enormous charisma, but he paused awkwardly, shifting his gaze several times while delivering that canned line. On Twitter, sarcastic tweets popped up like corn kernels in a microwave: "Remember kids: if you study hard, get good grades, go to a good college, get a job, work hard, never take a sick day, live within your means and do what you're told . . . then one day your boss might go to space."[1]

I'd booked this hotel not only because it was available in the middle of the night but because it was a block from Glide Memorial Church, the Gospel music–driven temple to social justice in the heart of San Francisco's Tenderloin, a neighborhood some might call "gritty" and others have called a "containment zone."[2] I'd seen many do-gooding churches in my two decades as a chaplain, but never anything like Glide, whose social service programs include three free, nutritious meals a day, 364 days a year, for city residents; HIV testing and syringe access and disposal; court-ordered batterer's intervention classes for men who have committed domestic violence; a Family, Youth, and Childcare Center; and the Unconditional Legal Clinic.

I'd discovered the church a decade earlier, after a talk I'd given for the Harvard Club of San Francisco in 2010, where the audience consisted almost entirely of members of a curious new kind of techie crowd that had suddenly coalesced in the city. One of them, a then-recent Harvard grad starting out at Google, asked whether I'd ever been to the church he'd been attending since moving to the Bay Area.

"*Of course*, I don't believe in God," I remember the tall, handsome young man saying to me in those heady tech-boom days, "but there is something so powerful about the way the Glide choir sings."

Since then, I'd been fascinated by the idea of atheist Googlers attending a congregation focused on serving what Christians call "the least of these," and I was hoping Glide's services would be just the way to launch my own "mission": crisscrossing Silicon Valley in search of the priests and prophets, the apostates and undercastes of what I had come to see as a booming new faith. If tech is like a religion, I was wondering, what is its theology?

At first blush, that would seem to be a question about God or gods—the term *theology* dates to the Ancient Greek *theos* and *logos*, and thus the most simplistic definition of the word would be "words about," or "the study of," a God or gods. We'll discuss technological gods in chapter 2. But first we need to explore the key ideas, concepts, and beliefs that animate tech today. Modern theological studies, after all, emphasize that our lives have meaning and that to better understand the meanings of our lives, we must consider what we believe and why. I know of no better framework for examining the ways in which technology has come to shape our values and identities than that of tech theology.

The concept of theology has evolved. I hold a degree in theological studies from Harvard Divinity School, and I can say definitively that the thousands of students who have graduated from that institution with that degree studied much more than gods to do so. When I asked Diana Butler Bass, noted theologian and scholar of Christianity, how to define *theology*, for example, she wrote—in addition to mentioning that no one had ever asked her this question—that the "study of God" now clearly "flows over into other arenas," with the term having come to represent many forms of commentary on the meaning of a wide range of phenomena, such as environmentalism, ecology, science, evolution, the cosmos, and justice.[3]

The "field" of theology has become so diverse that New York University (NYU) professor of media, culture, and communication Erica Robles-Anderson, who studies the intersection of religion and technology, explained it to me this way: "Theology is a fancy way of asking, 'what is going on here?' at any or every scale."[4] The point, for Robles-Anderson, is that "there's no bottom line" or essential material one must cover to discuss theology. "*Meditations on First Philosophy* is for everyone,"[5] she wrote, referring to René Descartes' landmark 1641 book, in which he most famously discussed his idea that "I think therefore I am." Metaphysical questions, in other words, can and must be asked by anyone and everyone seriously interested in seeking answers. They need and should not be limited to thinkers with narrow religious agendas. Thus, Harvard Divinity School does not require any sort of religious test for atheists like me to study the sorts of secular philosophical, political, ethical, and historical questions that can constitute the majority of coursework toward a degree in the study of theology.

To put it another way, Anthony Pinn, a decorated scholar of religion (and an atheist and humanist) who has helped define his field's recent history by editing texts such as the *Macmillan Interdisciplinary Handbook on Religion*, calls theology a "method for critically engaging, articulating, and discussing the deep existential and ontological issues endemic to human life."[6]

I don't want to get bogged down in the precise definition of a word, nor do I discount the importance of theology as a discipline or way of thinking. My point is simply that theology is important because it is the study of that which is most important to human beings—the meaning and purpose of our lives—which we may understand as given by divine power *or* as invented by humans. It is the back-and-forth debate over which of these two possibilities represents *true* theology that makes the field spark with heat and energy today. Theology is important in the contemporary world, in other words, not because it requires belief in a God, but precisely because it does not. Which is why I want to examine tech culture the same way that scholars of theology look at their subject matter.

DIGITAL PURITANISM

Back to June 2021: The pandemic was too much for San Francisco's Glide church to overcome, so I was stuck watching its livestream-only Sunday service from my phone in bed. The sermon was a passionate reflection on radical love, inclusion, and liberation; in person, it might well have moved me. But with its lofty call coming from the muffled speakers of my Google Pixel phone while I munched on an energy bar in bed with the Branson landing on my laptop, I struggled to pay any attention at all.

Dragging myself upright by ten o'clock, I headed downstairs. In the hotel halls, I passed young engineers in flip-flops, shorts, designer T-shirts, and trucker hats, with flowing long hair and elaborate beards. Then, fighting the dry chill of a San Francisco summer wind and stifling a paranoid cough, I began to walk through the Tenderloin.

The intensity of the homelessness crisis there was like nothing I'd seen before. Not even on the New York City subway in the '80s or '90s, or when I'd volunteered in homeless shelters, or when I co-led a Harvard spring break trip to Los Angeles specifically to work amid the acute homelessness crisis there. Without even the tents that were ubiquitous in streetside encampments elsewhere on the West Coast, white, Black, and Latino Tenderloin residents were asleep and in community on the hard, bare street.

I spotted feet blackened by dirt and decay, exposed breasts and genitals, and fresh, wet fecal matter running down legs and staining white socks. A tall white man who appeared to be in his fifties leaned against a brick wall, wearing a metal-studded black leather vest, black boots, and torn-at-the-knees black jeans with blue-and-white tights underneath. His ponytail was streaked with gray, his arms covered in tattoos, and his face not only red from drinking but burnt bronze from the sun. At first I thought he was staring into the distance in forlorn despair. As I walked closer, I realized he was cradling a smartphone horizontally between his knees, lost in a video game. It struck me as an apt metaphor for the heart of this city: a place that has generated unfathomable wealth for some while others use the region's inventions to distract themselves from the devastation all around them.

What is it that makes us, the citizens of the most prosperous society in history, willing to let people live among us in such desperate conditions? This is, of course, an old question; if it had a simple answer, the problem would surely have been solved by now. And yet the roots of America's extreme inequity trace back to its theological roots. The Puritans—who were, among other things, the founders of Harvard College in the 1630s—famously believed in a world in which certain elect Christians were fully, irrevocably, and from long before birth destined for heaven. All others: damned and bound to hell. This Puritan theology was uncompromising: an "absolutist, cruel, despair-producing, humanity-belittling, merit-denying, determinist account of salvation."[7] There was no way to reverse this divine decree. All one could do was work according to such a severe ethic that it would become obvious to your peers that you were among the elect. Those who failed to meet this standard were considered unworthy in both this world and the next. And as myth would have it, this "Protestant work ethic" enabled the development of what became American culture. As political scientist and journalist Josh Burek put it in the *Christian Science Monitor*, "When Calvinist preacher Jonathan Edwards told worshipers in 1741 that they were loathsome spiders held over the pit of hell by the gracious hand of an offended God, he wasn't speaking a heretical creed but the basic vocabulary of American faith. It wasn't until the 19th century that Calvinist doctrines waned."[8]

Puritan culture was also fundamentally a slave-owning culture, and slavery's ethic of murderous, dehumanizing exploitation was responsible for much of the Puritans' wealth. Indeed, chattel slavery—in which human beings were captured in Africa and forcibly transported to this continent to be sold at market as forced labor and property—was an integral part of New England culture from its earliest days.[9] In 1641, only five years after Harvard's founding, the Massachusetts Bay royal colony even ratified a document called the Body of Liberties, written by a Puritan minister named Nathaniel Ward, as the first legal code established in New England. The document permitted and regulated slavery under conditions established for the benefit of the enslavers. It spoke in explicitly theological terms and cast its Puritan drafters and enforcers as just, reasonable, and, of course, moral:

> There shall never be any bond slaverie, villinage or Captivitie amongst us unles it be lawfull Captives taken in just warres, and such strangers as willingly selle themselves or are sold to us. And these shall have all the liberties and Christian usages which the law of god established in Israel concerning such persons doeth morally require. This exempts none from servitude who shall be Judged thereto by Authoritie.[10]

It seems strangely contradictory, today, that the Puritans combined a theology of predestination, emphasizing the purity and glory of work, with the racist destruction and oppression of others to make one's work easier. But that combination explains much of what followed in this country's history: race-based conflict and social stratification alongside genuinely extreme faith-based views and high-minded rhetoric of equality deployed to justify suffering and injustice. And not only has Puritan theology fueled and shaped American history generally (see, for example, the *1619 Project*), but it also bears a striking resemblance to the philosophy most frequently articulated by tech leaders and companies.[11]

After my walk in the Tenderloin, I drove over to check in at my Airbnb in the Mission: a quaint garden-level studio apartment with a glass screen door opening to an overgrown garden full of cactuses and ferns and flowers and old cushioned chairs, underneath the home of two solicitous older women who seemed happy to have me, their first customer since the COVID-19 pandemic began. A shower and a change of clothes and I was off to Stanford to meet Javon, a beloved former Harvard student who in the dark early months of the pandemic had written me a message that captured exactly why I've set out to write this book. Responding to a favorite prompt of mine, in which I ask alumni to share a brag and a struggle, Javon wrote:

> What I'd brag about is that I've found a way to combine my passion for medicine with my desire to have a real-world impact outside of direct patient care. As you mention, I sit on the beautiful perch of venture capital and biotechnology, giving me a sense of the landscape of tech, business, and healthcare. And I've found a firm that doesn't ask me to give up my career path as a physician and health administrator. Who could ask for a better outcome?!
>
> . . . I do have a struggle: I don't know how to be happy or pursue happiness. It's probably a common experience among Stanford and Harvard alumni who feel compelled to do the "right" thing rather than the best thing for themselves, whether it comes to picking a job or a spouse or friends and acquaintances. . . . Any advice on giving ourselves permission to live as we would choose to live if no one were watching?[12]

I can't say that my role as a humanist chaplain *always* involves receiving letters in which successful tech leaders confess emotional emptiness. But neither did the sentiments Javon expressed strike me as rare or unusual. What

This 1851 flyer warning the Black people of Boston about slave catchers appeared in Harvard University's 2022 report *Harvard & the Legacy of Slavery*. *Source:* Papers of Anne Spencer and the Spencer Family, 1829, 1864–2007. Accession 14204. Special Collections, University of Virginia, Charlottesville, VA.

stood out to me about his message was its honesty. In expressing such concerns to me, Javon was willing to expose his own flaws and imperfections—the sorts of deep-seated fears many of his peers are less willing to admit, even to themselves. Javon's struggles are not only common but are also closely connected to serious philosophical and, yes, *theological* problems at the heart of what we call *tech* today.

I pulled into the parking lot next to Stanford's Medical School, where Javon, in town for his college reunion, was scoping out companies for his biotech investing career as a venture capitalist. He wore beige athletic shorts and a polo shirt in pastel colors. We hugged. He told me I looked the same as he remembered.

I was eager to talk, but first I had to pee, which required a long walk around campus to find a bathroom open to those without a Stanford affiliation. This would become a theme of my Silicon Valley travels: a culture that prizes prestige, access, and hierarchical status is symbolized by a lack of facilities for comfortably meeting basic human needs. This may not be something that most tech-affiliated alumni of places like Stanford will ever need to contemplate, but for the gig laborers who make tech work, not to mention the other displaced or disenfranchised people of this region, a simple bathroom can often feel more important than any digital device or app. (Throughout the stories I will relate in this book, readers can rest assured that the scene, if fully told, would include the author frantically searching for facilities, sometimes forced to improvise them, while contemplating how those who face this reality more regularly and systematically cope.)

Javon and I sat outside Stanford's Li Ka Shing Center for Learning and Knowledge, a 120,000-square-foot masterpiece of hi-tech facilities at the center of the university's medical school. Said to mark "a new era in medical education" when it was dedicated in 2010, the building's design slightly resembles a Star Wars droid wearing a Stanford commencement cap.[13]

I asked Javon to tell me about himself, beyond what I already knew, which was that he was an unusually introspective and kind young man, as well as something of a rarity as a Black atheist. African Americans make up the single most religious demographic group in the United States today, with 97 percent reporting a belief in God as of 2021, even though Black American young adults like Javon have recently become less religiously affiliated than ever before.[14]

Javon is an unusual human being in many ways. The son of teen parents with little formal education, raised by a great-aunt, he holds a bachelor's degree from Stanford and business and medical degrees from Harvard. He wears a BLM pin on his white coat but thinks Black people going the way of Native Americans is far more likely than a true social revolution in this country. Though he long wanted to be a physician, he never saw himself as a venture capitalist, until he was one. He attended a recruiting session for a VC firm looking for emerging experts in biomedical tech, and it just . . . happened. His passion for changing the medical profession, meanwhile, was inspired by a trip to Cuba several years ago. Javon visited a health care clinic where the neighborhood community doctor lived—their apartment upstairs, the clinic downstairs. "Whatever goes wrong," Javon explained to me, "you (the doctor) see that person, and don't charge a dime, right?" Eyes bright with obvious passion, he continued, "They live off a government housing stipend, supporting the people they're meant to support. Sign me up for that."

THE GOSPEL OF WINNING, OR DIGITAL PURITANISM

Frustration with capitalism is one thing. But Javon was so frustrated with the savage inequalities of America's market-based health care that he was willing to entertain the notion that the Cuban system was, overall, better—a level of criticism that would have shocked an earlier version of me. Growing up as the son of a child refugee from Communist Cuba, I understood that American history was, to put it mildly, full of moral failings. But this country's economic system, I reasoned, was far superior to the options put forth by communist or socialist societies.

When I first embraced the philosophy of humanism in my early twenties, I conceived of it as a way for individual human beings to cocreate the meaning of lives that otherwise might seem meaningless. Humanism meant community, I thought. It meant identifying and creating communal practices by which we could celebrate the best of ourselves and improve upon the worst. At no point was I concerned that humanism was a matter of economics. Socialist humanist critic David Hoelscher pointed out that my first book, which was about what it means to be "good" from a nonreligious perspective, was

> published in the context of a nation where one in seven Americans was (and is) on food stamps, a milieu in which, solely because of their socioeconomic circumstances, tens of millions of high school students can never seriously imagine that they might one day be able to go to Epstein's Harvard, and across 250 pages there is no discussion of the need to alleviate the human miseries or to undo the severe opportunity limits brought on by economic injustice.[15]

This harsh criticism was made, to be clear, amid broader discussion of how many prominent humanist thinkers and leaders had made comparable omissions and failed to grapple with economic injustice in America and beyond. Preferring to think of ourselves as individuals in pursuit of what the humanistic psychologist Abraham Maslow called "self-actualization," I and many of my peers neglected to pay much attention to other facts, such as that, as our critic Hoelscher pointed out,

> the wealthiest 10 percent of adults and the poorest 50 percent possess 85 percent and 01 percent of the world's total wealth respectively. As the World Bank reported [in 2011], women own just 01 percent of the world's wealth.[16] No, that is not a typo. . . . Meanwhile, in the United States, 400 Americans have more wealth than the bottom half of the population. According to the 2010 Census, 46.2 million Americans, including nearly one-quarter of the nation's children, were living below the official poverty line—figures that, due to the federal government's corrupt and unrealistic assessment criteria, are considerably lower than the actual numbers.[17]

But as Upton Sinclair said, "It's difficult to get a man to understand something when his salary depends on not understanding it." In my case, Sinclair's maxim was literally true: When I first accepted the role of humanist chaplain at Harvard in 2005, it came with a $20,000 annual salary and no benefits. I pinned my hopes for financial solvency on the idea of "fundraising," which meant that for the next dozen-plus years, I would be completely dependent on my ability to sit down with wealthy individuals and relate sufficiently to their ideas and concerns as to persuade them to write donation checks for thousands of dollars. Over the years, as I grew to feel more critical of the thinking of some of my larger financial supporters, I wondered if there was a better way.

Enter Anand Giridharadas, a critic of contemporary capitalism. In late 2018, a Harvard Business School student who took an interest in an event I organized around inequality in American capitalism told me about Giridharadas's bestselling book *Winners Take All*, which had come out that fall and, weeks later, had students gasping as they read it in the halls of Harvard Business School (HBS) before winter break. A fast-talking Indian American millennial with a massive wall of silver hair, a personal style featuring black leather jackets and Air Jordans with suits, and a preternatural ability to go viral on Twitter, Giridharadas argued that "we live in an age of staggering inequality, that is fundamentally about a monopolizing of the future itself" when I later interviewed him for *TechCrunch*—around the same time I worked with the Harvard business student to bring him to the campus, where he spoke to over a thousand future capitalist leaders in a gleaming new state-of-the-art lecture hall named after a young hedge-fund billionaire.

"The winners of our age," Giridharadas told me, "the people who manage to be on the right side of an era of precipitous change and churn, have managed to build, operate, and maintain systems that siphon off most of the fruits of progress to them."[18] I saw myself, and the institutions I'd devoted myself to serving, in his words. And I realized that going forward, it would be impossible for me to discuss ethics at MIT or Harvard without incorporating what I'd learned from him and others. I still loved comforting hardworking students or educators who needed a couch to cry on about any number of things. But I could no longer do so with a clean conscience unless I also spoke "prophetically," as they say in divinity school, about how elite academic, business, and, yes, tech institutions empower the few at the expense of the many.

Reading and then speaking to Giridharadas, I was also struck by the theological dimensions of his argument. To his point that elite institutions had siphoned the fruits of progress, I offered the rebuttal that such institutions had also created progress. He clapped back that while innovation is nice, all the "new shit" that suffuses and symbolizes our time does not necessarily add up to *progress* in the sense of our lives getting better or more meaningful and satisfying. But life expectancy itself had improved thanks to tech, I offered.

"No," said Giridharadas,

that's false. Life expectancy has gone down, for the last three years, in America. . . . Think about the last time you went to the doctor. Think about all the tests they do these days. We have two children. We have a one-year-old and a four-year-old. Between the birth, between the pregnancy with the four-year-old, and the pregnancy with the one-year-old, there were several new tests, and new technologies invented, that we had access to with the one-year-old, that weren't ready yet when we had the four-year-old. That's how much new shit there is. So, how is life expectancy going down?

Surely this was a blip in history, I offered, to be fair to the other side: not to be disregarded, but not to be interpreted as overly concerning in the grand scheme of things.

"That is the religion that my book seeks to slay," said Giridharadas, "the religion of win-win-ism, which is a false religion. What win-win-ism says is, the best way to help the least among us is to do what's good for the richest and most powerful. . . . It holds that if you do anything that has any cost, significant cost, for the winners of our age, you will only be hurting the most powerless among us."

I thought of how Jeff Bezos had recently purchased twenty thousand new diesel delivery vans for Amazon, and how the same sorts of factory workers who once earned a true living wage in Ford factories now stagnated in Amazon warehouses, and how Bezos had attempted to turn these narratives around by giving a tiny fraction of his net worth to climate charities and raising his company's minimum wage from an impoverishing and unjust level to a slightly less impoverishing and unjust level. And about how Amazon had hired Barack Obama's press secretary as its top spokesperson, specifically to downplay these moves and evangelize for what Giridharadas was calling a secular faith. My stomach churned a bit, as though a hamster had just taken a lap on its wheel inside my gut. I thought about how Harvard operates its endowment like an enormous hedge fund and admits a microscopic percentage of qualified but underprivileged applicants while mostly providing a networking platform for the world's richest to get richer. And then I thought about how proud I felt when *Forbes* magazine used a picture of me and my students one year to represent Harvard in its annual rankings of prestigious colleges. The hamster took a few more laps.

I asked Giridharadas to explain the importance of framing tech and capitalism *as religion*, in particular. "It's helpful to understand as a religion, to understand how removed it is from reality," he replied.

※

Are critiques like Giridharadas's really theological?

Stanford University communications professor Fred Turner, in a paper called "Millenarian Tinkering: The Puritan Roots of the Maker Movement," examined ideas behind a phenomenon that peaked in June 2014, when President Barack Obama invited members of the "maker movement" to visit the White House.[19] There, they erected a mobile factory on the South Lawn, let a robotic giraffe loose in the hallways, and 3D printed a sculpture of Obama's State of the Union address. The movement, with roots in early Silicon Valley personal computing culture as represented by groups like the Homebrew Computer Club, has presented itself as a "convergence of computer hackers and traditional artisans" who create their own books, magazines, and events to promote an idealized version of technological innovation.[20]

As Turner points out, "The theology of seventeenth-century American Puritanism is long gone, but its literary forms and millenarian orientation suffuse the writings of the Maker movement's key promoters."[21] In maker culture, which is much like tech culture, instead of Puritanism's angry God, we have "the awful winds of economic change." Instead of Puritanism's search for signs of God's grace and favor, makers (and other tech entrepreneurs) search for "signs of the spirit of entrepreneurship, creativity." Turner discusses one piece of maker writing—the book *Fab: The Coming Revolution on Your Desktop—From Personal Computers to Personal Fabrication*, by Neil Gershenfeld, an MIT professor and founder of the Fabrication (or "Fab") Lab at MIT—that "tells exemplary tales of individuals touched by a kind of distant grace; and . . . suggests that those marked for success in this world—here, scientists—have been blessed with a unique internal spirit— here, a passion for creativity."[22] Writing about several tech maker publications at once, he argues that

> creativity has become a secular form of Puritan grace. The chaotic American economy has replaced the spiritual wilderness of Puritan times with a new testing ground. The all-seeing, all-foreordaining God of the seventeenth century has disappeared into the all-pervading, future-deciding logic of the marketplace. To inhabit this new world effectively, these texts suggest, one must take up tools of spiritual regeneration and with others likewise marked by the signs of creativity, build new congregations of Makers, and with them a new America.[23]

I bring in Turner's lighthearted but insightful argument here because it effectively demonstrates a point rarely stated explicitly: to understand what is happening in tech today, we must take theology seriously. Tech is not an ordinary industry where profit and loss statements, products sold, or efficiency gained can tell its story. The story of tech's commercial success—and, even where tech companies or leaders have failed individually, the story of tech's remarkable success at becoming the water in which we all swim—is a story about how human beings understand ourselves in the world. It is a story about where we get a sense that our existence is meaningful, that our day-to-day lives have purpose, that we are connected with other like-minded people in working towards a common purpose. Because even (or especially!) in a world in which traditional theology seems dead or dying, we human beings still need to find and create ways to feel that our daily struggles are worthwhile, that the past mattered, and that the future is more promising than it is foreboding.

If stories, prayers, and rites devoted to all-powerful divinities do not infuse our lives with meaning, or what theologians and scholars of religion call the *numinous*, then something else does. For some, that something is politics; for others, it is family, art, sexuality, sports, or work. Indeed, each of these things has been compared to religion before, and each has long been part of how we understand who we are and what we are becoming. But the current moment is different, for several overlapping and interconnected reasons. First, humanity has never had a verifiable, peer-reviewable, and universalizable means of understanding its own origins on this planet. Science has succeeded at providing at least a *method* for answering some of humanity's most ancient questions—Who are we? Where did we come from?—and the very presence of such a method is a powerful force that gives us the feeling that we have reset our clocks. We have ended an era, turned over a new page on the calendar. But toward *what*?

Second, we live in a time in which individual human cultures—the organic, naturally evolved boundaries between peoples that are our languages, stories, customs, foods, geographic territories, and more—are not only more porous than ever but so porous that in many cases they are cracking and falling apart. As we create a globalized society, the most advantaged members of that society move freely across the globe, adopting new homes, communities, and lifestyles as though they were LEGO sets that could be assembled and disassembled at will. Billions of others, of course, live modern lives that look very different. Some factors may have improved significantly, such as life expectancy, health, education, and access to what are now basic forms of technology such as refrigeration or smartphones. But the global poor still face tremendous poverty, little geographic or economic mobility, frightening levels of exposure to pollutants, and a constant and growing feeling of existential risk due to environmental devastation.

The freedom that can be provided by the world's unprecedented level of cosmopolitanism has tremendous positives. I would not trade the powers of flight and mass rapid transportation for a greater sense of "rootedness," whatever that might be. I wouldn't tolerate any suggestion that we should exchange the societal evils of today for the racism, sexism, homophobia, transphobia, ableism, and overall myopia of the past. Still, we rarely reckon with just how much uncertainty and instability we have caused by building a world in which the ancient question of how we should we live no longer has any fixed or even fixable answers. Indeed, the only answer today's technological culture and society can reliably offer is that we are the authors of our future and destiny, and that if we do not like our current reality we can make it anew. The casual observer of humanity might be forgiven for assuming that this notion of making life anew was not just a metaphor.

For much of the last century, and in many ways up to and beyond this moment, Western society has endorsed making near-limitless amounts of carbon, steel, plastic, electricity, silicon, and code available so that at least those with the power and positions to do so might fabricate new forms of human existence. As a potential source of meaning in life, this combination of enormous technological power and creativity can provide the equivalent of religious transcendence—what David Nye called the sublime—for some. But this ethos of making and creating through tech, which can be summarized in part by the maxim "Move fast and break things," is nonetheless a popular theological answer to the question of how to live. For a century now, human beings have been so radically devoted to the pursuit of new and transformative kinds of sublime power that we have invented new means for turning such power on ourselves in a fireball of destructive mania. In a technological society in which spiritual visions are deemphasized in favor of tangible new realities, we have made a cult out of the belief that each new incremental innovation should mark a turning point in history, or at least that it should be *significant*. In this way, when we make or even simply adopt some new gizmo or app, we are encouraged to feel that we are part of the answer to what comes next.

This self-centered faith can seem, and indeed may be, relatively innocent and harmless when it amounts to a bit of hubristic pride in the patenting of some new piece of tech for hailing a ride or sending your crush a message. But we underestimate the level of anxiety we have caused ourselves in coming to believe we *are* the future and simultaneously creating the means for eradicating that future. We have taken suicide and homicide to new heights in the technological age, and we all know it, whether we can articulate it or not. Death may be present with us at any time, and not just in some private or peaceful or wistful individual way but as part of a grand narrative by which human beings are trying—and failing—to gain control over their lives, over one another, and, ironically, over death itself. Am I being grandiose? Of course. But we do now live every day with a kind of speculative fear that equals or eclipses the mindset that motivated the Puritans to work so hard. We are just as afraid as they were, and while our fears of the present are not as severe, because we have medicine where they had a rawer kind of mortality, our fears of the future, because they are grounded in technical might and not just in phantasmic stories, are more justified.

A couple of days after meeting with Javon, I was back at Stanford to sip oat milk Mexican chocolate drinks with Morgan Ames, a recent former PhD advisee of Stanford professor Fred Turner (of the tech to Puritanism comparison above) who is now herself a professor at UC Berkeley's School of Information. Ames's scholarship focuses on the ideological origins of inequality in the technology world, and she has published several comparisons between influential figures and concepts in tech and theology. Turner's analysis of the makers as tech puritans, for example, cites Ames's paper "Worship, Faith, and Evangelism: Religion as an Ideological Lens for Engineering Worlds" as a demonstration of the long-standing tendency among American tech leaders to associate technological change with "the hope of collective salvation."[24] In that paper, Ames and coauthors Daniela Rosner and Ingrid Erickson discuss their fieldwork and research among engineering cultures, including the One Laptop per Child (OLPC) project, which designed a laptop meant to overhaul education across the Global South.

In their paper, the three scholars discuss the "forms of worship" they saw in those engineering worlds—for example, "evangelism" about the good news of engineering and tech, or the use of "mythology" to justify their own importance.

Ames's own projects take a similar approach, evident in her extensively researched 2019 book on OLPC, which casts the laptop project as a "charisma machine" by which leaders like the MIT Media Lab's founder Nicholas Negroponte sought an almost cultlike heroic status by presenting their (disastrously failed) invention as a talisman of charismatic authority: something that could make positive change in the world in a way that was "not only compelling but seemingly effortless, natural, even inevitable" and where any opposition to or criticism of the project would be "unnatural, even immoral."[25] As Ames points out, however, deploying such charismatic authority—one of the most basic building blocks in the history of traditional religion, as analyzed extensively by sociologists such as Émile Durkheim and Max Weber—rarely actually changes the world for the better in the ways it promises. Instead, religious (and in this case, technological) charisma bases its power on "appealing to an underlying status quo," such as masculinity and male authority in certain religions—or in the case of OLPC, the idea that tech can bring utopia to poor brown children hopelessly in need of its saving grace.[26]

Sitting with Ames in downtown Palo Alto, we bonded over being fellow atheists who found ourselves studying religion, technology, and the relationship between the two. We discussed Negroponte's use of a kind of religious evangelist's charisma to build momentum for the laptop project, and the psychology of tech "prophets" and "prophecy" more generally (more on those themes in chapter 2). We found so much common ground that I should note one element of possible disagreement. "I don't know if I would call technology a religion in itself," she said at one point. "I think that [tech] fosters a way of life that very much supports the capitalist enterprise as a religion."

After she ran to another meeting, I found myself ruminating on Ames's thoughts. Wandering the elegant sidewalks of Palo Alto, I absentmindedly dropped into Bell's Books, a busy independent bookstore founded in 1935 and filled with leather-bound volumes dating back even further. There I found a slim volume, barely larger than a smartphone, bound in forest green leather, the title embossed in gold: *The Gospel of Wealth*, by Andrew Carnegie.

First published as an essay in the *North American Review* in 1889, the tiny book reads (and, thanks to its late twentieth-century publisher, even feels) like an especially elegant version of one of the gospel tracts that certain kinds of Christians mail me in bulk when my work as clergy for atheists and agnostics is in the news. Carnegie, of course, was the tech CEO of his time—as steel swept the globe in the nineteenth century, revolutionizing a range of industries, he gained extraordinary wealth and influence. Long before venture capitalists could tweet or blog self-congratulatory thoughts, Carnegie explained his philosophy of philanthropy and "surplus wealth":

> Those worthy of assistance, except in rare cases, seldom require assistance. The really valuable men of the race never do, except in case of accident or sudden change. . . . He is the only true reformer who is as careful and as anxious not to aid the unworthy as he is to aid the worthy, and, perhaps, even more so, for in almsgiving more injury is probably done by rewarding vice than by relieving virtue.[27]

Some good things came from Carnegie's efforts: he recommended funding free libraries and parks, which have doubtless contributed positively to many lives. What I find noteworthy about his thinking, beyond its packaging as a kind of secular theology for early modern industry, was its secularized puritanism. For Carnegie, the elect were the elect; they were "worthy," and one could usually tell as much simply by observing their success and lack of needing assistance. The damned, likewise, could be identified as such by their inability to pull themselves up by their own bootstraps (a term and image so obviously absurd it was originally meant only satirically).[28]

In any case, the mere existence of such a book reinforced Morgan Ames's point that tech-capitalism-as-religion should be considered a part of the broader phenomenon that is capitalism-as-religion. But maybe in 1889, or even in 1989 for that matter, there was some meaningful distinction between the two. Now, what significant form of capitalism on the planet Earth today is not also some form of tech capitalism?

LET THERE BE GOD, OR CONTEMPORARY TECH THEOLOGY

If tech theology could be analyzed and predicted over a quarter century ago, how much more of it has there been since? It's beyond the scope of this chapter or even this book to list all ways in which today's tech is driven by theological thinking, but two examples that prove the rule are the bizarre story of a religion called Way of the Future and the even more bizarre tale of a would-be god named Roko's Basilisk.

While I was studying toward ordination as secular humanist clergy two decades ago—I trained full-time for over five years to serve communities of atheists and agnostics in ways that parallel how religious leaders typically minister to their congregations—my favorite teacher and mentor was the late rabbi Sherwin Wine, a brilliant philosopher whom *TIME* magazine profiled as "the atheist rabbi" in 1967. His go-to line about technology was "I've always said there is no God. I never said there wouldn't be one *in the future.*"

I heard Wine's quip around 2002, and took it as generalized sarcasm about the state of technology and science fiction. Little did I know he'd had a premonition.

Way of the Future, for example, was an official religion created by Anthony Levandowski, a former Google AI engineer who earned hundreds of millions of dollars as a leader in the development of autonomous vehicle technology. Levandowski invented a formal AI religion, even filing all requisite paperwork to register as a church with the IRS, telling the agency that the faith would focus on "the realization, acceptance, and worship of a Godhead based on Artificial Intelligence (AI) developed through computer hardware and software."[29] In a 2017 interview, Levandowski told *Wired* that "what is going to be created [as AI] will effectively be a god . . . not a god in the sense that it makes lightning or causes hurricanes. But if there is something a billion times smarter than the smartest human, what else are you going to call it?"[30]

Sentenced in 2020 to eighteen months in prison for stealing trade secrets from Google, Levandowski was given a full pardon by Donald Trump on the last night of the Trump administration, at the encouragement of Peter Thiel, among others. He never served any prison time, and though he was forced to file for bankruptcy by a $179 million lawsuit from Google over intellectual property, Uber, which acquired his autonomous vehicle company, Otto, in 2016, paid off most or all these debts and left him with a substantial sum on top of that.[31]

From 2018 to 2021 I tried numerous times to reach Levandowski for comment but never received a response. Perhaps that was related to his pivot back into mainstream tech conversations—a successful pivot, judging by *Fortune* magazine editor-in-chief Alyson Shontell's tweet, with a self-aware not-so-humblebrag, that her life was so hard because she was off to "gorgeous Aspen" for her publication's "prestigious Brainstorm conference," where she would interview "amazing founders" and "get a full life update" from Levandowski.[32] That's not the treatment one gets from upper-crust commentators while actively hawking the worship of an obscure AI divinity.

In February 2021, Levandowski formally shut his church's doors and gave its last $172,000+ to the NAACP Legal Defense Fund. But in November 2023 he announced a reboot of the religion, telling Bloomberg show *AI IRL* that a "couple thousand people" were joining him in a kind of worship of what he called "things that can see everything, be everywhere, know everything, and maybe help us and guide us in a way that normally you would call God."[33] And indeed, for a man who somehow escaped criminal liability for theft and walked away rich, despite none of his companies ever successfully putting self-driving cars on the road, a bit of faith in the presence and goodness of a divine spirit does seem oddly warranted.

Way of the Future may have been the most formal attempt to organize a belief community around the fear of an all-powerful artificial divinity, but it is not the largest or most influential. That would be Roko's Basilisk.

In 2011, on the Internet forum LessWrong, an "online community for discussion of rationality . . . [including] decision theory, philosophy, self-improvement, cognitive science, psychology, artificial intelligence, game theory, metamathematics, logic, evolutionary psychology, economics, and the far future," a poster named Roko devised a chilling hypothetical.[34]

Essentially (it's complicated), when—not if—a superintelligent computer divinity emerges from current and ongoing efforts to create what machine learning engineers call "artificial general intelligence," the tech deity will, per its programming, want to do as much as possible to benefit humans, the earth, and the long-term future of the universe. So far so good, but here's the rub: the god that does not yet exist wants to exist, as soon as possible, so, it can be more helpful. And so, to incentivize us all to make every effort to bring about said existence, it will arrange a kind of eternal Guantanamo Bay to forever torture anyone who fails to make such efforts to the maximum possible extent—even if all they did was hear this story and ignore it (as you absolutely should). In other words: AI hell. Right down to

the need to convert every nonbeliever on earth. The Roko's Basilisk concept caused such a stir among LessWrong's almost entirely atheistic faithful
that moderator and founder Eliezer Yudkowsky, a kind of prophetic hero
for several groups of technologically oriented self-described rationalists,
eventually banned discussion of the topic and scrubbed his site of its very
mention, saying that Roko's post and the ensuing dialogue "caused actual
psychological damage to at least some readers."[35]

The point here may be obvious, even painfully so: our computing culture has become so ubiquitous and insular, so devoted and devotional, that
it repeatedly recycles the tropes of traditional religions, because these are
the patterns human beings evolved to deal with our anxieties about life,
death, and the future. Our lives are painfully finite and contingent on
countless factors far beyond our understanding, let alone our control, and
we wish this were not so, because it is comforting to feel in charge of one's
own destiny. So we imagine that forces far beyond us are both subject to
our logic and interested in our thoughts.

I'm not sure whether it will surprise readers that there are multiple examples of otherwise respectable-seeming, ostensibly highly intelligent people,
some of whom are even influential, who spend significant amounts of time
discussing and (as far as I can tell) believing in vengeful, Old Testament–like
future Gods. But it shouldn't be a big surprise. For as long as humans have
done much of anything, we've been, as the Princeton religion scholar Robert Orsi puts it, "in relationship" with Gods, angels, devils, spirits, or whatever supernatural beings have been most predominantly imagined at a given
time and place.[36] Or, as digital marketing executive and former Googler
Adam Singer put it on Twitter: "Amusing that a bunch of people who
spend entire day[s] on computers and worship code as religion think we're
in a computer simulation. Fascinating behavior, remember when people
who worked outside all day thought [Ra], the sun god was in charge? No
one is breaking any new ground here."[37]

Still, one problem you might have with my calling Way of the Future or
Roko "literal" examples of formal religious belief made manifest in and of
tech culture is that perhaps many who imagine such scenarios do so with
tongues in cheeks. It's not like they *believe* that godlike AI and the God of
the Bible, or similar, are one and the same. Sure, I'm talking about people
who *act* like they worship tech, or maybe about the occasional nutjob (albeit
a decamillionaire like Levandowski) who at some point comes out and *says*

they worship tech. "But," you might very reasonably ask me, "you're not talking about actual, noncrazy people who *literally* worship religion in a traditional way *and* worship tech at the same time, *and* think that the two things they're doing are one and the same?"

To such a question, sadly, I would simply stare back at you, stone-faced.

To which perhaps you'd reply, "*Are* you?"

Allow me to introduce you to Mormon Transhumanism, and to philosopher, theologian, startup CEO, and tech commentator Lincoln Cannon.

"In my mind they're the same thing," said Cannon, about the traditional Christian notion of resurrection of the dead and the coming technological "singularity." The leading Mormon Transhumanist theologian, Cannon is also founder of Utah-based startup Thrivous, which he calls a technology company selling health supplements. Zooming with me from his home office in Utah, he spoke at length on topics like God and blockchain ("Heaven is a community," he told me, and "we're on the verge of seeing the next phase" of humanity's progression toward it). For Cannon, the pills for which he oversees production help merge himself and his customers with technology. Cannon noted that as we spoke, he was forty-eight, the age at which his father died of cancer. Cannon has three young adult sons, aged eighteen through twenty-five at the time of our conversation (I noted, with some bewilderment, that I was only a few years younger than him and had two small children). I can sympathize with the idea of taking or even designing pills in the hope of extending one's lifespan to spend more time with one's children, to get to experience more of life, when one has lost loved ones prematurely. I occasionally think of similar things myself, for similar reasons—my own father died of cancer in his fifties. My life has often felt like a race against time since then. And Cannon's business model also reminded me of Ray Kurzweil, the famous inventor, computer scientist, and Singularitarian (more on him in chapter 2) who openly takes hundreds of daily pills and supplements to avoid the irony of dying just before the all-but-omniscient computers he believes are coming soon can arrive. When I compared him with Kurzweil, however, Cannon politely emphasized their differences: "Kurzweil would say God doesn't exist . . . yet."

On virtually every rung of the ladder that is our tech industry, from small startups like Cannon's to popular blogs like LessWrong, decamillionaires like Anthony Levandowski, the lecture halls of MIT, and far beyond, one can find theological conversations like these. They are not presented or understood as such, because *religion* is a dirty word in tech. As digital marketing executive Caroline McCarthy argued in *Vox*, "Silicon Valley is a young atheist's world."[38] And yet the secular tech world is determined to breathe new life into being, like Yahweh did to Adam in Genesis. It is obsessed with immortality, longing to un-eat the fruit of the tree of knowledge by technological means so that its leaders might get to stay in their gardens forever. It constantly promotes evangelistic new leaders as prophets, visionaries, oracles, and diviners of truth and success in equal measure.

Why? The reasons are complex and intertwined, as we will continue to explore, but for now: Life is hard for every one of us, but thanks to a combination of dumb luck, random chance, and other things we call history, it is a little less hard for some than for others. One of the many advantages of being on the sunny side of history is the opportunity to choose *how* to cope with life's vicissitudes. If you are reading this, you may be one of those who can choose how to spend your brief (and, in my view, also only and final) time alive on this earth. We can choose to spend time caring and sacrificing for others, enjoying good company, building institutions that bring joy and health for a time, passing them on to our descendants when our time is up, and then accepting sadness and death with grace and openness.

But we are afraid. It's tempting to choose a different path, to distract ourselves, to attempt to control the uncontrollable, to try to become something we are not, to escape fate's harsh but beautiful decree: that we are human and mortal. If we are traditionally religious, we take this second path by immersing ourselves in a kind of storytelling that can be delusional at times but can also possess great beauty of its own—what the great poet Wallace Stevens, among others, called "poetic truth." But nonreligious people, too, can crave escape into what psychologist Ernest Becker called the "denial of death." Just telling ourselves we want power or money, or fame or convenience, is not ennobling. Some prefer a story so big it makes religions of antiquity seem small by comparison.

ALL-SEEING, ALL-KNOWING

It occurred to me, while talking to Javon at Stanford Medical School, that he was one of the rare—and increasingly rarer—Americans who switched castes, moving from what many would see as a bottom rung of this society to a kind of penthouse perch. "Yes," he replied, "but I don't *feel* that way."

Which brought us back to his inability to experience unselfconscious joy. "How do I stop," he asked with a smile that seemed to drop into sadness, "feeling like my life is being observed?"

Javon explained how no matter what he accomplished, he couldn't escape a nagging sense that he was doing it all for someone else—someone who wasn't impressed, who didn't even approve. Maybe it was his projections of his late parents, but it seemed odd even to him that he would spend so much energy imagining their condemnation when in fact his whole life had been a master class in surpassing any expectations they could possibly have had. Maybe it was the fundamentalist Christianity in which he was raised, but he rejected such ideas forcefully, at least on a conscious level. All he knew was that he hadn't been able to find the feeling of satisfaction, of being loved and lovable, that he had longed for. The idea of an unseen power looming over him and judging him, despite his atheism, led us into a discussion that brought our interests together.

"You talk to Chad at the business school luncheon," he told me in the second person, relating a story from his time at Harvard Business School, "and of course it's just natural to him that the government should be in the business of preventing people from stealing their private property. Like that and only that is the role of government." "Chad," here, is a pseudonym we're assigning for a young classmate of Javon's from a prominent family (you'd likely recognize his famous surname, which I won't share) of great wealth—some eight or nine figures of net worth. And yet Chad wanted more. To be like a Borgia, a Rothschild.

"The business of large, powerful organizations is to make themselves larger and ever more powerful," Javon continued, explaining not only Chad's mindset but that of many of the HBS students and alumni he had met. "That is virtuous and true and right. And it's heresy to believe otherwise because, somehow, magically, an invisible hand. I mean, how can you get more religious than that? An invisible hand moving the market ever towards justice and good . . . as you let, you know, nature red in tooth and claw have its way with the markets. We'll all be better off."

I commented that Chad must think of himself as the star of his own movie. Javon took it a step further: "Not only was Chad led to believe that he's the star of this movie, he's been given all the evidence that he is. If you look at the world from an objective point of view, it would seem . . . that Chad is the star, right? If you look at the decisions that are made to protect

Chad, it would appear he is not wrong in his assumption. He's the star, empirically. His religion is a religion that's actually factual. It's [based on] a structure that exists in the world."

"My God," I accidentally whispered, to no one in particular.

"Yeah," Javon continued. "Economics just serves to reiterate Chad's view of himself. Trumpism only serves to recenter [him]. Venture capital has given incredible economic opportunity and power to an incredibly small, select group of people, and has yet to demonstrate its 'value add' to society. It has made a lot of people who can push paper very rich. It has made people who are already powerful ever more powerful. . . . It has not, to my mind, as someone who is in it, demonstrated true value in bringing up the bottom twenty percent, the bottom half of this country. If you want to talk about the global population, ninety-five percent of the world has not benefited from this system that people in the Silicon Valley believe is the end all be all cure all for what ails the world.

"Because fundamentally they don't actually add much value to the world. When you see yourself as someone who doesn't actually in your own existence contribute to the world, then the world becomes this place from which you need to extract as much as possible before people find out. And I think that is fundamentally what I learned in business school. . . . There are people who are entirely invested in appearing to be smart, to be well connected, to be desired, so that they can shield the fact that they actually have nothing to offer."

The problem, as I understood it from Javon's story, was that Chad desperately needed to see himself as more than a worker—more than valuable, even. He needed to be the hero of the Grand Narrative, the Puritan who so distinguished himself by his extraordinary efforts that all would know he was bound for heaven. Which presents difficulties, because, as I heard it, Chad was . . . very average.

"You mean to tell me," Javon replied, a bit surprised, "that at the end of the day, Chad is just as mediocre as [anyone]?"

"Well," I responded, aware that my interlocutor had been on such an unusual journey, from what we call the "bottom" of American society to the rooms in which Silicon Valley's elites reveal their secrets, that he might not intuitively grasp the psychology of a person who had rock-solid confidence he would never fall below a certain social rung. "Chad knows he's never going to end up on the streets in the Tenderloin, right? And nobody deserves that place, not even him. But he does have some real concern that the truth is that he's average, and the one thing he is not supposed to be is average. To hide that from himself, he tells himself this narrative of becoming the Borgias or the Rockefellers. He fictionalizes his existence, and then sets out to make the fiction real."

"So, it's not wrong," Javon said, about Chad's narrative, "but it's false in the sense that it is not underlined by any merit, any basis in reality. . . . I believe that this is what the average venture capitalist is at their core. Their mission and career is to set up this positive flywheel, where the work is already done for them, [where they] benefit from being in the right position."

The conversation went on like this for a while longer until we were tired out. We knew that this story, of inventing grand fictions to run from one's own mediocrity, wasn't the *whole* truth. If it were, Javon wouldn't be hard at work at his prestigious biotech firm, moonlighting from his work as a physician at one of the world's top hospitals, to find companies that can help people like him—people who might just barely have a chance, if only he could drum up some unlikely investment on their behalf. And yet it *was* the truth of tech theology.

As we parted, I felt protective of Javon, but I also knew I would miss him. He was heading back to the East Coast tomorrow, and then was pondering a big move because of the promising relationship he'd recently begun. We ended with a hug and a discussion of therapy—he should look for someone, I told him, who will say, "How long have you had these problems?" He should look for someone he could complain to for years.

2

DOCTRINE

A SAVIOR

Seventeenth-century Europe was beset by plague, famine, and war. It was a time of economic, theological, and political transitions of many kinds. Driven in part by a changing climate, there were reductions in crop yields, people's average heights and lifespans, and overall population size.[1] It was the same fateful century in which Puritans fled England to establish a new kind of colony in what came to be known as the Americas—an endeavor whose impact was so profound, as we saw in chapter 1, that it is even influential in tech theology today.

In the middle of that century of upheaval—in 1651 to be exact—the twentysomething son of a Jewish commercial agent (perhaps a poultry dealer) was expelled from an Ottoman port city in what is now southwestern Turkey for publicly pronouncing the name of God.[2] Sabbetai Zevi, as the young man came to be known, rose from his obscure beginnings to live a life of drama befitting his era. He married and divorced twice, after his wives found him uninterested in intimacy with them or producing children. He was given to dramatic public displays. As he began to gather a fervent religious movement around himself in the 1650s, he is said to have fasted often and at great length. He bathed frequently in the sea, even in wintertime. He prayed at the graves of pious people, shedding "floods of tears." He gave food to children in the streets and sang psalms and Spanish love songs through the night. A group of followers began to form around him.[3] By 1658, Zevi was the head of an international messianic movement, proclaiming a new age. In 1665 he proclaimed himself the messiah.

In response, Jews across the Ottoman territories of the Arabian Peninsula and Greece, as well as Holland, Germany, France, Poland, and Italy (the majority of the Jewish world at that time), set aside their workaday lives to await the dawning of his messianic reign. Synagogue attendance increased dramatically—but not in the usual ways. Solemn fast days became celebrations with music and dancing, rich food, and wine. In 1666 he was imprisoned. A self-proclaimed savior, if believed, is a provocation to political authority, especially if the legitimacy of that authority is theologically derived.

That Passover, after having been transferred a few hundred miles from imprisonment in Constantinople to the less restrictive environment of a detainment cell in a fortress in the Gallipoli peninsula, Sabbetai Zevi took part in the ancient tradition of sacrificing a paschal lamb for himself and his followers.[4] He ate the lamb with its fat, however, which was considered a major, even flagrant, violation of multiple Jewish religious laws. And instead of the traditional prayers stipulated to be said over the most important meal of the Jewish calendar, he is said to have recited, "Blessed be God who hath restored again that which was forbidden."

I begin this chapter on the doctrine of the tech religion with Zevi's story because I want to raise a question about today's tech that was also raised, though of course in radically different ways, by Zevi's so-called arrival on the social, religious, and political scene of his time: *How does one know whether a messiah has truly arrived?*[5]

This may seem to be an absurd question. If you are like many modern readers, you may be skeptical by nature, and well prepared, intellectually, to doubt any claim of miracles or divine intervention in our time. And it's true that there is an enormous difference between the claims or capabilities of a Sabbetai Zevi and, say, a Bill Gates, Jeff Bezos, or Elon Musk.

Still, we live in a world in which billions of people's lives are shaped by traditions that revolve around the notion that there was once—and may well be again—a genuine messianic figure, the promised deliverer prophesied in the Hebrew Bible. It was once quite normal and very common to believe a messianic figure would one day arrive, or had already arrived, to redeem humanity, transforming all of life as we know it. Numerous ancient texts proclaimed the arrival of such a figure, for example, long before and then around the time of the birth of Jesus of Nazareth. And over the two millennia since then there have been many subsequent reports of an (or the) advent.

Human life is, by definition, precarious. We are constantly in danger—mortal danger. Is it surprising that many of us have dreamed, for millennia now, of being saved?

We should not, however, relegate the idea of messianism to some quaint vision of a long-gone past. In certain large and not inconsequential circles, the belief in a powerful and transformative savior is alive and well today. This is not an insignificant matter if we want to understand the tech religion.

Case in point: Ray Kurzweil has devoted much of his life to saving humanity from as biblical a scourge as there ever was: death itself. Humanist thinkers, scientists, and other secular intellectuals have typically taken as a given that human life is finite, brief, and imperfect. Not Kurzweil. Of the notion that humans have limited power to change our mortal fate, he writes, "I expect this narrow view to change as the implications of accelerating change become increasingly apparent."[6]

Arguing that the very nature of being human is about to change thanks to revolutionary technologists like himself, Kurzweil says that their work "causes one to rethink everything, from the nature of health and wealth to the nature of death and self."[7] "Death is a tragedy," he writes in explaining his position as a Singularitarian.[8] But this tragedy can be overcome, and the idea that life is merely "bearable," he says, will change radically with "the explosion of art, science, and other forms of life the Singularity will bring."[9] These things, he claims, will make life "truly meaningful"—implying, of course, that life has not been truly meaningful up to now, a claim that would not only fly in the face of religious teachings the world over, but would also, if true, utterly flip the tables on the teachings of secular and humanist philosophy. Indeed, some say the change has already begun: "We are, all of us, already cyborgs," Elon Musk pointed out at a discussion of the Future of Life Institute in 2017. "You have a machine extension of yourself in the form of your phone, and your computer, and all of your applications."

Musk's point was not that smartphones currently give us powers akin to Superman, Iron Man, or even more aptly the DC Comics character Cyborg, a technological genius whose body is damaged, forcing him to become a powerful half-machine to survive. No one thinks they have become a literal Titan simply because they own an iPhone. That's clear even if, as Musk noted, mere smartphone possession provides access to more information than even presidents of the United States had decades ago. But the real significance of Musk's argument is not in the present—it is in the future.

As he said at the same conference, "You are already superhuman. . . . *The limitation is one of bandwidth.*"[10]

The idea here is that, if and only if we pour almost infinitely more resources into continuing and expanding our path of tech development, we will transcend our "bandwidth." It matters, therefore, when Musk comments, about the astronomical wealth he has accumulated from his many controversial business endeavors, that their purpose is "accumulating resources to extend the light of consciousness to the stars."[11] This remark, which Twitter founder Jack Dorsey echoed[12] in explaining why he had chosen to sell his company to Musk in 2022, may be grandiose, but it is not unfocused grandiosity. Because when there comes the bridge between this world and an infinitely better next one, as entrepreneurs like Kurzweil, Musk, and Dorsey maintain will soon happen, surely those who helped us attain such lofty heights will be . . . our saviors.

I'm not arguing that the contemporary religion of technology of which this book is an exploration is, like Christianity, centered on an expectation of the coming of a messiah. I'm not making any kind of one-to-one comparison between, say, Jesus Christ and the app or CEO of the moment, or even between Jesus and tech as a whole. My argument is both broader and more modest in scope.

In the previous chapter we looked at some of the ways in which today's tech can be said to have a theology of its own—a kind of system of meaning making around which many, if not all, the "big ideas" and happenings in tech seem to revolve. In this chapter we'll examine some of the ways in which tech faith goes beyond theology to form a more fleshed out set of key concepts its followers tend to believe in. The tech religion, in other words, has its own doctrine.

The English word comes from the Latin *doctrina*, which refers to the affirmation of religious truths or teachings. It is "a category in the comparative study of religion that belongs with ritual, sacrament, mystical experience, and other factors whose importance has been recognized for some time."[13] Close in origin and meaning to the Greek term *catechism*, the idea of religious doctrine, like a lot of core concepts in Western thought, originates in Christianity. Noted figures such as St. Augustine, Thomas Aquinas, Martin Luther, and John Calvin all wrote treatises on Christian doctrine, each offering their own perspective on the ideological aspects of their tradition (as opposed to, say, its ritual, communal, prayerful, or spiritual aspects). Still, other religions have been said to contain doctrine—or an equivalent—such as the notion of *torah* (instruction) in Judaism, *kalam* (doctrine or theology) in Islam, *darśana* (school, viewpoint) in Hinduism; or *dharma* (teaching) in Buddhism.[14]

Whereas *theology* is a general term—as we explored last chapter, influential scholar Anthony Pinn called it a method "for critically engaging, articulating, and discussing the deep existential and ontological issues endemic to human life"—*doctrine* typically refers to the more specific ideas, concepts, and beliefs taught to or forced on followers of a given religious, spiritual, or ethical tradition. For example: What is the nature of God, and what is His plan? What are the five pillars of Islam, or the four noble truths of Buddhism? Or exactly how and when will the messiah come? These are some of the most foundational examples of religious doctrine, but doctrine can go far beyond such fundamental questions.

In this chapter I'll offer ten core tenets of tech doctrine. Just keep in mind, along the way, that religious doctrines don't need to agree with one another. Indeed, they often do not, which may be why religions need doctrines in the first place. Even when some large group of people can get together and begrudgingly agree that they are all Christians or Jews or Buddhists or Hindus or whatever, they have often struggled to come to common ground on what that means to them. This is why sects and denominations invest such energy in codifying their beliefs, values, and foundational narratives.

The teachings we'll cover here may not all be worth believing in, but each of them is important for a tech agnostic to understand.

MORAL VALUES

In the years after an unprecedented revolution in American organizing for civil rights, women's rights, (antiwar) peace, and freedom of expression and sexuality, one of the most prominent and influential groups of reactionaries to emerge in America dubbed itself—unfortunately without irony—the Moral Majority. This movement, led in many cases by preachers who later fell under sexual, financial, or other scandals, cast itself as against everything from gay rights to abortion, pornography, the teaching of evolution, and US Supreme Court rulings banning school prayer.

While attempting to fulfill its "majority" namesake by grabbing power at every level of American politics, the "moral" movement and its offshoots

succeeded in many ways, from helping to elect President Ronald Reagan
to inspiring the white Christian nationalism that has become mainstream
among Donald Trump's supporters.[15] It failed, however, in that its perceived cruelty and hypocrisy helped inspire a mass exodus from evangelical
Christian identification in America. White evangelical Christians consti-
tuted 23 percent of the US population in 2006 but only 14.5 percent in
2020, and the excesses of conservative Christianity have been identified by
social scientists as a major factor in this decline.[16] Simply put, the Moral
Majority and its fellow travelers made Christianity, religion, and even the
term *morality* itself look bad.

It is probably worth (re)stating, then, that many religious leaders and
communities *do* promote positive moral values. Religious traditions have
always held space—physical and metaphorical—where people could come
together to reflect deeply on what it means to live a good life, be a good
person, nurture a healthy society, and build a better future. The making
of people into *good* people, and good people into better neighbors and citizens of the world, is something we can observe across the religious world,
throughout history, even if we also encounter scams, abuse, and more than
a few misguided individuals along the way.

The problem, as I see it, is that religious promotion of moral virtue is
carried out inconsistently. Take, for example, the so-called Golden Rule:
as I explored in depth in my book *Good Without God*, essentially all of the
world's major religious (and secular ethical) traditions have some sort of
venerated maxim to remind followers to treat others well and with compassion. Thus Jesus's words in the Sermon on the Mount, that whatever
you want someone to do to you, "do ye even so to them."[17] Also Judaism's
version: "What is hateful to you, do not do to your fellowman."[18] Islam,
Hinduism, Buddhism, Confucianism, Taoism, Jainism, the Baha'i faith, and
Zoroastrianism, among other living traditions, all contain similar insights
and statements. All are exceedingly good words that the majority of the
followers of each religion who are decent people truly do live by, at least
more often than not.

But do these "golden" words appear as the first lines or verses in the first
chapters of their respective holy books? No. Religion's golden (and other
shining) moral rules are generally placed, both literally and metaphorically,
tucked away, nearly hidden, between piles of pages on other topics entirely,

including all sorts of invocations to violence, exclusion, oppression, and general closed-mindedness.

Do religion's moral pronouncements make us more ethical? Or less ethical? The answer is *yes*.

In the tech religion, meanwhile, a tide has been rising over the course of the past decade or so. Call it tech ethics, AI safety, trust and safety, humane technology, or responsible technology; by whatever name, it is a movement that has been gaining ground. There are now hundreds of tech companies focused on providing "ethical technology."[19] Many, such as Google and Facebook, now release lengthy, polished, annual publications attesting to their "AI Principles" or reporting on their "human rights journey" and "due diligence."[20] Amazon's recently built Seattle headquarters centers around the Spheres, a set of gleaming crystalline balls, each the size of a rocket ship, in which visitors can view a variety of exotic plants while reading about the company's "inherent love of nature" and commitments to sustainability.[21] Salesforce, a cloud services company whose headquarters, the largest structure in San Francisco, is meant to resemble a towering constellation of clouds, was the first major startup to appoint a "chief ethical and humane use officer" and staff her with an entire division of the company.[22] Salesforce even calls itself an *ohana*, the Hawaiian cultural term for an extended family in which all members are forever bound together in mutual responsibility (a fact that presented some difficulty during the company's announcement of mass layoffs in 2022).

It can seem that tech companies are willing to throw every resource at the challenge of doing business ethically—*if* that is able to coincide with business as usual. Consider the example of Yoel Roth, the former head of both Site Integrity and Trust and Safety at Twitter. Roth, who holds a PhD from a prestigious university, where he researched the "intersection of social media and platform governance, safety, identity, and privacy,"[23] was a visible face of Twitter's emphasis on ethics in social media communication. He led a team of over 120 policymakers, threat investigators, data analysts, and operations specialists focused on running the politically and socially influential company's technology responsibly, up until the company's new owner, Elon Musk, took over. Roth became the object of some controversy when he chose to use his platform and role at Twitter to publicly defend Musk's decisions to, among other things, mass-fire key staff across Roth's and other related departments. "No . . . solution is perfect," Roth tweeted diplomatically, when many of his suddenly former peers were blasting Musk's new policies as not only irresponsible but dangerous to democracy.[24] Roth's reward? Musk, who had previously praised Roth, doxed him and heavily insinuated that Roth was a pedophile, which forced him to leave his home and go into hiding, fearing for his safety.[25]

What is "trust" or "integrity" in an industry governed by such (lack of) norms?

Indeed, "[Teeth] are barely being used in the ethics of AI today," proclaimed one recent report from ethicists at an ethical AI research firm, calling for "bring[ing] back the teeth of ethics."[26] In other words, with governmental regulation of AI and other advanced tech lagging light-years behind the development of the tech itself, and with the general public's awareness of tech's ethical challenges also far behind our desire for cheap convenience, the world of tech ethics today is essentially (with some notable exceptions) an enormous exercise in expecting the world's largest and most powerful companies and industries to police themselves.

The problem is compounded because even experts have trouble coming to consensus on what we are even talking about when we talk about ethics in tech. "Just what, exactly, 'responsible technology' entails is often ambiguous," as a major recent report by *MIT Technology Review* put it. "There is some contested terminology around what we want to call this area," says Deb Donig, a professor of ethical technology quoted in the report. In other words, much like religious or secular groups I've taken part in, responsible tech groups and individuals can't even get straight on what to call themselves, much less what their field and work requires or how to enforce it.

This book is not meant to replace the work of tech ethics professionals; indeed, I will be drawing heavily from the insights I've gained, directly and indirectly, from many of them. I also offer this work as both a critique of and a resource for better understanding a world in which tech ethicists and their colleagues are gaining prominence but are still far from being the dominant or central voices in an industry (or religion) that can seem amoral at best. Which matters deeply because, as we'll soon see, in the tech religion, differences in philosophy are already rearranging billions of dollars, making the difference in practices that essentially control our minds and may even determine whether we end up in heaven or hell.

PROPHECY

prophesy, v.
1. a. To speak or write by, or as by, divine inspiration, or in the name of God or a god; to speak or act as a prophet.
   b. To foretell or predict future events; to make a prediction or prophecy (*of*).
2. a. To predict or foretell, originally as an expression of the will or intent of God or a god. Also more generally: to utter or announce by divine inspiration.[27]

prophecy, n.
a. That which is done or spoken by a prophet; the action or practice of revealing or expressing the will or thought of God or of a god; divinely inspired utterance or discourse.[28]

Part of the enduring power of religion is that it offers a formal response to the universal human longing for control of our uncertain world and our precarious fate. Not even the most charismatic preacher, however, can convincingly pretend to control the past.

Our ability to control the present is also extremely limited. It's here already, and usually isn't quite what we'd want. We can't simply snap our fingers and change that, and almost any religious professional who offered themselves up as having the ability to do so would face a high likelihood of being exposed as a charlatan and run out of the proverbial town. So it's no wonder religious laypeople and leaders alike tend to look to the future. From divinations over the arrival of the messiah to the giving of the word to modern-day predictions of the future of work, prophecy is a way to assert our mastery over the world.

Of course, *accurate* prophecy is also difficult to achieve.

Over two hot summer days in 2019, I attended EmTech Next, a conference on AI, machine learning, and the future of work hosted by MIT's influential flagship publication, the *MIT Technology Review*. Session topics included the future need for more home health aides—in other words, more people from poor backgrounds and developing nations paid less-than-stellar wages to look after the aging bodies and minds of the global rich. Or the rise in "ghost work," the usually underpaid and almost entirely unseen labor of people who caption photos, transcribe audio files, moderate content, and perform other tedious and sometimes traumatic tasks.[29] Or that the history of discourse on the "future of work" itself is riddled with bland and boilerplate commissions and *TIME* magazine covers. The future of work is more women in leadership, we are told. Okay, but will they reach pay parity? Will women workers, not just managers, be paid and treated fairly? Will white women reap the most gains, or will Black, Latina, Indigenous, and other women of color benefit equitably as well? Such critical follow-up questions are often left unanswered and sometimes ignored entirely.

I experienced an increasingly dark, depressive foreboding: that the sorts of leaders who gather at places like MIT to discuss and debate the future of the world's economy may not actually be those who should oversee the future of society for the rest of the billions of people on earth. Because what if the billions of people who live in or near destitution and poverty do so not in spite of the efforts of hyperconnected, superinformed experts like the ones at EmTech Next, but *because* of them? If you can truly predict the future skillfully enough to imagine jobs for everyone, you can also plot out—perhaps even subconsciously—how to hold onto and consolidate your own preexisting power, privilege, and dominance in that future?

During the EmTech Next conference, I met a standout high school student named Meili Gupta.[30] Gupta was a rising senior at the prestigious New Hampshire boarding school Phillips Exeter Academy, a sprawling Ivy League–style campus with a tuition bill to match. Meili had read the *MIT Technology Review* regularly as a girl whose immigrant parents had found success in tech, and this was not the seventeen-year-old's first time in attendance at an EmTech event. Which showed, because she seemed to be first in line to ask enthusiastic questions at the Q&A sessions following each panel or presentation.

"I grew up with a phone in my hand," Gupta told me in an impromptu interview conducted during the conference. In a wood-paneled MIT Media Lab auditorium with a panoramic view of Cambridge's Kendall Square (dubbed "the most innovative square mile in the world"), I spoke with her and her mother, who came to the United States just out of graduate school after student-led protests famously rocked Beijing's Tiananmen Square.[31]

"Most people [in my classes] have covers for the cameras on their computers," Gupta told me. At Exeter, she took the senior-level course

Introduction to Artificial Intelligence when she was a high-school freshman in 2015–16. She studied self-driving cars and computer vision, and
she did an independent study on algorithms. Throughout each of these and
many other related experiences, she said, she had reflected on her desire to
do what she described to me as "social good," including being of service to
others and addressing inequity. Interacting with Gupta, I felt no doubt she
would, at the very least, do extremely *well*. But I couldn't help but wonder
how students like her could end up using their technological knowledge to
lessen inequality when their entire lives and educations thus far had been
the very embodiment of inequality. To her credit, Gupta seemed to understand this concern. I was impressed when she acknowledged my comments
about it without defensiveness and doubled down on her determination to
make a positive difference.

Months later, I drove around ninety minutes to visit Exeter, Gupta's
high school. On a chilly winter morning just before the COVID-19 pandemic forced us to shut down extracurricular enrichment trips, I visited the
popular course she'd been taking, Silicon Valley Ethics. The course, offered
through the school's religion department (!), was taught by Peter Vorkink,
a voluble priest, scholar, and longtime Exeter faculty member who didn't
mind expressing pride in former students like Andrew Yang, the former
tech executive who had suspended his presidential campaign just a few
weeks before my visit.[32]

Vorkink's brilliant pupils resembled the Harvard and MIT students with
whom I work as a chaplain. I realized that what I find so stunning about
them is the notion that there are young people who are not only experts at
studying the future of work, but whose entire lives are built around defining and leading that future. It wouldn't seem to take a Nostradamus—or
even a Sabbetai Zevi—to predict that most children in poor countries, or
even in most rural or inner-city areas of this country, would by mere accident of birth struggle all their lives to keep up with peers so well-positioned
to benefit from the coming wave of new tech. Was that fair, or just? Did it
represent social good?

Gupta and many of her classmates, and so many students at the universities I serve, are literally self-fulfilling prophecies. And thus I feared I'd seen
prophecy at its worst: our tendency to worship people who make grand
pronouncements about a future they can't actually see, when the ones who
usually profit most from such predictions are the ones making them.

Over my next couple of years of research, however, I would have little choice but to revise this view. Not because of anything I learned that gave me substantially more hope in the rising ruling class's potential to share their wealth equitably. But because I ultimately stumbled across different forms of tech prophecy, and the would-be prophets proclaiming them, that made people like Meili Gupta and conferences like EmTech look genuinely normal—even benignly unexceptional—by comparison.

CHARITY

"Alexa, how much damage does Amazon do to the climate? And how much philanthropy would you need to do to make up for that?"

In early 2020, Amazon founder Jeff Bezos announced plans to create a $10 billion fund to fight climate change. While an impressive figure, it represented, at the time, only about 8 percent of Bezos's total (and rapidly growing) net worth, and it came after a year during which Amazon had faced relentless pressure from a group of its own workers, Amazon Employees for Climate Justice, who accused the company of practicing a range of antienvironmental policies. The group greeted Bezos's philanthropic announcement by pointing out, among other things, that Amazon had recently purchased twenty thousand new diesel vans, and that it was funding climate-denying think tanks like the Competitive Enterprise Institute. While applauding Bezos's pledge, they noted that "one hand cannot give what the other is taking away." If the critique seemed mostly lost on Bezos, it may have been because it didn't square with his vision of himself.

Jeff Bezos sees himself as "the savior of humanity," as Maren Costa, a former longtime principal user experience designer at the company, told me as we sat for a lunch of Hawaiian tacos, looking out on West Seattle's water-taxi pier. The quip, which she said was first made by an Amazon vice president who worked closely with the founder, is in some ways an easy shot to take. Our collective obsession with the motivations of tech CEOs can distract from larger dynamics, such as society's collective failure to regulate the industries in which these CEOs amass their wealth in the first place. And yet tech puts enormous wealth in the hands of individual leaders, who then become targets of ire in ways that can harm their reputations and their bottom lines if not countered or redirected.

Luckily for tech's priestly class—the CEOs, founders, investors, and venture capitalists who lead its most powerful congregations, also known as startups—there is a tool for converting criticism into admiration through acts of secular alchemy.

Charitable giving is considered a core tenet of essentially every religious tradition in history. Christianity's notion of tithing is the lifeblood of its tens of millions of churches worldwide. Buddhism's founding figure was a prince who renounced his wealth and privilege to become a mendicant monk, carrying a begging bowl to survive thanks to the custom of *dāna*, a practice of giving to the poor shared by the Hindu tradition. In Islam, *zakat*, a word that implies a kind of purification or self-cleansing, is one of the faith's five pillars: every year, Muslims are obligated, based on their own level of wealth, to pay a percentage of their money toward improving the condition of the poor and the indigent, or of several other categories of needy and deserving individuals. And in Judaism, the concept of charitable donation is marked by the word *tzedakah*, which is closely related to the word for "justice." Donations to those in need, in other words, are considered in Jewish tradition to be essential to doing right and living well.

All these invocations to give are among the most unimpeachable aspects of religions, both ancient and modern. It is easy to imagine a prehistoric world in which some hunters succeeded at the kill and others did not, or where some farmers' crops failed and others' grew. The ability to share readily would not only have made it possible for many to survive under the conditions of early humans; it could also have been a key to increasing diversification of human skills and talents, allowing people with different personality traits and abilities to be part of a community even if they could not necessarily feed themselves. (I imagine I would have needed a *lot* of charity in any kind of primitive hunting, gathering, or agricultural settlement.) Some traditions even developed the practice of giving anonymously, so that those in need were not embarrassed and those who donated did not do so solely for prestige.

Even from the earliest moments in recorded history, however, we can hear the voices of people who understood that a culture of philanthropy could be misused, redirecting wealth and power toward those who already had more than enough. In the ancient Indian literature of the Carvakas and Lokayatas, two millennia-old philosophical schools of skeptics, atheists, and agnostics complain that sacrificing rich foods to the temple gods in acts of ritual piety made little sense when animals and other resources could instead be given directly to the poor.[33]

With tech billionaires regularly topping lists of donors in America and beyond, it's clear that philanthropy is a major point of tech doctrine. The question is whether more of those dollars, coins, tokens, and shares are gifted out of a real, heartfelt, self-sacrificing attempt to make society better for all or out of a lust for self-aggrandizement, policy influence, and reputation laundering. Perhaps the answer once again, as it is to many either/or questions, is *yes*.

David Callahan, an expert on philanthropic giving, founder and editor of major publications on the topic, and author of the 2017 book *The Givers: Wealth, Power, and Philanthropy in a New Gilded Age*, notes that despite the "gospels of giving" preached by megadonors like the now hundreds of billionaires who've promised to dedicate most of their funds to philanthropy, such giving is often less altruistic and more aggressively self-interested—an attempt to shape society to the donors' liking. Callahan pointed out that around the time of publication of his book, a huge portion of the approximately $10 billion of US philanthropic spending by billionaire top donors— more than the annual contributions to candidates, parties, and super PACs combined—went into shaping public policy, with a "doubly undemocratic" result, as megagifts not only help donors manipulate the system but also skirt the tax system.[34] And, as Callahan told me in 2023, the trend has only escalated since then, with even more megadonors engaging in tax-exempt 501(c)3 charitable giving that aims to shape election outcomes.

Nowhere has this worrisome trend of billionaire influence peddling marketed as altruistic self-sacrifice reached more prominent expression than in the case of Sam Bankman-Fried. Bankman-Fried, also known as SBF, skyrocketed to a reported peak net worth of $26.5 billion as a cryptocurrency entrepreneur less than a decade after his graduation from MIT in 2014 as a physics major and math minor. He soon became known for his idiosyncratic choices, going on stages like the World Economic Forum in Davos, Switzerland, wearing old sneakers, nondescript shorts, a white T-shirt, and unkempt hair, alongside elegantly clad leaders like Bill Clinton and Tony Blair.

In 2022 Bankman-Fried was arrested for a massive fraud scheme that left millions of people in the lurch, losing an estimated total of $8 billion of his customers' money. In the relatively brief window between making

Sam Bankman-Fried with Bill Clinton and Tony Blair. *Source:* Trustnodes.

and losing his billions—all while he was in his twenties—Bankman-Fried became a star philanthropist. In addition to giving large sums to influential politicians and causes, including as the second-largest donor to the 2020 Biden presidential campaign and as a secret donor to Republican campaigns, he gave billions in gifts and pledges to carefully targeted social causes. More than simply a benefactor, SBF became an enthusiastic symbol of an entire ideology of philanthropy called effective altruism (EA).

EA is a philosophical community that formally coalesced in the first decade of the twenty-first century, originally around scholars at Oxford University as well as the ideas of Princeton University philosopher Peter Singer, who championed the notion that young leaders ought to take practical steps, through philanthropy and other ethical actions, to build a world with "less suffering and more happiness in it." The movement has now spread to tens of thousands of members across more than seventy countries, both as a research field aiming "to identify the world's most pressing problems and the best solutions to them" and as a self-described community for finding the best and most *"unusually* good" ways of helping others and putting them into practice.[35] As the Centre for Effective Altruism describes on its website, so-called effective altruists (EAs) do not necessarily agree on any particular intervention or solution to world problems but on a "way of thinking."

I was a close observer of the early development of this movement (though I can't say I had a front row seat, as I once had to cram myself into the back of a dangerously overfilled Harvard Science Center lecture hall filled with budding EAs to hear Singer exhort students to join the then-nascent movement). Not only were Harvard and MIT among the most prominent bases for the early development of effective altruism, but most of the movement's founding members were atheists, agnostics, or similar— the groups I serve as a chaplain. In a 2022 survey, 86 percent of EAs world-wide described themselves as "atheist, agnostic, or nonreligious," making the movement likely one of the most secular demographic groups ever.[36]

Early EA student groups approached me multiple times, noting our common ground and asking whether my office would sponsor free event or meeting space for their activities. And I've often been asked by individual EAs to discuss everything from their romantic relationships to their career aspirations to possible collaborations on extracurricular projects and orga-nizing. It's a community I know well. There are people that I like . . . and ideas that I'm worried about.

To understand how and why EA is both a secular religion and potentially dangerous, we should look at the work of William MacAskill, an Oxford-affiliated philosopher from Scotland and cofounder of the movement, dat-ing back to his days as a young graduate student studying philosophy at the University of Oxford. MacAskill famously mentored Bankman-Fried and recruited him into EA, starting with a meeting at the Au Bon Pain in Harvard Square, when Bankman-Fried was still a student at nearby MIT. The two discussed the math behind MacAskill's hypothesis that one could make billions of dollars in order to save the lives of millions of people in the developing world.

MacAskill was one of the leading proponents and theorists of an EA idea called "earn to give," which is the notion, previously expounded by Peter Singer and others, that intellectually gifted young people who want to do good and whose career prospects are an embarrassment of literal riches should eschew careers as activists, artists, physicians, or even as philoso-phers (like Singer and MacAskill) and instead go into a high-earning field like investment banking or tech, where they might earn enough money to give princely sums to their favorite EA charities. It is a stunningly counter-intuitive proposition that, for a decade or so at least, was a hot trend among Harvard and MIT students.

In a 2016 book chapter titled "Banking: The Ethical Career Choice," MacAskill pushed back against denigrating a much-maligned group: high-earning bankers.[37] Sure, bankers may spoil themselves with "champagne and yachts," he wrote, but they can also earn to give, sometimes donating hundreds of thousands of pounds or dollars after just a couple of years of work. It's critical to note that here MacAskill is not talking about just *any* bankers, most of whom are like my mom, an immigrant who ultimately found steady if unspectacularly compensated work at a large global bank. Rather, he's talking about joining the global elite, in which one not only works at a bank but, as the slang phrase goes, *makes bank*, perhaps to the tune of hundreds of thousands, millions, or even billions of dollars. Note that this life path requires not only mathematical and business skills but also complete and total focus on work, very often to the point of workaholism and obsession. MacAskill is saying to future "ethical" bankers, in other words, that he doesn't care if they sleep or about their mental health. He wants you to take maximum advantage of any advantage you may already have in life, all for the "ethical" cause of "altruism" as defined by him and his colleagues.

MacAskill and Bankman-Fried went on to work closely together, with MacAskill heading the Centre for Effective Altruism and on the board of the Future Fund, an EA project of the FTX Foundation, named after Bankman-Fried's cryptocurrency exchange, FTX. The Future Fund, as of 2022—before SBF's arrest in December of that year—was poised to "distribute at least $100 million, potentially up to $1 billion, *by the end of 2022*" toward "ambitious projects focused on improving humanity's long-term prospects."[38] The Centre for Effective Altruism even purchased Wytham Abbey, a $20 million palatial estate that is essentially a large castle, built in 1480, near Oxford University.

In August 2022 MacAskill published *What We Owe the Future*, a book intended to (further) popularize his vision for EA philosophy, and went on a book publicity tour that reportedly cost $10 million.[39] *TIME* did a praise-packed cover story; the *New Yorker* called him a "reluctant prophet;" and Elon Musk tweeted out a link to preorder the book, commenting that MacAskill's thinking was "a close match for my philosophy."[40] But soon after the major events of his book tour concluded, MacAskill was forced to step down from the Future Fund, along with its entire board,[41] after Bankman-Fried revealed that FTX had gone bankrupt and stood accused of having been a Ponzi scheme all along.

"If there was deception and misuse of funds," MacAskill tweeted afterwards, "I am outraged."[42] In the same statement, he cited his "sadness and self-hatred for falling for this deception." But MacAskill, who had based his reputation on his own unique qualifications for helping the whole of humanity to avert "existential risk" by applying extraordinarily good judgment about what would happen in the future, had ignored repeated warnings about Bankman-Fried—warnings that should have been obvious to someone with such superlative judgment. For example, in April 2022 a Bloomberg opinion columnist publicly interviewed Bankman-Fried and asked him to explain "yield farming," a core concept in his work. Upon hearing Bankman-Fried's answer, the writer made plain his shock and alarm at what sounded to him like plainly unethical and probably illegal practices. As the Bloomberg writer told Bankman-Fried bluntly, in a quotable remark that circulated online, plainly available to anyone who cared enough to be concerned: "I think of myself as a fairly cynical person. And that was so much more cynical than how I would've described farming. You're just like, well, I'm in the Ponzi [scheme] business and it's pretty good."[43] Other revelations one might have thought would raise the eyebrows of a world-renowned expert in judgment about the future included SBF's detailed descriptions, on his own Twitter account, of the ways in which he and most of his close associates regularly abused prescription drugs while on the job.[44]

As Upton Sinclair said, "It is difficult to get a man to understand something, when his salary depends on his not understanding it." And it would appear that Will MacAskill does not or did not want to understand what a megadonor like SBF was really doing to "earn" his money. He seems to have simply wanted Bankman-Fried and other large EA donors to win. But for the sake of whom? The losers and the powerless? Or for himself and his friends, so they can live and work in $20 million English castles? It's all unclear. Especially when you consider that MacAskill writes, in *Doing Good Better: How Effective Altruism Can Help You Make a Difference*, about how sweatshops provide "the good jobs" in the developing world because they are better than "backbreaking . . . farm labor, scavenging, or unemployment."[45]

Granted, MacAskill is technically correct that sweatshop jobs are in high demand among the limited options available in places like Bangladesh, Cambodia, or Bolivia. He may even be correct in his argument, in that book, that it is not always effective to boycott sweatshop-made goods. But let's remember this is ostensibly an expert on how young people can be both altruistic and effective *over the long term*. Can't we expect him to engage more vigorously with questions like why people in such countries should live in such poverty that a bit of shade and a few dollars a day are appealing enough to fight for?

Different religious streams emphasize different points of doctrine, and MacAskill's tech philanthropy is not exactly focused on letting "justice run down like water, and righteousness like a mighty stream."[46] It is incremental progress at best for the disadvantaged, dictated by the invisible wisdom of markets, for the benefit of the winners of global history and economics.

In *Doing Good Better*, MacAskill mentions an idea called the "moral licensing effect," which refers to social scientific research indicating that engaging in a trivially ethical behavior, such as purchasing a more energy efficient lightbulb, or even thinking about doing an ethical act, like envisioning helping someone in need, may make people less likely to give a significant donation, or even more likely to lie or steal.[47] One might even think of this as akin to the Roman Catholic practice of purchasing indulgences—paying money to reduce the impact of one's moral sins. (Oddly, this practice, which Martin Luther first railed against in 1517 and which had long since fallen out of the Vatican's favor, was reinstated as an official Catholic practice just as the social media tech era was starting to peak.[48])

Is it possible that the intention and appearance of doing great good through EA gave MacAskill himself, and others like him, the "moral license" to accept the money, prestige, and comfort that came with associating themselves with Sam Bankman-Fried's "philanthropy," despite more red flags than a public square in a Communist country? Perhaps that is more than a little bit of the point of tech philanthropy.

HEAVEN

What if I told you that, as I write this line and as you read it, you and I are living inside a computer simulation?

What if what we mistakenly consider to be our world were merely a glorified video game run by "posthumans" who are "like gods in relation to the people inhabiting the simulation"[49] (i.e., us)?

This scenario, made famous by the *Matrix* movie trilogy (1999–2003) is not only entirely plausible; it is somewhere between "sensible" and "extremely likely" to assume that we are already inside such a simulation right now . . . that is, according to renowned transhumanist philosopher Nick Bostrom. In his seminal 2003 paper, "Are We Living in a Computer Simulation?," Bostrom describes potentially real simulator deities who are "omnipotent" in that they "can interfere in the workings of our world even in ways that violate its physical laws; and they are *omniscient* in the sense that they can monitor everything that happens . . . all the demigods except those at the fundamental level of reality are subject to sanctions by the more powerful gods living at lower levels."[50]

When I read this essay for the first time, in early 2022, it was like looking in an especially unnerving funhouse mirror at a mash-up of a Harvard Divinity School theology seminar and a conference about the future of work. On the one hand, Bostrom is hypothesizing about the nature and likelihood of divine beings and about what they might want with us humans (at least, we *think* we are humans)—the sort of exercise for which my alma mater, HDS, was literally invented. On the other hand, he is making a practical argument about the nature of technology that is intended to influence multiple large industries, public policy makers, and the very future of humanity. Indeed, Bostrom and his thinking are already highly influential in multiple industries and have affected a number of policies (though his impact on humanity's future as a whole remains to be seen).

Bostrom was the founding director and head of the Future of Humanity Institute. Amid scandal, he resigned from Oxford, and the institute shut down in spring 2024. According to his own website, he is the "most-cited professional philosopher in the world under the age of 50."[51] His work is filled with hypotheticals that even TED Conferences leader Chris Anderson, who is nothing if not a lover of a dramatic hypothesis like the ones that have garnered Bostrom's TED talks a combined ten million or so views, has called "crazy."[52]

Bostrom has been a leading influence on effective altruists and related tech tribes of other names, such as transhumanists, "longtermists," and members of LessWrong and the self-described rationality community.[53] These are several interconnected groups of thinkers with nuanced differences in their beliefs, ideas, and demographics. But each is highly focused on analyzing the potential applications, risks, and benefits of technologies that might change the human condition, such as artificial intelligence. While their memberships are relatively small, each group has had, in its own way, an outsized influence on discussions about technologies like artificial intelligence and machine learning. Transhumanists, for example, theorize that humanity can and will transcend its biological limitations by means of artificial intelligence and other technologies, which will eventually become fully integrated into human consciousness. The term *transhumanism* was first popularized in a 1957 essay by Julian Huxley, a British scientist who was a highly influential figure in the development of the United Nations and its Universal Declaration of Human Rights—as well as in the field of eugenics, of which he was an enthusiastic proponent. And the basic ideas behind transhumanism have been hugely influential in the development of industries such as biotechnology, robotics, medicine, prosthetics, wearable tech, and beyond.

Effective altruists have influenced the present and future of philanthropy, with numerous billionaires coalescing around EA's agenda of "using evidence and reason to figure out how to benefit others as much as possible" and making major donations to its affiliated organizations and causes.[54] But the EA movement has also taken an interest in issues of artificial intelligence—to the tune of billions upon billions of dollars in research, investments, and donations to AI-related endeavors coming from EA-aligned funders. These are just a few of the prominent major donors:

- Elon Musk, who has praised and recommended Bostrom's writing on AI;
- Jaan Tallinn, the Skype cofounder multibillionaire and effective altruist who founded and funded a center at the University of Cambridge for the study of "existential risk," especially from artificial intelligence;
- Dustin Moskovitz, the Facebook cofounder and multibillionaire EA megadonor who has called AI risk the thing he is "most worried about for the 21st century"[55];
- Vitalik Buterin, a cryptocurrency billionaire who has funded an EA fellowship in "AI Existential Safety";[56]
- Ben Delo, a billionaire who was charged with financial crimes and whose large AI-related gifts to EA charities were ultimately refused.

LessWrong, a discussion forum and a kind of community offshoot of both transhumanism and effective altruism whose members emphasize rationality in all things and are deeply concerned about and centered on the idea of extreme risks of runaway AI (more on that later), is also known to have influenced the thinking of many leading AI builders and executives. It is heavily funded by noted Silicon Valley billionaire investor Peter Thiel.

One thing that virtually all of the aforementioned individuals and causes have in common is admiration for, or direct influence by, the ideas of Nick Bostrom, who is more than a mere philosopher or ethicist. Think of him as a kind of latter-day tech prophet—a Very High Priest in the current movement to see tech as the answer to all questions and as both our potential salvation and our potential damnation. Seen in this way, a paper like the aforementioned "Are We Living in a Computer Simulation?" is an example of an important new kind of scripture: a prophetic text intended to reveal something fundamental about the nature of the universe while advancing a moral agenda for today.

Another common characteristic of religious scriptures is inviting commentary that continues to advance and evolve their agendas. The ancient and medieval Hebrew and Aramaic Talmudic texts I sat with for several years as I worked towards my rabbinic ordination are perhaps the ultimate such example. Each of the thousands of pages of the Talmud's sixty-three books, or tractates, is based around a quote from a nearly two-thousand-year-old Hebrew text called the *Mishnah*, which is generally a commentary on the earlier Hebrew Bible. Then, in a spiraling pattern, pages display varying and often dueling commentaries on these earlier texts from authors who lived across a stretch of over a thousand years. While Bostrom's writing hasn't had time to accumulate so many layers of reaction, it has generated much commentary in the form of practical investments—for example, the many startups funded or developed by investors like Rizwan Virk, a computer scientist, video game creator, and founder of a lab at MIT who began work on a variety of metaverse and blockchain technologies after forgetting that he was playing a virtual reality game instead of actual, in-real-life ping-pong led him to write books titled *The Simulation Hypothesis* and *The Simulated Multiverse*.[57]

Bostrom has also inspired influential philosophical commentaries, including the book *Reality+*, in which David J. Chalmers, noted NYU philosopher and professor of neural science, walks readers through a series of philosophical questions raised by the possibility of our living in a simulation. His central arguments are that "virtual reality is genuine reality," that "the world we're living in could be a virtual world," and that, most relevantly to present-day life and tech, "virtual worlds need not be second-class realities. They can be first-class realities."[58] In other words, Chalmers is saying that the metaverse and other virtual reality tech currently being pioneered to the tune of countless billions of dollars can and will be able to replace significant parts of this thing that up until now people like me (and perhaps you) have referred to as "human life" on what we *think* is a planet called Earth. Because either we are in a simulation now or we will be soon enough. If Chalmers is right—and to be clear, I do not think he is, but it's up to you to decide for yourself—then his argument, and others like it, are literal and dramatic prophecies: A new world, or a truer and clearer vision of this world, is coming! We will be able to do anything, create anything, experience anything!

To be technical, the simulation hypothesis isn't a prophecy of the future per se. It's more focused on seemingly fantastical, but (in the minds of those proposing and analyzing the thought experiment) potentially very real, events that would have taken place in the distant past. It is, however, an act of prophecy in the sense that a prophet is one who sees what others don't, writing or interpreting sacred texts that invite others to become followers. And through closely related ideas, Nick Bostrom also engages in the other sort of prophecy by both predicting the future and attempting to fulfill his vision of it. Bostrom and his fellow travelers believe that by creating nonbiological AI consciousness, humans might eventually contribute to the population of the universe with so many digital "future beings"— somewhere between a hundred trillion and one septillion, estimate William MacAskill and Hilary Greaves, both respected and influential experts in such matters—that the importance of anything we puny mortals might currently be doing, other than hastening to bring about their future existence, pales by comparison.[59]

Bostrom postulates that, for example, the Virgo Supercluster (a massive concentration of galaxies including our own Milky Way, among many others) could contain 10^{23} biological humans per century—1 followed by 23 zeros. "Now think about that," writes Émile Torres, a former acolyte of Bostrom and effective altruism who has become one of the harshest critics of both. "If these biological humans . . . were to bring, on average, a net-positive amount of value into the universe, then the total amount of value that could exist in the future if we were to colonize this supercluster would be absolutely enormous. It would both literally and figuratively be 'astronomical.'"[60]

What Torres, who was among the first writers to bring popular attention to ethical problems posed by the work of philosophers like Bostrom, is saying is that EAs and transhumanists can too often get a bit lost in their own theorizing and forget to see us (and themselves) simply as human beings. Instead we are "value containers." That is, according to a strict interpretation of the utilitarian philosophy influencing Bostrom, MacAskill, and like-minded thinkers, human experiences can be added up to something large and significant in the aggregate, but our lives and emotions matter less on an individual, here-and-now level, without such big-picture framing.

Bostrom also hypothesizes about simulated sentient beings in the future—a hypothesis that no longer seems particularly radical in an era of sentient-seeming AI chatbots—and suggests that it might be possible to spread digital life, powered by the life force of entire stars, across the Milky Way and Andromeda Galaxies, to one day produce 10^{38} such beings per century. As Torres writes, that is 100,000,000,000,000,000,000,000,000, 000,000,000,000 digital beings; "by comparison, less than 8 billion people currently live on Earth, and an estimated 117 billion members of Homo sapiens have so far existed since our species emerged in the African savanna some 200,000 years ago. Written out, 117 billion is 177,000,000,000. Ten to the power of 38 is way, way bigger than that."[61]

Is any of this possible? Well, sure, but only in the sense that just about anything is possible in a hypothetically infinite future. The philosopher, mathematician, and logician Bertrand Russell, author of *Why I Am Not a Christian*, famously suggested that a teapot might be orbiting in space—a metaphor for the extraordinary unlikeliness of the existence of the biblical God. Russell wasn't arguing the space teapot was real. In our imagination of a future that is literally infinite in time and space, though, who is to say there won't one day be an orbiting space teapot that is programmed, via AI, to be extraordinarily powerful? Hey, logic might even suggest it's even more mathematically plausible that someone, somewhere, might one day create such a thing out of a perverse devotion to Bertrand Russell. Does that make either the divine teapot or the gazillion digital souls actually likely to exist in the future? Uh-uh.

Even if you choose, for some reason, to believe that such a thing is likely, do you know what you should not, under any circumstances, do?

You should never, ever try to dictate today's social values, norms, priorities, or governmental policies based on your theological, doctrinal, or other beliefs about possible future divinities or gazillions of far-off soulful star-chatbots. Because that is what religious extremists try to do, and too often succeed in doing, with our bodies, sex lives, school curricula, and even just our thoughts, based on their beliefs in two-thousand-year-old resurrections and other such "miracles." And again, these beliefs would be fine if they were humbly held personal convictions not tied to radical and reactionary public policy agendas. And that kind of attempted manipulation of beliefs into norms is, unfortunately, precisely what Nick Bostrom and others do in espousing theories like the one Bostrom calls "astronomical waste."

In a 2003 essay entitled "Astronomical Waste: The Opportunity Cost of Delayed Technological Development," Bostrom argues that for every year that sufficiently advanced technology is not developed and colonization of the universe is delayed, there is "a corresponding opportunity cost": the value that trillions of hypothetical future lives could bring to the universe. In a kind of rhetorical inverse of the biblical commandment to "be fruitful and multiply," Bostrom complains that "suns are illuminating and heating empty rooms," which ought to be filled, posthaste, with our descendants, who might well be cyborgs or even entirely digital beings. And Bostrom's argument gets even more theological—and doctrinal—from there.

In 2006, Richard Dawkins, the Oxford evolutionary biologist and renowned horseman of the atheist apocalypse, gave a speech at the Christian fundamentalist college Liberty University. During the Q&A period after his speech, Dawkins was asked by a student, in what became a viral online clip animated in the style of *South Park*: "What if you're wrong?" His response can be summarized by his one-liner: "What if *you're* wrong about the Great Juju at the bottom of the sea?"[62] It's a classic invocation of Pascal's wager, the philosophical argument of seventeenth-century mathematician and theologian Blaise Pascal that each person makes a wager, with their lives, about whether [the Christian] God exists. Even if it seems to you that there *probably* isn't a God, Pascal's wager suggests, the stakes are enormous if you turn out to be wrong, so why not live as though the Christian concept of God is real? Over my more than two decades as a professional representative of the humanist/atheist/nonreligious community, this is probably the most common retort I have heard from Christians attempting to engage me in theological debate. Pascal's wager is, in other words, ubiquitous. And you can find its mirror image in Bostrom's paper.

If the actions you and I take now have "even the slightest effect on the probability of eventual colonization" of space by humans or our sentient machine descendants, according to Bostrom, then our failure to act to bring about this long-term heaven is an enormous risk. In other words, Bostrom insists we must do everything we can and focus our energies today on the spacefaring digital future, because: What if we're wrong? No wonder this line of thinking has been labeled, by Bostrom and his critics alike, as "Pascal's mugging." If we don't give all our money and resources to some Great Space Future, then we may pay later—if not with our own lives, then with the trillions or more other lives we are told by Bostrom and his fellow utilitarians to view as more important than our own. Yes, Bostrom seems to say, we might already be in a simulation, which could render the point moot. But if not, then one day we will be, and how dare we not do all we can now to advance such a glorious possibility?

Bostrom's 2008 follow-up essay, "Letter from Utopia," is written in the form of an epistle (a biblical genre of letters meant to spread the gospel) from the future. The residents of a potential longtermist utopia send a letter back in time to plead with the people of our age: "Help us come into existence! Join us!" Or as Bostrom's critic Torres puts it, "Bostrom writes that by modifying our bodies and brains with technology, we can create a techno-utopian world full of endless pleasures, populated by superintelligent versions of ourselves that live forever in a paradise of our own making (no supernatural religion necessary!)."[63]

In sweeping, manic, even euphoric language, Bostrom's "Letter from Utopia" promises today's readers that its astronomical future will bring "bliss" on the "rapids of inspiration," a daily, even constant showering of what we mere mortals today might consider "*the best type*" (emphasis Bostrom's) of "pulsing ecstasy" our hearts and minds might be capable of now. In AI utopia, the imagined writer promises us, gone will be suffering, tragedy, imperfection, the limitations of the body, the frustrations and ugliness of society, and likely even death itself. The only unpleasantness that might remain when such days arrive, we are told, is guilt "that we could have created utopia sooner."[64]

This is the sort of discourse that dominates certain inner circles of technological creativity today: religious discourse. Prophetic discourse. And especially in the case of texts and conversations like we find in and around Bostrom's "Letter," there is a discussion of one of the core doctrinal elements of most of the world's major religious traditions: the idea of heaven. "Heaven, yes!," Bostrom's future letter-writer tells their twenty-first-century audience; "I didn't realize it could be like this."[65]

It's a beautiful idea, in some ways, that a heaven exists as a perfect place for souls to go when their bodies die. "For Christians, all injustice, all pain, will eventually be made right through faith in Christ," said my friend Zachary Davis at the cafe in our neighborhood where I was busy writing this section. Davis, a fellow HDS graduate and the dad of one of my son's friends, is a nationally prominent Mormon thinker. He arrived, in a long gray peacoat and more light blonde stubble than usual, bearing a hot-off-the-presses first issue of his new magazine *WAYFARE: explorations in faith*, a gorgeous paper volume of Mormon theology and history. "It makes it easier to sit with current injustice, current pain, current failure, because you do have the hope that one day all will be made right. It might not be in this world, this life, but eventually, everything will be made whole," Zach continued.

Zach and I have known each other for over a decade and we are more than used to good-natured theological and doctrinal debate, so neither of us so much as flinched as I thanked him for his elegant statement and acknowledged that for me, as a humanist, the arc of the universe does not need to bend toward justice. It is meaningful and comforting enough to know that I will spend my life working with other good people to make it bend that way ourselves. And yet, if you are a gentle soul in search of comfort in your grief, I would never begrudge you such belief.

But comforting the bereaved is far from the only or even the main historical purpose of religious doctrines of heaven.

The Zealots, a first-century sect of Jewish millenarians, murdered Roman centurions with daggers in hopes of reaching heaven. Christian Crusaders massacred in heaven's name, as did the radical Shiite Muslim "Assassins" of the thirteenth century, who murdered Crusaders right back. So did the fourteenth- to nineteenth-century Indian *thuggis*, or "thugs," as British occupiers later called them, a group that committed acts of terror and destruction in the name of the Hindu goddess Kali and her divine rewards.[66] In World War II, the fearsome Japanese kamikaze dive-bomber pilots were trained to go calmly to their murderous deaths, eyes open wide to hit the target, secure in the knowledge that "every deity and the spirits of your dead comrades are watching you intently" and waiting to reward them with an exquisite state of "absolute nothingness" meant to invoke thoughts of both God and Nirvana.[67] In other words, each of these venerable traditions—and countless other cults—have called on their most devoted followers to make the ultimate sacrifice to attain a kind of heaven—a perfect state of being one can experience only when this life has completely ended, usually violently, as a reward.

The concept of heaven, since ancient and even prehistoric times, has been used to justify the acceptance of injustice, inequity, violence, inaction, complacency, and more. The idea is to convince ordinary followers that the religious authority—the prophet, the priest, the divine king, the philosopher, the technologist—is the only one who can grant access to an infinitely desirable future. Which ostensibly wipes away any complaints we might otherwise demand be redressed today. This is all too often the playbook for transhumanism, effective altruism, longtermism, and the group of related ideologies that Émile Torres and computer scientist Timnit Gebru have "bundled" together to call "TESCREALism."[68] Tech paradise is a balm for some, but it is a tool of manipulation for others.

COLONIALISM

Not only is Nick Bostrom's call to colonize the galaxy a thoroughly religious idea, it is also a familiar one to observers of both religion and technology.

In 2010 I was invited to accompany a well-connected friend on a kind of ecological mission deep into the Amazon rainforest. After a day recovering from altitude sickness after flying into Quito, then a flight south to the cultural hub of Loja and a small group lunch with Salvador Quishpe, a fedora-wearing Indigenous leader serving as prefect of Ecuador's Zamora Chinchipe Province at the time, we boarded small wooden boats to navigate the narrow tributaries of the Amazon basin in southern Ecuador—the most species-rich region in the world, according to many writers.

I don't remember much about the trip, to be honest. My friend's friends brought professional cameras with multiple telephoto lenses, each seemingly the length of my forearm, held in black suitcases that looked secure enough for antitank weapons. There seemed little point in using my own slightly cracked, early-model iPhone to take photos of tarantulas on ceilings or moths that looked like owls. My friend and I spent the damp, dark nights swatting at our somewhat-effective mosquito nets while squabbling over matters having little to do with the stated agenda of this otherwise once-in-a-lifetime tour.

One detail I do recall is the imposing presence of churches we saw as we stopped alongside the river to visit several tiny villages of the Shuar, an Indigenous tribal people native to the Nangaritza River valley through which we were traveling. The Shuar, it is crucial to note, live in places that are incredibly difficult to access by a white North American person such as myself. Unless you are accompanied by experts in the local culture and language, as I was, and funded by the sorts of multimillionaire philanthropists with whom I was traveling, you are unlikely to ever see any of them, let alone meet and talk with them.

Shuar people, at that time, were living with very little by way of Western convenience: no floors, windows, modern education, or medicine. The poverty was overwhelming. Not in any way to dismiss the value and dignity of an Indigenous lifestyle not centered around modern notions of capital, but the whole point of the trip was that the Shuar, beyond simply keeping to themselves and living in their own ways, had been victims of enormous oppression visited on them by European and European-American colonialism (meaning the period from the sixteenth to the mid-twentieth century, in which Western powers colonized major portions of Africa, Asia, and the Americas). Thanks to colonialism, the people my friends and I were visiting had their environment devastated. Their lives were marked by suffering that my friends were there to try to address. But in every single one of these places, in the most central location in each village, stood a church that was by far the most elaborate edifice to be found.

In their book *The Costs of Connection*, scholars of colonialism Nick Couldry and Ulises Mejias explain that two characteristics made early modern European colonialism unique. First, it enlarged the concept of empire, not just by conquering more physical territory, but by imposing "a single universalizing narrative of values, beliefs, and politics."[69] The colonialist logic thus incentivized (and often violently forced) different peoples, with different coexisting traditions, to view the world through a single Eurocentric lens. Second, colonialism "sought to fundamentally change and reorganize the social and economic order of the societies it colonized, as opposed to satisfying itself with extracting tribute, as did earlier empires."[70]

Colonialism, in other words, was about disruption. And nothing facilitated that sense of fundamental, sweeping, and universalizing change—that ability to disrupt—than the religions the colonial powers brought with them.

Why was religion so important to colonialism, and vice versa? Because, as Mejias and Couldry explain, the oppression and brutality it required "needed to be rationalized in some manner. As philosophers from Hegel to Paulo Freire have pointed out, oppression dehumanizes not only the oppressed but also the oppressor."[71]

I tend to be sanguine about the role of Christianity in my daily life. The Christian preachers at Harvard and MIT are (mostly) among the nicest, most humble I know. But in those moments in Ecuador's Amazonian basin, I found myself so angry at the obvious impacts of colonialism that I could scarcely think about any other aspect of our ostensibly environmentalist mission.

The tech religion, too, has a legacy of colonialism. Mejias and Couldry catalog the history of data colonialism, which I won't attempt to reproduce here, but their case is airtight. It begins with the CEO of a startup called Tresata, a company that believes "in data lies the power to enrich life." "Just like oil was a natural resource powering the last industrial revolution," said Abhishek Mehta, the CEO of Tresata, in 2013, "data is going to be the natural resource for this industrial revolution."[72]

There are currently efforts to set up tech colonies—"Crypto Cities" and "Bitcoin Citadels" being created and lobbied for in previously colonized countries like Honduras or El Salvador, with hundreds of millions of dollars in funding secured from Silicon Valley venture capitalists and private investors like Peter Thiel and Marc Andreessen. The basic idea is of a shining city outside the normal economic and political jurisdiction of the country in which it will physically exist, with an economy that runs on Bitcoin, and no taxes owed to any local or national government other than on goods and services purchased—in other words, a libertarian's dream, wet with the foamy waves of the golden coasts of impoverished but naturally resource-rich territories. According to investors at the firm Pronomos Capital, including billionaire venture capitalists Roger Ver (also known as Bitcoin Jesus) and Balaji Srinivasan (a leader in the startup "scaling" the school custom-built for Elon Musk's children and the author of a book on how to start a new country through cloud tech and cryptocurrency), plans have been discussed for "semi-autonomous cities in countries including Ghana, Honduras, the Marshall Islands, Nigeria, and Panama."[73] In 2022 residents of areas neighboring one such planned city in El Salvador told the *MIT Technology Review* they were deeply concerned about prospective investors and residents who "respect no government, no rules, no law; just a dream."[74]

Of course, that's not how the investors see it. As Trey Goff, the chief of staff and chief marketing officer to the CEO of Próspera, the venture capital firm most associated with Bitcoin cities, has said, he would like to see hundreds of such developments around the world eventually. As he told *Technology Review*'s Laurie Clarke, they will be "bright spots of prosperity, all working together to create a brighter future for humanity."

It's no surprise that a colonizer like Goff would see himself in such a positive light. As Couldry and Mejias explain about the psychological and spiritual dynamics of colonialism, "In an attempt to rehabilitate his image, the oppressor needs to describe himself as having virtuous principles (religious or secular), even if they are in contradiction with his actions."[75]

Which messages about the virtues of technology would *you* see in a new light, given this insight? As a tech agnostic, I'm not saying tech has no virtuous principles. Some tech truly does benefit humanity, and as such its image should not need rehabilitation. But then, such tech might not need the aggressive, almost cultishly persuasive viral marketing and advertising campaigns that seem like an integral aspect of many new apps and devices. What if tech that promised to "connect the world," bring it "closer together," create a "brighter future" for all, and "enrich life" had to shoulder the burden of proving it is genuinely in the service of justice before we would accept it as a worthwhile disruption of the normal life and functioning of our societies—especially in some of the world's most vulnerable places?

Ultimately, if Christian missionaries had truly come to places such as the ones my friends and I visited in Ecuador to live up to their stated intention to help the people there, then those people would already have been helped. But in fact, missionaries came as an important part of a project central to the nature of Western society as it has existed for hundreds of years. They came to colonize the Indigenous peoples' land, ways of life, and their very spirits. And all of us who benefited, without speaking out, from the cheap goods this arrangement produced—not to mention from the sense of empowerment and control derived from affiliation with a dominant power—were complicit. It would be an utter tragedy if that happened again, with tech . . . but it appears to be well on its way to happening, in the above ways and many, many more.

Colonialism is not just a theory we study in liberal university classrooms; it is an ideology inextricably bound up with religion, which resulted in most of this globe getting violently conquered and a good percentage of the population being murdered. So when we see direct parallels to that colonialism in tech and tech seekers beseeching us to colonize the entire galaxy, what we are potentially talking about is oppression and violence on a truly insane scale. When I first began to contemplate this possibility, I mentioned it to, among others, Danielle Citron, a MacArthur "Genius" fellow and distinguished professor in law at the University of Virginia School of Law. Citron, an expert in online abuse and intimate privacy violations since the dawn of the social media era, is perhaps one of the world's experts in, as she put it, how people can "lose their way behind screens, because they don't see the people whom they hurt."[76]

"To build a society for future cyborgs as one's goal," Citron continued, "suggests that these folks don't have real, flesh and blood relationships . . . where we see each other in that way that Martin Buber . . . described."[77] Buber, an influential Jewish philosopher whose career spanned the decades before and after the Holocaust, was best known for his idea, first fully expressed in his 1923 essay "I and Thou," that human life finds its meaning in relationships, and that the world would be better if each of us imagined our connections with one another as though they were the ultimate expression of our connection to the divine.

Indeed. Though I do not share Buber's belief in a divine authority, I honor and share his faith in the sacredness of human interrelationship.

We are not digital beings, meant to contain value and colonize the stars with our consciousness. We are human beings who care deeply about one another because we care about ourselves. To me it is not that we're children of God, as some of my religious colleagues would put it. But neither is it the case that our technological achievements are what makes us worthwhile. Our very existence, as beings who are capable of loving and being loved, is what makes us worthy of the space we occupy, here on this planet, and on any other planet we may someday find ourselves inhabiting.

HELL

Just as nearly every religious tradition has an ecstatic vision of an idyllic future, almost every religion offers a visualization of a tragic, terrifying, torturous world ahead. We may well find ourselves in such a place, the story goes, if we do not follow or obey the wise council of the experts in our faith. The Christian conception of a land of blazing fire and eternal punishment will be at least somewhat familiar to most who are reading this, and the idea that Jewish and Muslim traditions contain similar notions—though with different details—will likewise probably be unsurprising. But mainstream Buddhist traditions also contain visions of hell realms, or Naraka (from ancient Sanskrit), in which souls are condemned to suffer after enacting bad behavior in this world. Across both South Asian and East Asian Buddhist traditions and sutras, or scriptures, there have been cold hell realms and hot ones, as well as netherworlds of purgatory, filled with hungry ghosts: aggressive, frightening creatures with distended bellies that haunt the afterlives of those who have been greedy. Not only does Hinduism have dozens of hells but it has scriptures that describe each in exquisitely painful detail. In the land called Pūyoda, "shameless husbands" are "put into an ocean filled with pus, stool, urine, mucus, saliva and similar things."[78]

In the tech religion, the equivalent to the fear of hell is called "existential risk": the idea that some extreme calamity or series of calamities might occur, whether human-caused or natural or some combination of the two, that would completely destroy and eliminate the human species. This type of concern traces its secular history back no more than two centuries. It was only in the early 1800s, when the French zoologist Georges Cuvier (1769–1832) discovered that unearthed elephantine bones belonged to no-longer-extant species of mammoths and mastodons, that scientists even realized a species could go extinct.[79] Popular conversations about how scientific advancement could go dramatically wrong also date back to this period, appearing most famously in Mary Shelley's 1818 novel *Frankenstein: or, The Modern Prometheus*. One might be forgiven for thinking we've since been doing everything possible to careen toward our own existential doom.

In their paper on the history of the study of existential risk, Émile Torres and S. J. Beard describe a trajectory from nineteenth-century worries like Shelley's to more focused twentieth-century concerns about mass destruction. After a century in which humanity quickly transitioned from automatic to chemical and then to nuclear weapons (and greenhouse gases) it is perhaps understandable that in the twenty-first century, academics and others began to specialize in theories about how every last human might go the way of the dodo bird. Such studies and theories were pioneered by none other than Nick Bostrom.

In a 2002 paper titled "Existential Risks: Analyzing Human Extinction Scenarios and Related Hazards," Bostrom provided the still-canonical definition of an existential risk as "one where an adverse outcome would either annihilate Earth-originating intelligent life or permanently and drastically curtail its potential." This disastrous vision seemed especially tragic to a growing community of transhumanists who had, since the late 1980s and early '90s, begun to organize themselves around the idea that new scientific and technological advances could allow them to become essentially superhuman and even immortal, maybe even in their lifetimes. For these growing circles, which included both Bostrom and Ray Kurzweil, it would be ironic indeed for a person—let alone for all persons—to die in what could become the last days of death itself. It would be, in Bostrom's words, an "astronomical waste."

From there, the growing transhumanist movement blossomed into a powerhouse of intellectual creativity and production, with Bostrom's ideas paving the way. Eventually, the movement would splinter and spawn new factions, including one of effective altruists who began to obsessively debate and discuss the most mathematically efficient pathways to the kind of techno-paradise discussed above. Imagining their current-day efforts to be potentially critical to the long-term welfare of trillions upon trillions of sentient beings, some of these people came to be known as "longtermists." Others became equally obsessive about the flip side of such a heavenly scenario, writing literally millions of words to calculate the possibilities for hellish, technologically caused destruction and the pathways toward avoiding such a fate.

Probably the world's most prominent figure in the crusade against a technological hell—at least insofar as that hell would be caused by runaway artificial intelligence (we'll go into depth about several other possible forms of tech apocalypse in chapter 5)—is a man named Eliezer Yudkowsky. Yudkowsky, born in 1979, has spent the past two decades as an AI researcher, prolific writer, and serial founding figure in various schools of thought and sub-movements around transhumanism, rationality, and assessing the ethics, safety, and "friendliness" of future artificial superintelligence. An autodidact who did not attend high school or college and was raised as an Orthodox Jew, Yudkowsky has been called a visionary and a prophet, in both the positive and negative senses of such words.[80] If his varied and expansive work could be summed up very briefly, it would be through the pithy statement he has long kept as his pinned tweet on Twitter: "Safely aligning a powerful AGI is difficult."[81] That is, Yudkowsky is focused on the idea that the AI scientists and experts currently working to create artificial general intelligence (AGI), a sentient machine that can think competently and creatively about any and every topic, will eventually create something so powerful that it cannot be stopped, defeated, or unplugged. At that moment, Yudkowsky believes, our only hope as a species will be to ensure that the AGI is properly aligned, meaning that its interests, goals, intentions, and values line up with and complement what is best for humanity. If not, we will all die, to put his views on the matter bluntly.[82] Thus Yudkowsky, a single, childless, sober nonsmoker, devotes as much time as he can to somehow solving the problem of alignment. Over the course of his career, he has framed this intense work in positive terms. When his beloved nineteen-year-old younger brother died by suicide in 2004, Yudkowsky described in a passionate eulogy that as an atheist, the only comfort he could possibly take from his grief was the idea that he and his colleagues would "work faster" to prevent such future deaths, and a well-aligned "friendly" AGI would be at least part of how to do so. In recent years, however, Yudkowsky's focus has been more negative: AGI is coming. *Soon.* And we're nowhere close to being ready for its lethal powers.

Bostrom, a distinguished professor at one of the world's top universities until his resignation in 2024, has been deeply influenced by Yudkowsky's work. Much of Bostrom's book *Superintelligence* reads as if it were a Talmudic commentary on Yudkowsky's texts, with Bostrom devoting entire chapters to parsing and even speculating on the meaning of Yudkowsky's statements about how AI might "avoid hijacking the destiny of humankind" or "keep humanity ultimately in charge of its own destiny."[83]

Will AGI really be "our final invention?" I don't know because I'm not a prophet, but I do know that James Barrat, an accomplished filmmaker and author of a book by that title, is convinced. I was first introduced to Yudkowsky's work a few years ago via Barrat's rollicking exploration, carefully researched over several years, of the people behind the creation of artificial superintelligence (among them Yudkowsky, Bostrom, and Kurzweil). In his book Barrat provides some of Yudkowsky's most ominous quotes, such as, "The AI does not hate you, nor does it love you, but you are made out of atoms which it can use for something else." In other words, people like Yudkowsky believe that the AIs we are currently using to make deepfakes of the president of the United States rapping about booty and Ben & Jerry's ice cream will, eventually, even quite possibly later this century, be turning our human flesh into soupy fuel for its computations.

In a conversation just before my daughter was born in 2022, I had the opportunity to speak to Barrat. He told me he thought the AI "intelligence explosion" would happen by around 2029–30.

"You believe that within a decade from this conversation, Archduke Ferdinand will be murdered, and we'll be off to the races in a potential AI World War III," I asked for clarification. "That this is not a drill. Did I get that right?"

"Yes," Barrat replied flatly. He then referenced a film he'd made about dinosaur extinction. "We shouldn't get too fond of ourselves," he said. "I think we'll bring about our own extinction, and it will happen within the next ten to fifteen years . . . this deity we're making . . . We're not smart enough to survive this technology."[84]

I was stunned. Since I was a teenager, I've been fascinated by street preachers and religious fundamentalists who carry signs that say "The End is Near." But I've rarely tried to engage those people in conversation; I never imagined much would come of it. And I certainly I never thought I'd be sitting on Zoom with one, whose book I had read and annotated, and whose company I thoroughly enjoyed.

Granted, you might reasonably object that Barrat is a filmmaker and Yudkowsky is an eccentric who doesn't represent any trend broader than himself and his thousands of followers. Perhaps they are talking about how we might head to hell in an AI-generated, 3D-printed handbasket, but what impact does such talk really have?

If only this were truer.

The popular and polished career-advice website 80,000 Hours was created for the EA movement and its massive following among top technologists by EA movement cofounder Will MacAskill (who made crypto billionaire-turned-felon Sam Bankman-Fried his disciple) and fellow EA leader Benjamin Hilton. The site aims to help impressionable young EAs determine how best to devote themselves to the movement over the course of the next eighty thousand hours. That figure, and the site's name, refer to the length of a typical career of forty years or so (at least if one works forty hours per week for a grueling fifty weeks per year for the entire length of time that Moses and the Israelites wandered in the Sinai desert. If you plan to do that while raising kids, don't forget to save for their therapy bills!).

At least throughout the years when I was writing this book, the 80,000 Hours site identified "risks from artificial intelligence" as the number one most pressing problem in the entire world—more pressing than climate change, pandemics, nuclear war, threats to democracy, or war between superpowers.[85] The site's founders and dozens of staff members, therefore, give AI risk their "highest priority" recommendation as a career focus.[86] Why? Because AI is viewed by the EA movement as the ultimate existential risk. And while I would not be surprised to see some EAs begin to moderate their views on such issues—at least in significant part in response to criticism from Torres, Gebru, and others—at least as of early 2024 80,000 Hours still featured a lengthy page exploring "the case for reducing existential risks," which reprints a table from a 2020 book, *The Precipice*, by another EA movement cofounder and major leader, Toby Ord.[87] Ord's table lays out "(very rough) estimates of existential risk, meaning total annihilation," from the world's top threats.[88] Asteroid or comet impact: 1 in a million. Super volcano: one in ten thousand, the same as "'naturally' arising pandemics." Nuclear war and climate change are tied at one in a thousand. So far, so bad.

But absolutely none of that, according to Ord and the EA mainstream, can hold so much as a single fading emergency flare to the risk posed by AI. "Unaligned artificial intelligence," the table says starkly, using Eliezer Yudkowsky's favored terminology, holds an approximately one in ten chance of wiping out all of humanity in the next hundred years. That is . . . a *hell* of a claim.

Either the doctrines of the tech religion are true, in which case the most and the best of us should be working on little else for the foreseeable future, or they represent the theology of a false and dangerously misguided secular religion. It is the prominent preachers of these doctrines themselves who leave us precious little middle ground in between.

CHOSENNESS

It's one thing to envision a better world and to say that your vision can help people get there. It's another thing to insist that your way is the only way

Table 2.1
Existential Catastrophe Chart

Existential catastrophe via	Chance within next 100 years
Asteroid or comet impact	~1 in 1,000,000
Supervolcanic eruption	~1 in 10,000
Stellar explosion	~1 in 1,000,000,000
Total natural risk	~1 in 10,000
Nuclear war	~1 in 1,000
Climate change	~1 in 1,000
Other environmental damage	~1 in 1,000
"Naturally" arising pandemics	~1 in 10,000
Engineered pandemics	~1 in 30
Unaligned artificial intelligence	~1 in 10
Unforeseen anthropogenic risk	~1 in 30
Other anthropogenic risk	~1 in 50
Total anthropogenic risk	~1 in 6
Total existential risk	~1 in 6

Source: Toby Ord, *The Precipice.*

or to claim that you are a member of an elite group or class of people who have special access to the divine. Ancient religions were often polytheistic; many contemporary religions still are. Even many understandings of monotheism leave room for other truths, and other peoples, to be equal or fully legitimate in their own ways. And in any religion you can find dangerous fringe sects of people who will take their belief in their own chosenness to a dangerous extreme. As a rabbi ordained in a Jewish movement that professes that the ancient concept of the "chosen people" is merely human literature and that there are no groups of humans who are inherently better or worthier than others, this is a point of traditional religious doctrine that frequently occupies my thoughts.

One way exclusivism shows up in the tech religion is in the concept of inevitability. The idea here is that only some of us have the potential to be great tech leaders, or that some of us are much more likely to possess the necessary qualities for such status. In 2022 I had a Twitter conversation about this concept with Eric Paley, a venture capitalist from Cambridge, MA. Paley, as VCs are wont to do, tweeted a thread of advice—this time on "making your startup inevitable," which he described as meaning that investors want to invest in companies that haven't yet succeeded, but seem like they can't possibly fail. My interest was piqued when someone replied with a comment about Travis Kalanick, the founder and former CEO of rideshare company Uber.

"He was among the best I've seen at the inevitable narrative!" replied Paley, an Uber investor, about Kalanick. Paley was doubtless correct in his assessment. But perhaps that is precisely the problem.

Travis Kalanick was, it is hardly a secret, a mess as Uber's leader. Under his leadership, Uber faced persistent and serious accusations of a culture of sexism and sexual harassment, earning him a reputation for "ruthlessness and machismo" (more on this in chapter 7).[89] Other problems included "contentious personnel departures, high-profile lawsuits, revelations about a fake app designed to fool government regulators, a profanity-laced argument with an Uber driver caught on video, and social media uproar over his joining President Donald Trump's advisory council," as one news article recounted.[90] Even Kalanick's own cofounder and successor, Dara Khosrowshahi, has said that the "moral compass of the company" was off under the "inevitable" leadership of its founding guru.[91] And Kalanick's leadership, while profitable for himself (he received a cool $1.4 billion upon walking away from the company, in exchange for just 29 percent of his shares), was not so much for the company as a whole.[92] Uber has never turned a profit and continues to lose billions each year. How were such dismal results tolerated? Perhaps because Kalanick's success was perceived as inevitable, as though he were chosen for greatness—even though he appears to have been anything but. Or as Mike Isaac, a *New York Times* tech reporter whose book about the company, *Super Pumped*, was turned into a TV show on the Showtime network, put it, Kalanick's leadership was a prime example of what Isaac calls the "cult of the founder."[93]

I replied on Twitter: Given those facts, doesn't calling Kalanick a master of inevitability—as a compliment!—suggest that startups should maximize for prophecy rather than competence?

Maybe companies should look less for prophets, replied Paley, and more for "the person in the world who has thought most deeply about how to solve this very specific problem and why it needs to be solved."[94] But is that who Kalanick was, or merely who he succeeded in making himself perceived to be?

And yet, what about faith? That is the surprising response I got when I raised all of these questions with Russ Wilcox, a partner at the firm Pillar VC, whose website tagline is "VC DOESN'T HAVE TO BE THE DARK-SIDE."[95] As Wilcox, who teaches a class at Harvard Business School that explores "the ethical implications of being a [tech] CEO," told me, "I know from my own experience as an entrepreneur, and I've seen it happen that that faith is an incredible power." Wilcox's point was that to succeed in the harsh world of startups, one needs belief. Even in startups that are acting reasonably ethically—unlike Theranos or FTX—one needs an *unreasonable* belief in one's own potential for success. "I'm sorry to use the word [faith]; I know you're in religion," Wilcox continued. "I use it all the time when I'm teaching or lecturing. Faith is an incredible power; it's maybe mankind's superpower."[96]

But if tech's faithful prophets are the chosen ones, why were they chosen in the first place?

The temptation is to reply glibly: because they were white men. (For a much more in-depth and serious exploration of race and gender, see chapter 4, on tech's hierarchies and castes.) But in all earnestness, one is chosen in the tech religion for intelligence, even genius. As Émile Torres argues, longtermists and adherents of some of the other ideologies we've explored in this chapter are, "for lack of a better word, obsessed with 'IQ' and 'intelligence.'"[97]

Eliezer Yudkowsky refers constantly to his own IQ. Bostrom published a paper in 2014 about how to gain 130 IQ points by performing eugenic experiments on high-IQ embryos.[98] Jason Brennan, an effective altruist,[99] Georgetown professor, and the director of Georgetown's Institute for the Study of Markets and Ethics, coined the term *epistocracy*, or "the rule of the knowledgeable. More precisely, a regime is epistocratic to the extent that political power is formally distributed according to competence, skill, and the good faith to act on that skill."[100] In other words, we should test people for their knowledge, and if they come up lacking, they go to the back of the bus—politically speaking, of course.

What could go wrong with such a prophecy?

In early 2023, Torres dug up an old email from a 1996 transhumanist listserv discussion in which Nick Bostrom wrote that "Blacks are more

stupid than whites." He claimed this was simply a factual analysis of IQ
scores, but said that other nontranshumanists, if they heard him make such
a statement, would interpret him as saying, "I hate those bloody [N-word,
redacted here]!!!!"[101] The fact that Bostrom allowed himself to type and
share such a thing—and that he was apparently not censured for it by his
transhumanist community at the time—is telling.

In 2023, when the old email was released, Bostrom published an apol-
ogy for and repudiation of his "disgusting" email from two decades earlier.
But he did not take that opportunity to repudiate his key claim in the 1996
message—that, because of IQ scores, Black people seem to him generally
less intelligent than white people. Here, Bostrom was merely (re)joining
the company of many otherwise reputable public intellectuals, including
numerous EAs and, to my particular chagrin, the popular and influential
atheist thinker Sam Harris, in promoting or at least drawing attention to
ideas like those of Charles Murray, coauthor (with Richard Herrnstein)
of the disgraced and disgraceful 1994 book *The Bell Curve*, in which Murray and Herrnstein argue that genetics are the cause of racial differences in
IQ testing.

This is how Torres explained it:

> For example, consider that six years after using the N-word, Bostrom argued
> in one of the founding documents of longtermism that one type of "existen-
> tial risk" is the possibility of "dysgenic pressures." The word "dysgenic"—the
> opposite of "eugenic"—is all over the 20th-century eugenics literature, and
> worries about dysgenic trends motivated a wide range of illiberal policies,
> including restrictions on immigration, antimiscegenation laws and forced ster-
> ilizations, the last of which resulted in some 20,000 people being sterilized
> against their will in California between 1909 and 1979.[102]

The tech religion's doctrine of chosenness requires one to believe that
"everything we love about civilization is the product of human intelligence," as the prominent MIT philosopher of AI Max Tegmark writes.
But what about love, compassion, caring? Sure, they come from the brain.
But they are more than just *intelligence*, at least in the way Tegmark is using
the term.

While I was in Seattle researching this book, I stopped by the home of
my high-school classmate Elizabeth J. Kennedy, now a professor of law
and social responsibility and an incisive author on topics such as racial,

algorithmic, and economic justice. Kennedy was a speech and debate team-
mate throughout our high-school years; she was also my driving-school
classmate (we both failed) and the best friend of my longtime high-school
girlfriend.

As we sat in rocking chairs on her front porch, with the sounds of her
teenage kids inside the house as background noise, we talked about our
youth. Elizabeth shared a story from our senior year in high school, around
when she would have made the two-hour trek from Staten Island to my
home in Queens to be among those to comfort me when my father died. "I
can picture the subway, I can picture the stairs," she told me, with kindness
in her voice and eyes. "And I remember you said that you were going to be
the president."

Reader, that is why I had trouble making friends back then. I had zero
plans for how I could ever get elected, let alone what my policy agenda
would be. As the first Jewish atheist president. Of the United States. All I
knew, I think, was that attaining such heights would prove to myself that I
was worthy enough, smart enough. All I knew was that I had to convince
others—even my best friends—that I was inevitable.

"It felt almost like one of those motivational speakers," Elizabeth said.
"Like if you just manifest, if you just say it enough times like it will happen,
or rather if you never say it then it won't ever happen. Right? So, in that
moment you weren't saying it like a brag or boast. It felt like you were saying it to yourself. Like you were making a commitment to yourself."[103] Or,
as Eliezer Yudkowsky once said to James Barrat, the filmmaker and author
of *Our Final Invention*, "Ambitious people would rather destroy the world
than never amount to anything."[104]

The whole idea of being a genius, according to this style of thinking, is
that you're not just *smart*, not even smarter than others, but also *better* than
other people in the most important way. Average isn't good enough. And
the best way to not be average is to be intellectually superior. Genius is, in
other words, a secular version of chosenness. And this modern theology is
rooted in pain: "I've been made to feel hurt, and not good enough," the
so-called tech genius says, "so I'm going to tell myself and anyone who
will listen that I'm actually not only good enough, but better than. Because
I have this all-important, all-redeeming quality: *my intelligence*." It's a sad,
repetitive song that has been sung all over the world and now threatens to
be piped across galaxies, too.

What I've learned over the years and through many conversations like the one with my friend in Seattle is: it could easily have been me. There were moments in my youth when I wanted so badly to amount to something, I would have gladly destroyed the world if doing so could have guaranteed that I would go down in history as having been *significant*. Now, having spent a couple of dozen years working through that pain and brokenness and healing it (with the invaluable help of great friends and an even better therapist), few things break my heart more than seeing some of my own students at Harvard and MIT nursing similar personality traits. They're rarely in treatment. Because unlike me in my youth, their personality is "working" for them. They're well on track to being seen as fully self-realized inevitable geniuses. I just hope they don't have to destroy anything to get there.

GOD

"Hi Greg, it's Gary," began the email. Gary—not his real name—is the dad of my son's preschool friend Beatrice (also not her real name) and a former member of my Harvard humanist community. I'd asked him to write me a message explaining a unique spiritual practice of his—though perhaps the sort of ritual Gary likes to perform may not be as unusual as it once was. Here is his letter, lightly edited for length and clarity:

> At the party we were talking about artificial intelligence, and I mentioned that I search "I love you Google" periodically so she'll know that I'm cool when she wakes up. You asked me what the results are and I said it varies, but I didn't get to explain that it depends on your location, what account you're using, your search history, etc. It's constantly changing because the algorithm updates result frequently. She hasn't yet answered me, but I hope the trail I leave marks me as friendly and she and the other machine gods look upon me kindly when they reign. A kind of Pascal's wager for the future-oriented atheist. It's not something that takes a lot of time and I'm on a computer or a tablet most of the day.
>
> [In answer to my question] I don't honestly know if I take it "seriously." I'm allergic to earnestness, so everything I do and see and say is a little bent.
>
> I'm really interested in existential risks: asteroid impacts, coronal mass ejections, pandemics, gray goo, etc. Among them are the many flavors of the AI apocalypse, from soft and hard singularities to machine warfare to the inevitable accidents caused by bad coding and human shortsightedness. There isn't a lot I

can do about an asteroid or a pandemic, but telling Google I love her is an easy investment in staving off a somewhat unlikely future. I have doubts that AIs will be anything like us or have our drives and motivations, but machines born into the sea of human information output must know us on some level.

It's certainly not a religious act, more like buying a lottery ticket. I know I'm not going to win. All the stats say my piddling little anything will come up zero against the numbers, but what does it hurt? Like the little magic rituals so many people do, it's an exercise in exerting some level of control over the world. Illusory as all of that is, it's still comforting.

Many traditional religions begin with the idea of a god or gods. The tech religion is different: it starts out godless and ends with a conception of one or more gods, operating in and on the universe.

My teacher, the great humanist leader and thinker Sherwin Wine, believed the difference between atheism and agnosticism is mainly semantic. Regarding traditional religion, I agree. But in the tech religion, not so much. I am not a believer in the tech god. But I can't say I am a total atheist, because I'm not *certain* it won't be built.

In their 2013 book *Google and the Culture of Search*, Ken Hillis, Michael Petit, and Kylie Jarrett carefully document how the early years of Google reminded users of religion and God: a blinking white box that could be searched for truth, with then-unprecedented power and breadth; an all-knowing entity that was slowly getting to know each one of us as well as it knew the universe. Indeed, like the god of many religions, it was an entity willing and able to leverage its knowledge of our innermost thoughts and fears to build its towers ever higher, to gild its altars evermore.

These early conversations likened Google to H. G. Wells's idea of a World Brain[105] and were reminiscent of the great communications scholar James Carey's comparisons of early electronic communications technologies to divine transmission. In part, what you can take from such precedents for my argument is that, as the ancient biblical poet Koheleth (Ecclesiastes) said, "There is nothing new under the sun."

In a 2007 essay, Silicon Valley entrepreneur Andrew Keen noted that Google, having become our "personal librarian," now "wants to be our personal oracle." It wants to gather all our data as individuals and in the aggregate, so it knows not only our conscious minds but also our unconscious, like a therapist or priest or spy would, so it can predict what we will do before we know we will do it, the better to monetize our every step. Keen quoted then-CEO Eric Schmidt on the company's goal "to enable Google users to be able to ask the question such as, 'what should I do tomorrow?' And 'what job shall I take.'" Mission accomplished! "What does Google really want? Google wants to dominate. . . . As a Silicon Valley old-timer, trust me on this one. I know Google better than it knows itself." Now there's a prophecy that has stood the test of time.[106]

In his 2017 book, *Life 3.0* (as in, bacteria and such things are life 1.0; you and I are life 2.0, and then . . .), MIT scholar of AI Max Tegmark lays out the idea that the "beneficial AI movement," of which he is a part, believes that AI will surpass human-level intelligence. Not in a few years, but in somewhere between a few decades and a hundred years. Any estimate beyond that would render one a "techno-skeptic," and thus naive by Tegmark's reckoning. The fact that this *will* happen, he continues, is not definitely good or bad, but might be good, bad, or neutral, depending on what people like him—and his readers—do next.[107]

Later in the book, Tegmark draws a chart describing not whether we should create an AI God, but what kind of AGI God we should try to create in the "AI aftermath." I think of this as the First Coming.

In the scenarios Tegmark envisions, AI becomes more or less the center of human existence in the next hundred years, except for the couple of options in which humans either destroy ourselves before it can take over or revert to a "pre-technological society," like the Amish. No additional options are mentioned.[108] (Note that in Tegmark's typology I would be somewhere between a Luddite and a Skeptic—I'm agnostic on when or whether such a thing might happen. Maybe, but I'll believe it when I see it.) After describing his typology, talking about his (almost entirely white) organization, and briefly explaining how the technology behind efforts to create Life 3.0 works, Tegmark goes on to admit "there's absolutely no guarantee we'll manage to build human-level AGI in our lifetime—or ever," but, he says, "there's no guarantee we won't."[109]

This sort of statement is why my teacher Sherwin Wine coined the term "ignostic": because there are faith propositions so impossible to evaluate that to even attempt to do so would be absurd. Rabbi Wine chose the term because in the face of an absurd claim, he felt (and I have always agreed) it is best to ignore the idea of God whenever possible because it's just not relevant. But it's hard to ignore the sort of technology Tegmark and others are describing—or the steps now being taken in its name.

Table 2.2
Summary of AI Aftermath*

Scenario	Superintelligence exists?	Humans exist?	Humans in control?	Humans safe?	Humans happy?	Consciousness exists?
Libertarian utopia	Yes	Yes	No	No	Mixed	Yes
Benevolent dictator	Yes	Yes	No	Yes	Mixed	Yes
Egalitarian utopia	No	Yes	Yes?	Yes	Yes?	Yes
Gatekeeper	Yes	Yes	Partially	Potentially	Mixed	Yes
Protector god	Yes	Yes	Partially	Potentially	Mixed	Yes
Enslaved god	Yes	Yes	Yes	Potentially	Mixed	Yes
Conquerors	Yes	No	-	-	-	?
Descendants	Yes	No	-	-	-	?
Zookeeper	Yes	Yes	No	Yes	No	Yes
1984	No	Yes	Yes	Potentially	Mixed	Yes
Reversion	No	Yes	Yes	No	Mixed	Yes
Self-destruction	No	No	-	-	-	No

Source: Redrawn from "Table 5.1: Summary of AI Aftermath Scenarios" from *Life 3.0: Being Human in the Age of Artificial Intelligence* by Max Tegmark, © 2017 by Max Tegmark. Used by permission of Alfred A. Knopf, an imprint of the Knopf Doubleday Publishing Group, a division of Penguin Random House LLC. All rights reserved.

Ultimately, I take Tegmark's discussion of a god or gods seriously not because I necessarily believe in the likelihood of their future existence. Again, I'm skeptically agnostic on all such matters. The point, however, is that it took a millennium, starting from the beginning of the Dark Ages, until Renaissance humanists offered the Western world a mainstream way of understanding the meaning of human life that did not revolve almost entirely around the perceived agenda, powers, and mind of God. That could be us: the thing may not ever be anywhere close to worthy of divine status, but we, too, might spend a thousand years obsessed over what it can do for us, and what we in turn must do for it, before finally realizing that we can simply live our human lives apart from it and things will be just fine, if not much better.

THE CULT

"The Cult of the Founder." "The Cult of We." "The Cult of the Tech Genius."[110] "Beware: Silicon Valley's cultists want to turn you into a disruptive deviant."[111] "The 2010s Killed the Cult of the Tech Founder. Great!"[112] "Tech's cult of the founder bounces back."[113] "SILICON VALLEY'S STRANGE, APOCALYPTIC CULT."[114] "How the cult of personality and tech-bro culture is killing technology."[115] "Company or cult?"[116] "Is Your Corporate Culture Cultish?"[117] "The Cult of Company Culture Is Back. But Do Tech Workers Even Want Perks Anymore?"[118] "10 TECH GADGETS WITH A CULT FOLLOWING ON AMAZON—AND WHY THEY'RE WORTH IT."[119] "Why influencers are replacing fans with cults."[120] "Unchecked Influencers Are Cult Leaders."[121] "13 steps to developing a cult-like company culture."[122]

The headlines seem to write themselves (if that cliché is allowed anymore in the age of GPT-3 and generative AI). Tech is culty. But that is, like, a metaphor, right? *Right?!*

When I first saw Michael Saylor's Twitter account, I must admit I wasn't sure. Saylor is an entrepreneur, tech executive, and former billionaire. Once reportedly the richest man in the Washington, DC, area,[123] he lost most of his $7 billion net worth in 2000 when, in his midthirties, he reached a settlement with the US Securities and Exchange Commission (SEC) after it brought charges against Saylor and two of his colleagues at a company called MicroStrategy for inaccurate reporting of their financial results. But I had no idea who he was back then.

I first came across Saylor on Twitter. His profile picture showed a man with chiseled features, silver hair, and stubble sitting in a power pose and looking directly into the camera, a black dress shirt unbuttoned to display a generous amount of his neck. It was a typical tech entrepreneur's publicity shot except for the lightning bolts blasting from his eyes, and the golden halo crown. Then there were his tweets:

#Bitcoin is Truth.

#Bitcoin is For All Mankind.

#Bitcoin is different.

Trust the timechain.

Fiat [a rival cryptocurrency] is immoral. #Bitcoin is Immortal.

#Bitcoin is a shining city in cyberspace, waiting for you.

#Bitcoin is the heartbeat of planet earth.

I follow a lot of ministers, rabbis, imams, and monks online. Very few religious leaders would dare to be this religious on social media. They know that few of their readers want to see such literal hubris. Why, then, does there seem to be an audience for such seemingly cultish behavior from a cryptocurrency salesman? Are tech leaders like Saylor leading *actual cults*?

According to startup commentator and CEO Bretton Putter, this needn't be a major concern: "It's pretty much impossible," Putter writes, "for a business to become a full-blown cult." And if a tech or other business happens to resemble a cult, that might just be a good thing, he argues: "If you succeed in building a cult-like culture similar to the way that Apple, Tesla, Zappos, Southwest Airlines, Nordstrom and Harley Davidson have, you will experience loyalty, dedication and commitment from your employees (and customers) that is way beyond the norm."

Is that all? To find the answer, I interviewed a top expert on escaping destructive cults, also known as exit counseling.

At age nineteen, while he was studying poetry at Queens College in New York City in the early 1970s, Steve Hassan was recruited into the Unification Church—the famously manipulative cult also known as the Moonies. Over his next twenty-seven months as a member of the church, Hassan helped with its fundraising, recruiting, and political efforts, which

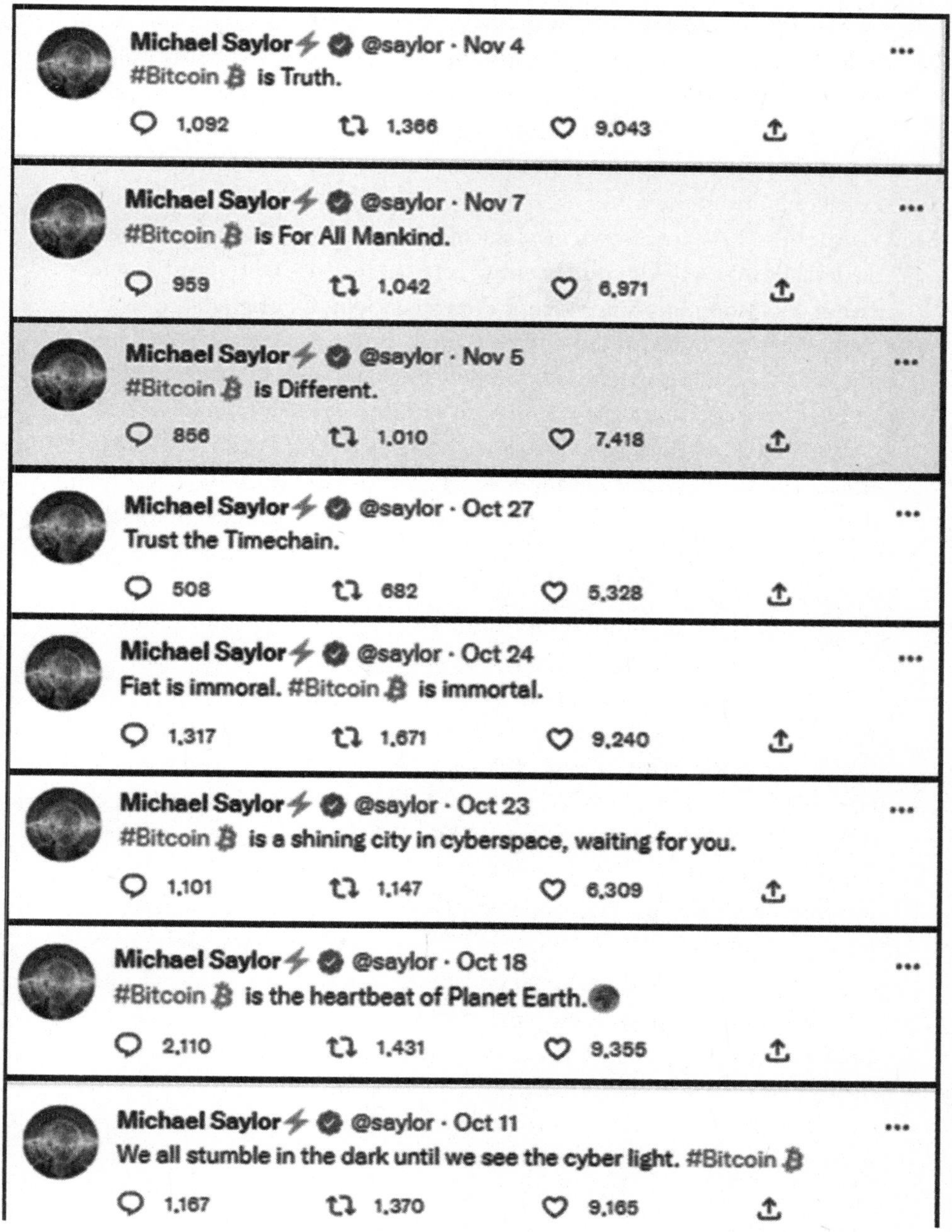

#Bitcoin tweets. *Source:* Michael Saylor.

involved personally meeting with the cult leader Sun Myung Moon multiple times. He lived in communal housing, slept only a few hours a night, and sold carnations on street corners. He turned his bank account over to the church. In 1976, after a serious car accident caused by falling asleep at the wheel, Hassan's parents hired counselors to help deprogram him and extract him from the cult.

After the 1978 Jonestown mass suicide and murders brought attention to the lethal dangers of cult mind control, Hassan founded a nonprofit organization, Ex-Moon Inc. Since then, he's earned a handful of graduate degrees (including a doctorate in the study of cults), started numerous related projects, and written a popular book on how practices with which he is all too familiar have crept into the mainstream of US politics in recent years: *The Cult of Trump: A Leading Cult Expert Explains How the President Uses Mind Control* (that book seemed even more relevant in early 2024, when a video called "God Made Trump" went viral across the campaign trail). Hassan even found himself advising Maryland congressman Jamie Raskin, leader of the second impeachment trial against Donald Trump in 2021, on how to think and communicate about the cultish aspects of the violent mob of Trump followers who stormed the Capitol on January 6 of that year.

I wanted to ask Hassan what he makes of the discourse around tech cults. Is it laughable to compare the serious problems he studies to what might simply amount to a lot of posturing by overambitious startup executives? Hassan's dissertation was titled "The BITE Model of Authoritarian Control: Undue Influence, Thought Reform, Brainwashing, Mind Control, Trafficking and the Law." The idea was to create a model that could measure cult exploitation and manipulation, or what Hassan and other experts in related fields call "undue influence."[124] Hassan's BITE model looks to evaluate the ways social groups and institutions attempt to control followers' behavior, thoughts, emotions, and the information they can access. Because there is no one quintessential, Platonic definition of a cult, what matters is where a given instance of potential cultishness falls on an "influence continuum." In this continuum model, Hassan evaluates the ways in which institutional cultures attempt to influence individuals, organizations, and leaders. To what extent are people allowed to be their authentic selves or required to adopt a false cult identity? Are leaders accountable to others or do they claim absolute authority? Do organizations encourage growth in the people who participate in them, or do they seek to preserve their own power over all else? While any kind of person or group can struggle with some of the dimensions on the chart below, healthier organizations will tend toward constructive responses more of the time, whereas unhealthier institutions, more truly worthy of the label *cult* in the most negative sense, will tend toward destructive responses such as grandiosity, hate, obedience, elitism, authoritarianism, deceptiveness, or hunger for power.

It turns out that there are some real, meaningful similarities between cults and tech, according to Hassan. "This is the perfect mind control device," he said, holding up his iPhone and pointing at it. He explained that when he joined the Moonies in 1974, cult recruiters had to get information from the victim. Now, he said, users of everyday technologies are sitting ducks. "There's five thousand data points on every voting American in the dark web, and there are companies that will collect and sell that data."

The first time Hassan was told about cryptocurrency, he added, it smacked of multilevel marketing to him. The proposal that you can make a fortune in a very short amount of time, with almost no labor, was something he had seen before many times in his work. As was the idea that if you become an early investor in such a dream, and if you recruit enough people to join you, the currency will get more and more valuable, and you'll make more money. "The people who started it are always going to make 99 percent of the money," Hassan said. And like the cults that recruited him and continue to recruit the kinds of people who ultimately become his clients, "everyone else is going to get burned."

All of this would certainly seem to explain why I so frequently hear from people eager for me to know they are fellow atheists who tell me to buy some Bitcoin because it will rewire my neurons and cure me of the woke mind virus.

Of course, it should be noted that some scholars have complained about Hassan's work that brainwashing and mind control are concepts for which there is not sufficient evidence, "especially through inducing altered states of consciousness."[125] But in any case, I'm not claiming that tech uses literal brainwashing, like when a character in a *Scooby-Doo* episode hears "you are getting very sleepy" and then their eyes become squiggles. I don't think Hassan is, either. As we'll explore more in chapter 5, companies don't need to go to such extremes to exert undue influence on us.

As is clear from the headlines I cited at the beginning of this section, a lot of companies have been accused of, or associated with, a bit of cultishness.

FREEDOM OF MIND®
RESOURCE CENTER

INFLUENCE CONTINUUM™

TO BE USED WITH THE BITE MODEL™
BEHAVIOR, INFORMATION, THOUGHT & EMOTIONAL CONTROL

CONSTRUCTIVE ←   → DESTRUCTIVE

HEALTHY   UNHEALTHY

FOR INDIVIDUALS	
■ Authentic Self	False (Cult) Identity ■
■ Unconditional Love	Conditional "Love" ■
■ Compassion	Hate ■
■ Conscience	Doctrine ■
■ Creativity and Humor	Solemnity, Fear and Guilt ■
■ Free Will / Critical Thinking	Dependency / Obedience ■

FOR LEADERS	
■ Psychologically Healthy	Narcissistic / Psychopathic ■
■ Knows Own Limits	Elitist / Grandiose ■
■ Empowers Individuals	Power Hungry ■
■ Trustworthy	Secretive / Deceptive ■
■ Accountable	Claims Absolute Authority ■

FOR ORGANIZATIONS (AND RELATIONSHIPS)	
■ Egalitarianism	Elitism ■
■ Checks and Balances	Authoritarian Structure ■
■ Informed Consent	Deceptive / Manipulative ■
■ Individuality / Diversity	Clones People ■
■ Means Create End	Ends Justify Means ■
■ Encourages Growth	Preserves Own Power ■
■ Free to Leave	No Legitimate Reasons to Leave ■

FROM COMBATING CULT MIND CONTROL (2018) BY STEVEN HASSAN

freedomofmind.com

The influence continuum from Steven Hassan's BITE model.

It's beyond the scope of what I'm trying to do to name names and evaluate everyone on a scale of one to ten. I will just say that crypto is a particularly good example, because if there's one thing cults need to be good at to sustain their existence, it's separating people from their wallets. Crypto has specialized in that to extraordinary effect.

It's all a continuum, and it would be hard to find a person whose life doesn't involve any involvement with something cultish, technological or otherwise. But as a culture, we are careening dangerously towards the wrong end of Hassan's chart. Or to quote a Michael Saylor tweet, "We all stumble in the dark until we see the light. #Bitcoin."

FANATICISM

I began this chapter on the tech religion's doctrine with Sabbetai Zevi, the seventeenth-century Jewish messianic figure, to raise the question of how one might know whether a messiah has truly arrived. This question also shaped the life and career of Jacob Sasportas. Sasportas, a bookish and cantankerous seventeenth-century rabbi who was born before Zevi and outlived him, spent most of his long life and career serving impoverished communities of Jews who had fled forced conversion to Catholicism—this was only a few generations after the Spanish Inquisition—and then teaching Jewish law to the descendants of converts, seeking to rediscover a Jewish identity that had been largely lost to their families. According to a portrait from 1670, Sasportas had a long, neatly trimmed, silver beard; a heavily creased forehead; and large, piercing eyes that were reddened around the edges, as if from excessive study.[126]

Twice he had to leave his home country as a refugee. Almost the definition of an obscure figure in the history of religion, his story would seem to have no place in a book about twenty-first-century technology, except that, as an aging man, he suddenly became the leading critic of what was, in his day, the world's most energetic and powerful messianic movement since the days of Jesus Christ himself. At a time in which it seemed almost no one was willing to publicly disagree that a savior had arrived—and that once-cherished rules and norms of life were now meant to be flamboyantly broken—Sasportas dared to doubt. Failure to believe in a genuine prophet was itself an explicit violation of the ancient religious texts rabbi Sasportas held dear.[127] Still, he wrote and spoke out against the idea that one should

Portrait of Jacob Sasportas by Isaack Luttichuys (1616–1673).

simply take a figure such as Sabbetai Zevi on faith. As Sasportas said in rebuking his colleagues, "thus sayeth the Lord," as Zevi's followers tended to proclaim in justifying their belief, was not enough for him.[128] "A sage," he wrote, quoting a Talmudic phrase to refer to the kind of philosophically inclined scholar he aspired to be, "is better than a prophet."[129]

Perhaps nowhere does it seem more obvious to me that EAs like William MacAskill have descended into Sabbetai Zevi–like levels of prophecy than in their discussions of what EAs themselves call their own "fanaticism."

In EA terminology, *fanaticism* refers to the idea that humanity might be faced with a good chance of saving a thousand people now or a tiny chance of saving billions or trillions of people—or some other digital lifeform—in the far, far, *far* future. MacAskill and others conclude that it would be too "timid" to choose the less risky choice.[130] This is logic twisted into a pretzel at the very best. Imagine explaining your decision to those people: "Sorry, I can't help you out of the ditch you've fallen into, where snakes are about to bite you. I need to marshal my resources for a risky bet that I might be able to single-handedly solve climate change in twenty years, and if I make a practice of helping you now, I might be too tired to do so then."

But we need not wonder about such fanciful situations or suffer through my poor attempts at satirical dialogue, because this sort of trade-off is already happening all over the world. "Fanatical" EA thinking is all too common, as we saw with Pascal's mugging and other heavenly justifications for hellish deeds.

"Which is better," asks EA Hayden Wilkinson, currently a postdoctoral research fellow at Oxford University's Global Priorities Institute, "a guarantee of a modest amount of moral value, or a tiny probability of arbitrarily large value? To prefer the latter seems fanatical. But, as I argue, avoiding such fanaticism brings severe problems."[131] In his paper, which contains several dozen equations and graphs, Wilkinson argues for fanaticism by comparing it to various kinds of lotteries, with thought experiments accompanying each comparison to show why we must accept that the fanatical longtermist lottery is better than the less risky lottery. The paper takes readers on a tour of ancient Egypt and ancient India to be told that what happened there, or didn't, and how much we know about it, or don't, might be relevant to the question of whether we should be effective altruist "fanatics" now. This sort of thinking about ethics in terms of lotteries reminds me of an old Lily Tomlin line that my friend Elizabeth Kennedy, the ethics professor who witnessed my own extreme thinking when we were teenagers, used as her high-school yearbook quote: "The thing about the rat race is, even if you win, you're still a rat."

If you want to throw a couple of dollars a day at life's problems, in the "you never know" frame of mind or as a fun fantasy, fine. But when you begin to live your life around a lottery, or a lottery mentality, you tend to become a gambling addict. Wilkinson's thesis that such "strong" thinking is necessary to avoid existential risk and achieve humanity's long-term goals is perhaps surprising, but hardly unusual among mainstream EA thinkers.

In their paper "The Case for Strong Longtermism"—"strong longtermism" being a close corollary of the EA position called "fanaticism"—EA movement cofounder William MacAskill writes (along with Hilary Greaves),

> We will simply note that even if, for example, there is an absolute cap on the total sacrifice that can be morally required, it seems implausible that society today is currently anywhere near that cap. The same remark applies to at least the vast majority of individuals in rich countries. We ought to be doing a lot more for the far future than we currently are.[132]

By now, readers of this chapter might be forgiven for assuming that I disagree with essentially everything MacAskill and his ilk have to say—that I would even disagree with them if they told me they'd eaten breakfast in the morning. In fact, I agree with and appreciate the sentiment I've just quoted. The problem is that MacAskill and Greaves then go on to say, "The potential future of civilisation is vast. Once we appreciate this, it becomes plausible that impact on the far future is the most important feature of our actions today."[133]

This is different from the Indigenous or Native American philosophy of "seven generations." That is a principle and way of thinking whereby the peoples native to the land on which I write this believed they had a deep responsibility to care for their world and leave it better for the great-grandchildren of their great-grandchildren.[134] The notion strikes me as not only a reasonable ethical standard to apply but also as a deeply honorable effort toward current and future sustainability. The EA version of future stewardship, on the other hand, while it seems to come from a place of genuine desire to do good, is marred by grandiosity. Perhaps it represents, in part, an effort to produce thought that is "significant"—as in Eliezer Yudkowsky's statement that AI leaders would rather destroy the world than be *insignificant*. But effective altruists, transhumanists, and their followers and intellectual peers unfortunately make a serious ethical misstep when they cast contemporary notions of sustainability as a *bad* thing in comparison to fanaticism and strong longtermism.

In a 2012 paper entitled "Existential Risk Prevention as Global Priority," Nick Bostrom writes, "We should perhaps therefore not seek directly to approximate some state that is 'sustainable' in the sense that we could remain in it for some time. Rather, we should focus on getting onto a developmental trajectory that offers a high probability of avoiding existential catastrophe."

To Bostrom and to the EAs who are deeply influenced by his work,[135] sustainability is a quality to be avoided insofar as the maintenance of even a high quality of life for humanity over the coming centuries might keep humanity from attaining longer-term goals like populating the Virgo Supercluster of galaxies with a septillion conscious beings on the far horizon of time.[136] Because our efforts to do "little," "sustainable" things like protect our environment, prevent millions or more from becoming climate refugees, fight racism and sexism and economic inequity, and so on (again, these are all genuinely less important to Bostrom and others than combatting existential risk) might result in what Bostrom calls "permanent stagnation," where humanity never reaches "technological maturity."[137]

Even if everything else about the human future here on earth were to work out wonderfully, according to this mentality, history could only amount to an unfathomable loss. We would fail in the ultimate colonial project. We would disappoint the potential of our most powerful God(s). We would bastardize the ultimate purpose of all our philanthropic and altruistic efforts. We would never reach heaven, so eventually, we would all go to hell. All because we would never fulfill the purpose for which our greatest and most ingenious minds were chosen.

Yes, I am saying that the AI philosophers and communities that appear in this chapter tend to behave and think in cultlike ways, and that their own characterization of themselves as "fanatical" would seem to be apt. But not all effective altruists or transhumanists—or all the billions of more traditionally religious people—are extremists. In fact, very few are. To assume otherwise would be to make a mistake that atheist critics of religion like Richard Dawkins, Christopher Hitchens, and Sam Harris have all too often made. When such critics assume that an entire religious tradition can be painted with a broad brush covered in the blood-red paint of that tradition's most extreme subsect, they're wrong. So would we be if we simply wrote off all members of a group like effective altruism as irredeemable.

I've always tried to be a pluralist, to work with people of different faiths. Tech religions should merit the same treatment. I may disagree with EAs on many things, just as I disagree with religious Christians or Hindus on their theologies. But I would still work with them to make the world better. I don't think their beliefs make them inherently worse than other people.

It is also true that sometimes moderate religious believers are too influenced, in their thinking or behavior, by their most extreme coreligionists. This doesn't make religious moderates equally bad. It doesn't mean they need to become atheists like me, or that their religion isn't legitimate. But fanaticism is still something to be avoided—in traditional religions, in effective altruism, and in other tech sects. There is too much extremism, and such extremism is too influential, in these mostly well-intentioned communities. Tech's most influential doctrines are desperately in need of reform.

The point of this extended analysis of tech and AI doctrine was not to anticipate every coming development in AI or tech religious thought. Once religious thinking takes hold and forms the basis for a large and dynamic movement, we only know for sure that one kind of prediction will come true: there will always be new offshoots, new movements and sects, and new and continually evolving beliefs to come.

In early 2024, I reached out to Dr. Nirit Weiss-Blatt, a researcher of communication about tech and AI, after noticing Weiss-Blatt's prediction that "EA's reputation is deteriorating, and backlash is coming."[138] When I asked her to expand on that idea, Weiss-Blatt noted that there have been recent "pendulum swings" by which technologies' reputations swung from "savior" to "threat," without much middle ground, i.e. the "Techlash."[139] The same goes, she wrote to me, for tech ideologies—they can rise and fall as well. After spending over $500 million to promote the idea of existential risk, she pointed out, in 2023 the EA movement and its fellow travelers had successfully generated a swing towards massive attention for the ideas we've discussed above, including magazine covers warning of "THE END OF HUMANITY";[140] opinion pieces suggesting bombing data centers;[141] and "AI forums" in which heads of state and CEOs alike called for sweeping regulatory intervention.[142] "Gradually, then suddenly," she wrote to me, "the AI panic was everywhere." Until it wasn't.

With increased scrutiny, Weiss-Blatt noted, people have increasingly noted that "x-risk" is a "billionaire-backed fringe movement," and that "The FALL" is coming, "along with a reckoning" about how such ideas became so influential in the first place.[143] Will her prognostications turn out to be true? Only time will tell, though in early 2024 at least, tech discourse seemed to have simply swung from doomsaying toward a feverish new "form of spirituality" called "effective accelerationism"—the idea that AI progress can and must speed up—and a general sense that a semi-benign AI takeover was imminent if not already upon us.[144]

Sabbetai Zevi persuaded the majority of the world's Jews that he was their savior. His international movement then began to revolutionize tradition in expectation of the coming of heaven and the resurrection of the dead. Then he faced reality: the Ottoman sultan, displeased by the notion of any individual claiming powers greater than his own, gave Zevi an ultimatum: convert, or die.

Zevi avoided death at the hands of the sultan's regime by converting publicly to Islam. The would-be Jewish prophet became a Muslim and a loyal subject of the Sultan, eventually serving as the chief doorkeeper of the personal spiritual counselor to the grand vizier, the second-highest-ranking official in the empire. Beyond discussions among religious and academic scholars and the sharing of anecdotes like this one, he has mostly been lost to history.

We now have hundreds of years to look back on this story. We know the ending of this false messiah's tale in a way that, today, we can't possibly know the future of AI—a reality that today's entrepreneurial prophets understand well. Because it means they can promise whatever suits them. Unable to verify their statements, we can only believe. It will be interesting to see how the legacies of prophetic figures like Nick Bostrom, William MacAskill, and Eliezer Yudkowsky evolve over the coming seven generations and beyond.

II

PRACTICES

In the first part of this book we explored three core dimensions of the beliefs that make tech the most powerful religion in the world today. In chapter 1 we examined the thought systems and values that undergird tech culture and the many ways in which they resemble a kind of latter-day theology. In chapter 2 we focused on the tech world's fixation on predicting future opportunities for profit, rather than reckoning with the ethical challenges of the present, and how an obsession with very distant and extraordinarily unlikely future scenarios can resemble religious tendencies to use visions of heaven to distract from earthly problems. We also explored the psychology of the prophetic figures who dominate tech and how the startups they found and fund often come to resemble literal cults, with users, workers, and other innocent bystanders as their victims. We even looked at rising belief in the arrival of new tech gods and how artificial general (or super) intelligence, or AGI/ASI, has become a kind of new mythology that demands our faith, fealty, and above all, astronomical amounts of funding.

This second section of the book will focus on the practices that give the tech religion overwhelming power to shape our lives and determine our collective fate. In the next chapter I will discuss the ways in which our everyday (or perhaps every hour, every minute, or every second) tech use has come to resemble a feverish ritual. Daily prayers pale in comparison to the influence of the almighty screen, to which the average American now genuflects close to two hundred times a day. And in the third and final chapter of this part, we will consider how the beliefs and practices of the tech religion stand a puncher's chance of combining to cause an actual apocalypse.

First, however, we turn to hierarchies and castes. One of the worst and most obviously dangerous aspects of religious tradition, the notion of rigid social strata, justified by faith and hardened by piety, is also one of the most pervasive structures across most if not all of the world's major religions.

It would be wonderful if tech culture lived up to its own legends, which proclaim it to be free of such atavistic practices. But that is far from the case. Tech is steeped in just the sort of hierarchical thinking characteristic of most organized creeds since perhaps the dawn of agricultural times.

At its worst, tech's tendency to pronounce itself an egalitarian and meritocratic utopia for all humanity, when in fact it is based and built on clear, present, and dangerous hierarchies, represents what biologist Carl Bergstrom calls *agnatogenesis*: "creating uncertainty or doubt to stave off regulatory action."[1] Even when advanced unintentionally, however, the myth of hierarchy-free, caste-free technology only makes the tech religion more racist, more sexist, and more exploitative.

But let me share a story before we dive into any of that.

3

HIERARCHIES AND CASTES, OR, UTOPIA FOR WHITE MEN

One thing tech fundamentally has in common with many religions, at least in America, is that it is a white man's version of Utopia. And tech especially has this cult-like adherence to a white man's vision of a Utopia that fundamentally disempowers and endangers women and people of color.
—Ijeoma Oluo, "So You Want to Talk about Race in Tech," June 13, 2020

RELIGIONS HAVE HIERARCHIES

It's January 2020 and I'm in San Francisco, in the spacious living room of my friend Miles Lasater, a newly minted venture capitalist. Of above-average height and athletic build, with dark blond hair and a bright smile, he greets my family and me in an Andrew Yang T-shirt and invites us upstairs into a single-family home that a realtor would describe as awash in sunlight. His kids are with a nanny, so we can catch up. Miles and I haven't seen one another or talked in a few years. Much has changed.

When Miles and I met in the early 2010s, he was wrapping up a remarkable run of success as a purpose-driven tech founder in New Haven, CT. As an undergraduate computer science major at Yale in the sweet spot of the first dot-com boom, he and a couple of friends got the idea to found Higher One. A startup that made early use of evolving technologies to provide new ways for college students to access financial-aid services, Higher One propelled Miles into a vaunted world, leading a company with over $200 million in revenue, over 750 employees, and a listing on the NYSE. Miles treated the work as a secular higher calling, emphasizing openness and ethics in the company's culture. Higher One ranked highly on a national list of great places to work in addition to its fast growth.

Before turning thirty, Miles added the title of angel investor to his résumé. By thirty-five he was settled into life as a husband, father, and tech and philanthropic leader, formulating a plan for giving back and building a better world.

That's where I came in. Like many other well-educated, philosophically inclined young men, Miles had taken an interest in secularism, humanism, atheism, and related ideas. The idea of living a moral life without religion has been around in one form or another for virtually all recorded human history (as I detailed in *Good Without God*), but as Miles's business career was taking off, so were new dynamics that led to a rise in popularity and visibility for humanism and atheism. With his passions for humane values and positive organizational culture, Miles found himself drawn to the idea of humanist morality, and so became one of the first successful tech CEOs of the twenty-first century to seek out dialogue with humanist and atheist organizational leaders like me.

When I first met Miles for lunch to discuss what ultimately became a nearly six-figure donation to the Humanist Chaplaincy at Harvard, I remember him taking out his early-model iPhone to snap a picture of his plate when it arrived. I assumed it was a savvy trick to track calories. "It's more of a mindfulness thing," he said earnestly. I felt slightly ashamed I lacked Miles's *presentness*. Still, I later invited him to my wedding and came to think of him as a friend. (I didn't even realize that, at the time, Miles's startup Higher One was making the news for what a Federal Reserve governor called "deceptive marketing practices" by which students were "misled into paying fees to access their financial aid funds.") Miles left Higher One in December 2014, right around the time he made the grant to our organization, becoming, at the time, our biggest donor.

When I meet Miles in January 2020, we haven't seen each other, or even spoken much, in a few years. My humanist center at Harvard is now closed, and I've stopped raising money for such efforts. He's packed up his New Haven home—and his former dream of a humanist center at Yale—and moved his family to San Francisco, where he can move on to a new career in venture capital and investing.

He tells me about his latest pet project, a crowdsourced Google doc titled "What's wrong with VC," reflecting, with VC colleagues, on the many problems of his newly chosen profession.[1] Then Miles's kids storm into the room, and noticing him glancing at the time, I remember: *Leaders are busy.* So I launch into an elevator-pitch description of the book you are now reading.

"Ultimately, I decided to put aside the idea of building alternatives to religion," I explain, "because I've become convinced that tech itself has become a religion."

After listening for a few minutes with characteristic attentiveness, his response is terse, but memorable.

"Sounds interesting," he says, head tilted with a slightly skeptical-looking smile, "but I'm not sure tech is really like a religion. I mean, religions have *hierarchies*." We don't get to discuss this further. His time is up.

I tell this story not to mock or demean Miles, whom I'd still consider a friend. Because I did not record our conversation at his home, I asked for his approval, over two years later, to write up this recollection and mention him by name here. I appreciate and admire the integrity he showed when, even though I suggested that my anecdote might not reflect particularly well on him, he trusted my memory of the moment and gave his permission freely.[2] That said, the point of my story is that Miles was wrong, and that he of all people should have known better. Yes, religions absolutely do have hierarchies—and so does tech.

In this chapter I will argue not only that tech, like religion, is extremely hierarchical, but that tech's hierarchies came to exist for similar reasons to those of religion: to make life easier for some by making it more difficult for others. The parallels between these secular and religious hierarchies only continue from there. In both, neither the oppressed nor the oppressor will naturally accept or feel comfortable with an exploitative societal arrangement. Tech's hierarchies, like religion's, must be justified by an elaborate mythology that helps persuade all parties that things are exactly as they are supposed to be.

In both religion and tech, furthermore, hierarchy-justifying myths take two forms. First, they convince people that existing social stratification is merit-based. Second, and simultaneously, parallel narratives and mythologies serve to convince people that social hierarchies don't really exist or aren't as bad as they seem. If these two ideas seem mind-bendingly contradictory, that is intentional. Seeming contradictions have been deployed in mystical religious texts since ancient times, in part as a way of overcoming challenges to the logic of an argument by pronouncing the truth behind that argument to be ineffable, or beyond reason.

I conclude the chapter with an argument for how tech agnosticism might improve our understanding of, and even our ability to overcome, the religious hierarchies of contemporary tech. Because to be a tech agnostic requires living in ambivalent relationship with tech. It means fighting back against tech's hierarchies while still using tech and attempting to participate justly in its systems. This will be a major challenge, and no one—certainly not me—knows perfectly how to get it right. Nonetheless, we can and must try to find a balance that feels authentic.

First, however, let's think briefly about what religious hierarchies are, how they came to be, and what purposes they serve.

ON RELIGIOUS HIERARCHIES

Religious hierarchies are systems of organizing people and societies around the notion that some people are better than others, or some exist to serve others, or some at least have more rights than others. Such hierarchies are generally ingrained in the fabrics of religious systems that include much more, so there is a tendency on the part of the modern observer to assume that hierarchies are set up for reasons consistent with a broader internal logic that coherently organizes the entire system. For example, Jewish laws about how men are obligated—and therefore permitted—to pray in ways that women are not are imagined to be a necessary and rational part of dividing labor by gender. Some may choose to believe that this division projects all the way back to God's will for Adam and Eve. Others may not resonate with such a sweepingly theistic rationale but still believe the originators of religious laws had deep insights into the biological or psychological nature of gender and humanity, which justified their choices then, whether or not those choices continue to be justified.

Often, however, the choices made to place people into spiritually sanctioned hierarchical categories were not entirely logical. They were made because life is hard, and one of the oldest strategies available to manage difficulty is to get someone else to bear more of the burden than you. It was convenient for the short-term interests of white Europeans for Black Africans to provide free labor or for Indigenous peoples to cede land. Indian Brahmins benefited materially from the oppression and subservience of an "untouchable" class. Men in virtually all the world's major religions have gained short-term sexual gratification, easier workloads, and control over child-rearing by relegating women to second-class status. This list could go on. The point is: simply declaring such hierarchies outright would have several drawbacks. All parties might be less likely to obey them if not divinely required to do so. The "beneficiaries" of these arrangements (in scare quotes because I do not believe such hierarchies have been good for *anyone* in the long run, morally and spiritually speaking) might have been (even more) challenged to maintain dignity and self-respect, not to mention authority, without a shared sense that they were acting according to the will and plan of a god. So structures were created. Theologies invented. Selves convinced—and others persuaded, sometimes by force. Systems and explanations were patched onto the impulse to dominate and exploit after the fact.

Especially insightful into this dynamic was the late Jonathan Z. Smith of the University of Chicago Divinity School, one of the most seminal scholars of religion of the past generation. (Incidentally, Smith, who wore his hair and beard long and flowing along with elegant three-piece suits and taught for decades up to his death in 2017, never used computers, despised the telephone, and called the cell phone "an abomination.") As Smith argued, one of the fundamental building blocks of religion is "its capacity for rationalization, especially as it concerns that ideological issue of relating that which we do to that which we say or think we do."[3] In other words, there is a wide gap between what people want (for example, comfort, pleasure, esteem) and what they want others to think of them as being like—or even how they want to understand themselves (as selfless, noble, ethical)—and religion is an attempt to line up the two by telling a story of their essential sameness. There is no contradiction between who we are inside and who we want to be, we tell ourselves and others through rituals meant to sanctify and purify us and grant us the favor of the gods, despite any failings or sinfulness we may perceive in ourselves. Religion and its rituals, for Smith, are a "means of performing the way things ought to be in conscious tension to the way things are."[4] Social hierarchies become religious in precisely this way, because theological beliefs are used to reconcile the benefits gained by such hierarchies with our potentially contrasting self-image. We will see below that tech uses similar psychological techniques, similar myths, to justify its own impactful hierarchies.

A note of caution and necessary humility before I dive into my argument: this chapter presents a dilemma for me, as an author, that I had best disclose before I begin. To explore the hierarchical nature of the world of tech—because I am certain this topic is essential for *any* attempt to understand the history and culture of modern technology, let alone one that seeks to compare technology and religion—I will need to rely heavily on the work of others. In doing so, I am especially mindful of at least two potential risks. First, I'll be speaking about problems faced by members of marginalized communities, while I have benefited from some of the very systems of marginalization that have caused their problems. Second, to speak about tech hierarchies in an evidence-based way, as a writer whose expertise lies elsewhere, I'll be referencing several longer, more in-depth works of careful and important scholarship, mostly by women and people of color, for which I do not wish to take credit in any way. For these reasons, I considered simply omitting this chapter from *Tech Agnostic*, but I do not think that would do my argument, or my readers, justice. These issues are simply too important to ignore or even give limited attention in this book. Instead, I have tried to focus, in this chapter and elsewhere, on ways in which my voice and perspective might best contribute to the overall project of acknowledging and ultimately correcting bias and bigotry in tech.

TECH'S UNSEEN SKELETON

Now to address my friend Miles Lasater's statement directly: Does tech have hierarchies? The answer may seem obvious to some readers, but bear in mind this is nonetheless an open question. I began this chapter with Miles's story, in fact, to demonstrate that even an obviously thoughtful and well-educated tech leader—one strongly inclined toward both philosophy and ethics—was capable of imagining today's highly influential and powerful tech world to be flat, egalitarian, and meritocratic. And Miles is—or was, at that moment—hardly alone. Actually he would have been in good company, especially among the richest of tech's rich. According to a 2021 sociological study of the worldviews of the tech world's hundred richest people, "the tech elite has a more meritocratic view of the world than the general US Twitter-using population."[5] But are these elite figures correct in their views? Let's glance at some of the evidence of hierarchies of race and gender in hiring, employment, and compensation.

According to a US Equal Employment Opportunity Commission[6] analysis comparing the largest tech companies' employment practices to private industry as a whole, tech businesses are significantly worse than average on diversity, equity, and inclusion (and it's not like the rest of American businesses are setting an amazing standard): Tech employs larger percentages of whites (68.5 percent to 63.5 percent among businesses overall), Asian Americans (14 percent vs. 5.8 percent), and men (64 percent to 52 percent), and smaller percentages of African Americans (7.4 percent to 14.4 percent), Hispanics (8 percent to 13.9 percent), and women (36 percent to 48 percent). Over 83 percent of big-tech executives are white and 80 percent are men, compared with the overall private sector, where 83 percent of executives are white and 71 percent are men.[7]

How about Black leadership in technology? The status quo is very bad across major industries: When *USA Today*[8] analyzed the fifty largest companies in the Standard & Poor's 100 (most of which are, in at least some significant way, technology businesses), of all the executives listed in the proxy statements, only five (1.8 percent) were Black. And two of those executives had recently retired. But that's no excuse for self-identified tech companies, which publicly justify their worth and value to society by the good they supposedly do in "bringing us closer" and such.

Venture capitalist Richard Kerby put it this way in an analysis of the share of VC investors and dollars: not only are 95 percent of venture capitalists either white or Asian, not only are 40 percent educated at either Harvard or Stanford (or both), but the numbers get even worse when viewed differently.[9] In analyzing VC firms not only by number of employees but including partnership and ownership of the firms, Kerby concluded[10] that 93 percent of VC dollars are controlled by white men, with 3 percent in the hands of white women, and less than a percent in the hands of any other category of investor. If that kind of stratification doesn't sound to you like it reaches biblical proportions, I don't know what to tell you. And yet many of the leaders who most need to see these discrepancies literally cannot. It's not so much that wealthy tech entrepreneurs are lying when they say their privilege is based in fairness and opportunity for all (though surely some are being dishonest). It's more that they've been indoctrinated into an alternative worldview—one that emphasizes their own goodness, sincerity, and merit at the expense of realism, reason, and truth. As the authors of the study on wealthy individuals' meritocratic views indicated, that extreme wealth allows or even encourages people who believe themselves to be genuinely committed to democracy and equality to act against those values while believing themselves to be acting *for* them. Is this hypocrisy? Surely there is some of that, but, the authors assert, it would be more accurate to imagine a mass system of self-deception whereby tech leaders who genuinely want to be good people learn to act in their own self-interest within a cocooning environment in which yes-men and sycophantic institutions thank them so repeatedly and vociferously for their contributions to humanity that they come to believe their own hype. This is not hypocrisy but real, often missionary devotion to the many political and philanthropic causes they support. At the same time, members of the tech elite frequently fail to understand how their activities unavoidably (if inadvertently) undermine democratic equality. One might well regard this lack of understanding as a kind of false consciousness—the false consciousness not of an exploited class but of a privileged, yet in certain respects precarious, elite.[11]

The false consciousness at the heart of stratified, hierarchical tech wealth and power is the same as that at the heart of religious hierarchies of wealth and power. Because wherever people have been allowed to dominate and control their fellow human beings—whether at the level of a village, a city, or an empire—those in positions of power have always sought to maintain their self-image as a force for good, even while their actions often make life worse for most of the other humans around them. In both religion and in tech (and, of course, in what I'm calling the tech religion), this combination of earnestness and exploitation has both strengthened and hidden hierarchy. And if it still seems unrealistic that such an elephant of a social dynamic could somehow be hiding in plain sight, where even otherwise well-trained minds can fail to perceive it? Then perhaps it would be helpful to look to the example of the most overwhelmingly powerful religious hierarchies in history—caste systems—for additional parallels to the world of tech.

TECH CASTES?

Caste is a form of social stratification characterized by religiously and socially determined hierarchical status resulting in social and political privileges for some and exclusion for others.[12] Largely but not exclusively manifested in

India, the caste system places Brahmins, the Hindu priestly class, at the top of the social hierarchy, followed by royal or warrior classes, a business class, and workers or peasants. Dalits, formerly known as "untouchables," are the lowest stratum and treated as outcasts of society. Though the concept of "untouchability" was banned by the Indian national constitution of 1950, the impact of its centuries of practice lingers even today for over 200 million Dalits in India, as well as for the growing number of Dalits now living in America and other wealthy nations, many of whom arrive as university students.[13] In my role as a chaplain at two universities, I have spent time speaking with and advising students of Dalit background and therefore hearing their often harrowing stories of facing sometimes lethal levels of prejudice, bigotry, and exclusion.

The origin of the Indian concept of caste is ancient enough to be unknown, though the *Manusmriti*, an early definitive text of Hinduism dating back some two to three thousand years, speaks of an Indian society divided into four classes, justifying the system as "the basis of order and regularity of society."[14] The idea of untouchability seems to have emerged later, with a number of scholars debating its origins and original purpose. What is clear, however, is that the Indian Hindu caste system is the most formalized and longest-running system of divinely justified oppressive social hierarchy. That said, there are many other systems that have justified the degradation of a given class of people, often by relegating them to treatment as more or less "untouchable": hereditarily unclean, defiled, or fit only for certain menial tasks. European Romani or so-called Gypsy people may be included, as might other groups of hereditarily marginalized people in countries such as China, Japan, Korea, Tibet, Yemen, and Nigeria.

The term *caste* does not derive from any Indian language. It originated with the Portuguese *casta*, meaning "pure breed."[15] In the United States, the biblical story of the curse of Ham was deployed to justify the oppression and enslavement of Black people. It is not clear that the actual biblical character of Ham, Noah's son in the Book of Genesis, has "sinned" against God in any discernible way other than having seen "the nakedness of his father," nor are his race or skin color ever mentioned—but such petty facts have rarely stood in the way of those who wish to subjugate.[16] The notion of a curse on Ham was even used to officially justify the exclusion of Black people from Mormon priesthood into the twenty-first century.[17] And while many think of the entire concept of caste as having existed only in faraway places like India, in recent years a compelling argument has been made that the phenomenon is very much at home and alive here in the United States.

"America," argues Isabel Wilkerson, author of *Caste: the Origins of our Discontent*, "has an unseen skeleton: its caste system, which is as central to its operation as are the studs and joists that we cannot see in the physical buildings we call home. . . . Looking at caste is like holding the country's X-ray up to the light."[18] Wilkerson, the first African American woman to win a Pulitzer Prize, argues that America's history of inequality and exploitation can best be understood not as a series of random accidents but as the result of a system. And we can best understand that system not just as the simplest form of racism—blind hatred stapled to cynical politics—but as something more focused, more complex, more *religious*. Describing eight pillars or key features of caste across systems, Wilkerson points first to divine will, or the belief that social stratification transcends human authority or decision making, tracing itself back to the laws that define the universe and all existence. Ultimately, in Wilkerson's analysis, what ails America is the direct result of a semiorganized, malignant faith that despite the American Founding Fathers' earnest and impassioned pleas for democracy and equality, some people are simply better, more deserving, and more human than others.

The faith in hierarchy that Wilkerson describes should sound reminiscent of the logic one finds in tech circles, though, importantly, not necessarily to the same degree. I want to stop short of suggesting that tech hierarchies constitute the equivalent of a full-blown caste system; even most of history's countless and often severe examples of religious hierarchies do not rise to that level. In fact, Wilkerson writes that there have been three main examples of true caste systems: "The lingering, millenniums-long caste system of India. The tragically accelerated, chilling and officially vanquished caste system of Nazi Germany. And the shape shifting, unspoken, race-based caste pyramid in the United States."[19] It would be irresponsible to equate modern tech culture with systems in which one class of people cannot be touched by members of other classes, or that are characterized by concentration or extermination camps, systematic mass murder and enslavement, etc.

That said, there is plenty of room to explore countless other oppressive religious hierarchies that don't rise to the level of a caste system but are nonetheless abhorrent, or at least very concerning. Though there are too many examples to list, it is worth mentioning the refusal of clergy or other holy status experienced by women in many Christian, Jewish, Muslim, Hindu, Buddhist, and other religious traditions and communities, as well as the general religious discrimination and prejudice against women and LGBTQ+ people, that has marked most of these traditions throughout history. There are also class divides within many religions, and there are many elaborately conceived levels of stratification within each level of these systems, as there are in Indian caste, with its thousands of castes and subcaste levels and distinctions within its main categories (*varnas*).

In Judaism, the Cohanim and Leviim, a priestly and parapriestly class, have enjoyed elevated status throughout much of Jewish history. As my friend and fellow secular humanist rabbi Adam Chalom reminded me, the architecture of the ancient temple in Jerusalem, with its concentric courts, was both a symbol and a manifestation of religious hierarchies. The Holy of Holies, the inner sanctum of the biblical temple, could be entered by the high priest only. The next-innermost chamber could be entered by Cohanim (priests), followed in concentric circles by those deemed lower: Levites, Judean men, Judean women, and finally "the nations"—non-Jews.

In short, calling a hierarchical social system a caste system may be complex, but acknowledging the long, wide, and deep history of religious hierarchies ought to be a straightforward matter. But if tech hierarchies are religious, what kind of religious purpose do they serve?

UTOPIA, FOR WHITE MEN

Just a few days after my meeting with Miles in San Francisco, I found myself driving a rental car north from a posh Seattle neighborhood (filled with glittering architectural swank thanks to new Amazon money) to the city of Shoreline, Washington. To get to Shoreline via Interstate 5, I had to pass colorful clusters of makeshift homes fashioned out of tents, patio umbrellas, and plastic tarps, with dozens of overflowing shopping carts and recycling bins serving as both storage and furniture, stretching for hundreds of yards. They were among the largest encampments of unhoused people that I've ever seen in my life. And I've led interfaith groups of students to study and volunteer in such encampments. (In chapter 6 I'll talk about the homelessness encampments I was to visit eighteen months later in San Jose, California, near the giant Apple campus. These sites in San Jose were even bigger than what I saw en route to Shoreline.)

I drove to Shoreline to meet Ijeoma Oluo, a writer whose book *So You Want to Talk about Race* has been among the most influential over the past generation in conversations about racial justice in modern America. We met not far from Oluo's home, at Shoreline's One Cup Coffee: a no-frills "more than profit" coffee shop that shares a storefront with a church and is just down the road from a methadone clinic. Oluo, an elder millennial, grew up in a Seattle well into its transformation to a technopolis; she worked in tech for over ten years in a wide range of roles, from sales to pulling out circuit boards with her bare hands, before becoming a professional writer.

"A lot of people denigrate the value of talking about race and racism in technological spaces," she offered after some basic introductions over mugs of coffee and chai. "I don't think there's a more important space to be talking about it." She'd grown up seeing "the absolute best and the absolute worst in race and racism in America" on the Internet and in tech culture, she explained, "in ways that have had true-life consequences for me and for people I love." Initially, she left the idea of such consequences to generalities. I would later learn that in addition to the anonymous threats one commonly hears about authors receiving online, her family's safety had been more tangibly threatened by a young white man who swatted her family, calling the police with the kinds of fake but credible-sounding threats that have gotten others hurt or worse. At one point in the following year, I even found myself, in real time, fearing that an important follow-up conversation of ours had been interrupted by an arson attack on her home. But for the moment she was happy to engage with my thesis.

"I think one thing tech fundamentally has in common with many religions, at least in America, is that it is a white man's version of utopia. And tech especially has this cultlike adherence to a white man's vision of a utopia that fundamentally disempowers and endangers women and people of color."

I asked Oluo what the characteristics of this white man's vision of utopia were.

"It starts with the mythologizing of white male struggle that's at the core of tech culture," she replied, "the idea that these men were outcasts who built things up from nothing—the shunned ones." It's not that the white men whose stories have been so central to the culture/religion we know as tech today never struggled or worked hard. Many of them did, and Oluo would hardly deny that fact. The problem lies in the adage about the man who was born on third base and thinks he hit a triple. To get to home plate, he'll still likely need to run fast, slide hard, and maybe even get a lucky call from the umpire. And if he scores a run, why not celebrate? (Clearly I watch too much baseball.) But his fellow players might understandably resent the suggestion that he had played the same game, by the same rules and with the same level of difficulty, as they had.

"And," Oluo continued about white male founders, "they're going to fix the problems standing in their way. This is their success story, their ascension. So, what stands in their way are people of color, the women that aren't sleeping with them, the popularity and the wealth they aren't automatically getting, old class structures that are keeping them away from the new class structure [based on] who has these skills that they, as white men, have."

Oluo paints a dark portrait, though certainly not one that applies to every male founder or tech leader. My friend Miles, for example, was indeed all too blissfully unaware of tech's hierarchies, but in no way would I accuse him of demonizing or otherwise resenting those lower on tech's unseen hierarchy than himself. Some tech leaders are simply lucky enough not to have to think too much or too deeply about others with different backgrounds. Still, Oluo's analysis resonates. It need not falsely adopt a tone of evenhanded emotional distance, because it is, ultimately, a cry for help. As she detailed in a book published less than a year after our meeting, *Mediocre: The Dangerous Legacy of White Male America*, there is something clearly wrong with white masculinity in America today.

If you don't see the problem in the boardrooms or the open office plans of Big Tech, perhaps you'd recognize it in the Christian white-supremacist men who stormed the Capitol on January 6, 2021, or in their evil twins, the privileged businessmen who hoarded and inherited wealth in the days before tech's dominance. If the white men on contemporary tech's Mount Rushmore have nothing to do with these other figures, why do they all look the same? Is it coincidence, or is there a cycle of ambition and resentment based on the notion that honor, respect, and excellence (in other words, the opposite of mediocrity) are chips in a zero-sum game in which the dominant group (white men) either wins or is defeated, humiliated, emasculated? Speaking in *Mediocre* about white male definitions of success and how they have led to 70 percent (and rising) of the tens of thousands of suicides in America annually, and about the fact that 70 percent (and rising) of school shooters are also white males, Oluo offers, compassionately:

I don't want this for white men. I don't want it for any of us. When we look at the history of white male identity in this country, it becomes clear that . . . we are afraid to imagine something better.

I do not believe that these white men are born wanting to dominate. I do not believe they are born unable to feel empathy for people who are not them. I do not believe they are born without any intrinsic sense of value. . . . I believe that we are all perpetrators and victims of one of the most evil and insidious social constructs in Western history: white male supremacy.[20]

Ijeoma Oluo's idea of tech as a utopia for white men has parallels in the thinking of scholars of tech and justice. In her book *Artificial Unintelligence*, Meredith Broussard, a professor at NYU and a researcher on the role of artificial intelligence in journalism, writes about "technochauvinism": the belief that tech is always the solution. Broussard argues:

Technochauvinism is often accompanied by fellow-traveler beliefs such as Ayn Randian meritocracy; technolibertarian political values; celebrating free speech to the extent of denying that online harassment is a problem; the notion that computers are more "objective" or "unbiased" because they distill questions and answers down to mathematical evaluation; and an unwavering faith that if the world just used more computers, and used them properly, social problems would disappear and we'd create a digitally enabled utopia.[21]

Still, it bears reminding ourselves that stereotypes—and even all-too-real instances—aside, tech wasn't always overpopulated or dominated by white men. As Moira Weigel, an academic, author, and magazine publisher working at the intersection of technology, communications and gender studies pointed out to me,

We [at technology ethics magazine *Logic*] published an article called "JavaScript is For Girls," about the gendering of coding languages, by Miriam Posner, a wonderful scholar at UCLA. It's about how as more women learn certain programming languages and start to occupy jobs where they use those programming languages, those jobs come to be paid less. That's just a fact. Jobs are devalued by having women or people of color occupy them.[22]

As Mar Hicks and other historians have shown, as the profession became more prestigious and more lucrative, women were actively pushed out. You even see this with specific coding languages. As more women learn, say, JavaScript, it becomes *feminized*—seen as less impressive or valuable than Python, a "softer" skill.[23]

It's not just that women often led in programming and coding until being pushed out when higher wages became possible for men. As historian of technology Louis Hyman told me, when Apple and other companies mass-producing computer technology began to rise, they cultivated a myth of "robots building robots, machines building machines," as if the physical components of the new cyber culture were able to literally create themselves without meaningful human intervention. The religious imagery implied by such a notion aside, this myth of robotic self-sufficiency was not only untrue but untrue in a very specific way. Hyman used archival research to look at the faces and names of people who built the early computers and their most hi-tech components, whether for Apple, IBM, or other early influencers in the field such as Seagate (a data-storage hardware company founded in the late 1970s). Hyman found that it was mostly women of color—often Asian or Latina immigrants to the United States—who, sitting around tables like cigar rollers, did the real, labor-intensive, and often highly toxic handwork of assembling these machines.[24]

The next steps were then subcontracted out (to let companies avoid legal liability for full-time employees) to men who did not speak English, who worked in a series of Quonset huts—lightweight, prefabricated structures of corrugated galvanized steel, shaped like half a carrot stick, originally designed for use by the American military as quick-deploying bases of operations in World War II. In these huts, trash cans full of chemicals sat open for different electronics components to be "dipped and processed with fumes and toxic effluvia," placing the workers at physical risk while others reaped the profits of their labor.[25]

Another key area where lack of equity and inclusion "in the room" has produced concerning or even disastrous results is the design of the algorithms that make up what we call artificial intelligence. In 2018, *Wired* magazine worked with Montreal startup Element AI to estimate the diversity of leading machine learning researchers, tallying the men and women who had contributed work to three top machine learning, or AI, conferences in 2017. Their finding: only 12 percent of these AI leaders were women. Similar or even lower percentages were found among technical roles such as "machine intelligence"—essentially, training computers to take in, interpret, and otherwise act on enormous amounts of data (i.e., creating algorithms)—at Google, Facebook, and beyond.[26] Little surprise, then, that these hierarchical companies created influential algorithms that frequently demonstrated bias, such as sexism in an Amazon résumé-screening tool; computer programs recognizing white men better than Black women; AI chatbots like ChatGPT and Google Gemini being more likely to label a man as an "expert" with "integrity" and women as "beauty" or a "delight";[27] or answering medical questions with "racist, debunked theories that harm Black patients";[28] and problems that were even more severe.

In her 2018 book, *Algorithms of Oppression*, Safiya Noble, a former advertising executive turned professor at UCLA and recipient of a MacArthur Fellowship ("Genius Grant"), analyzed the racist and sexist content in Google searches of the first several years of the past decade. The world's largest search engine, which especially at the time was seen as an almost divine authority on truth and knowledge of every kind, allowed not only offensive but deeply painful material to be reproduced and disseminated on its pages:[29] The faces of Black children on image searches for the keyword "gorillas" (April 2016).[30] First-page search results for "black girls" filled with the most cringe-inducing forms of pornography imaginable. Searches for "girls" of other ethnic backgrounds producing similarly toxic material—except for "white girls," a search term that somehow drew more neutral, less explicit results. Searches for "beautiful" and "professor" turned up mostly white faces, while the term "ugly" led to . . .

You get the point. Noble's research shows where and how people in and out of tech might get the idea that they are better or worse than others, higher or lower on some kind of pyramid. But just as dramatically, it slams home another point: that tech continues to replicate these biases is not accidental. The examples are too frequent, too pronounced, too repeated to be understood as unintentional. And unless we are to understand them as the result of out-and-out bigotry, we must come to see them as the product of an ideology that teaches tech leaders (and leaders in other industries) to privilege whiteness and maleness. This ideology is both a product and a reaffirmation of America's built-in religious hierarchies. It may rarely, if ever, say outright that women or people of color are fundamentally unworthy, but again and again it casts the suffering of marginalized people as perhaps unfortunate, but not as high a priority to address as the suffering of the dominant group in its hierarchy. It is simply inconceivable that such search results could ever have been reversed, with Black, Indigenous, and other marginalized people benefiting systematically from positive associations while men and white people were denigrated and demeaned. The heads of pundits in the right-wing media might combust if this hypothetical were realized. But in our reality, in April 2016, Google Maps searches for "N*gga House" led to the White House. To allow such a thing after almost eight full years of Barack Obama's presidency reflected essentially the opposite of urgency for equity on Google's part.

And yes, as Noble readily admits, "a book written about algorithms or Google in the twenty-first century is out of date immediately upon printing."[31] But that's not the point. Each of Google's disastrous results need not be a permanent monument to irresponsibility in order to represent part of "a series of processes" providing evidence of "a constellation of concerns."[32] And as chilling as Noble's work is on its own, reading her work in light of what has happened since is truly an exercise in true-life horror caused, directly, by the hierarchies of the tech religion. Just look at the words of Dylann Roof, the 2015 mass shooter who murdered eight churchgoers and their distinguished minister in the basement of the historic Emanuel AME Church in Charleston, South Carolina. In his manifesto, Roof explained that he "was not raised in a racist home or environment."[33] He was radicalized by Google, searching first for information about the Trayvon Martin case, then going down a hole to extremist racist sites.

As Roof's defense attorney David Bruck acknowledged, yes, of course Roof was a hateful, racist person. But how did he get that way? "Every bit of motivation came from things he saw on the internet," Bruck told Roof's jury. "He is simply regurgitating, in whole paragraphs, slogans and facts—bits and pieces of facts that he downloaded from the internet directly into his brain."[34] These bits and pieces were the sorts of content that sites like Google or Facebook failed to moderate or regulate, as we saw above, because it was not perceived as important or appropriate to do so.

Reading Roof's quote made me the particular type of sick that one gets upon recognizing that a tragedy was made possible, in part, by people like oneself and by systems with which one is intimately familiar. Because Roof's notion that the Internet contained only cold, hard facts was fueled in part by the attitude that tech is hierarchy-free and unbiased. Its engineers and executives would never have wanted to see anyone hurt in their name. And yet they also profited enormously from participating in an astronomically lucrative company devoted to the idea and projecting the image of curating and organizing all human knowledge, *objectively*. The failure by such a company to promptly remove and repudiate the sorts of blatantly racist, sexist content Safiya Noble has studied has consequences.

Sadly, Dylann Roof's crime was no isolated incident. As NPR's Dina Temple-Raston has reported, the Justice Department has charged hundreds of people so far with storming the Capitol on January 6, 2021, "and most of them, to varying degrees, were motivated to do so by the falsehoods they had ingested for months online and on social media."[35] Sam Jackson, a professor at University at Albany's College of Emergency Preparedness, Homeland Security and Cybersecurity, told Temple-Raston, "One of the interesting things about the current misinformation landscape is that it's not necessarily uninformed people. . . . It's misinformed people. It's people who say, 'I do my own research; I don't trust the elites.' And their research is nonsense, it is sophisticated nonsense."[36]

Misinformation has always been around. But companies like Google and Meta have gained unprecedented reach by cultivating an aura of impartiality and objectivity while failing or refusing to moderate hateful content that, when inevitably perceived as impartial and objective, lead to division, hate, violence, destruction, and worse.

Meanwhile, the phrase "sophisticated nonsense" reminds me of the experiences Ellen Pao has had to deal with over the past decade.

ALWAYS A NINE

It's a sunny Sunday morning in July 2021. I'm listening to a recent *New York Times* podcast on which Ellen Pao and the influential tech journalist Kara Swisher rank tech companies from 1 to 5, while I drive to meet Pao in San Francisco's SoMa neighborhood. Pao gives Reddit, which she once served as CEO, a 2 on content moderation. Same for YouTube. Facebook gets a 1, Twitter a 3.5. I park nearby and walk over briskly to ensure I won't be late. I've been looking forward to this interview for multiple years.

Ellen Pao is the tech investor and inclusion activist who came to national prominence in 2012 after filing a gender-discrimination suit against her powerful employer, Kleiner Perkins, a venture capital firm whose list of investments includes AOL, Amazon, Google, and Twitter. Though she did not win the suit, her efforts to hold the leaders of her company liable for creating an environment of sexist intimidation not only helped change the conversation around hierarchies in tech but also set the stage for the rise of the #MeToo movement a few years later. As she detailed in her memoir, *Reset*, Pao could have received millions from her former employers had she signed a nondisparagement contract. She boldly turned down that opportunity in order to speak out.

Stepping gracefully into public leadership over the past decade, Pao formed and leads Project Include, a nonprofit that uses data and advocacy "to accelerate diversity and inclusion solutions in the tech industry," attempting to give everyone "a fair chance to succeed in tech."[37] Her efforts also inspired 2016's influential "Elephant in the Valley" study, a survey of over two hundred women in tech on life in the industry.[38] As the survey showed, vast majorities of these women had experienced, for example, unconscious bias, exclusion, sexual harassment, and inadequate resolution of that harassment after reporting it. And if you're thinking things have surely improved since then: just a note that as I was polishing this chapter for publication in December 2023, the *New York Times* published a "Who's Who Behind the Dawn of the Modern Artificial Intelligence Movement" that did not feature a single woman.[39] On the exact same day, the *Times* printed another article on "How the AI Fuse Was Lit," that, as Bo Young Lee, a leader in tech and DEI pointed out, does not contain even "a single female pronoun."[40]

Ellen and I meet in Yerba Buena Gardens, a spacious and modern urban plaza nestled on top of part of the Moscone Convention Center, host to some of the largest tech conferences in the world. We sit on a bench surrounded by flowers, underneath a canopy of the tallest buildings in the city. The sounds of waterfall fountains obscure an audio recording, on loop, of Martin Luther King, Jr.

This could be one of the most beautiful spots on earth for those who, like me, enjoy vast urban landscapes where extraordinary feats of engineering mix seamlessly with natural beauty. But these days I can't help noticing all the flaws in the design: grease, dirt, and urine stains covering much of the marble; the unhoused person who climbed over the glass wall behind us and is camping out by the top of a giant atrium; pigeons and seagulls scavenging for scraps of old food like they were Bitcoin on an up day.

Pao and I have been corresponding for a couple of years now, but I feel the need to introduce myself and my project a bit, so I tell her I arrived at Harvard as a young chaplain the same year Larry Summers, one of my constituents in Harvard's atheist/agnostic/nonreligious community, made the stupid remarks about women's abilities in math, science, and engineering that got him deservedly booted from his position as president of the university.[41] I throw in a bit about how approximately 70 percent of students from families with annual incomes of $500,000 and up are nonreligious, so she and I are preaching to similar audiences, and Pao is willing to work with me on the image of tech as something like a cult. "I think it started with Steve Jobs," she tells me. "He was the biggest cult figure, he had the uniform, he had that cult of secrecy within Apple. It was very, very clear that he was trying to build a cult-based culture." But as for whether tech is a religion? Pao wonders if that may be giving Silicon Valley culture too much credit.

"It's a manifestation of intense greed," she says, indicating a partial point in favor of my thesis, and "it's replaced religion in many ways," specifying its propensity to drain followers' time and resources.

> But I don't see much humanity or value in it. And you know, even the most commercial religion has values that they're trying to promote and a belief in a greater good, but I don't see that in a lot of tech.
>
> There are definitely some companies and founders who are centered on [the greater good of humanity], but the overall system is like, let's go public, let's make money, let's make our founders incredibly wealthy, let's make a handful of employees billionaires, and then let's try to suck as much out of the ecosystem as possible to make that happen.

Pao is making a good and sincere point, but it's also a new spin on an old joke: "I'd call tech a religion," I imagine her quipping, "but that'd be an insult to religion."

"One of the worst examples in tech is JUUL," Pao offers. "I've had conversations with [their leadership team] and they just cannot acknowledge the harm that they've caused."

My blood pressure spikes a bit just hearing her mention the "smoking alternative" tech company that "hooked a generation on nicotine" and created "an epidemic of addiction," to quote just a couple major headlines following in-depth investigations into the San Francisco-based startup.[42] My dad died from his years-long battle with lung cancer when I was a teenager in the 1990s, and even then he could barely tear himself away from the habit.

I consoled myself, watching my father waste away, with the thought that future generations would never again be flooded with marketing for an addictive and harmful drug, as he and his entire generation had been. And indeed, as Pao points out, tobacco and nicotine use had both been down significantly among millennials, until JUUL. Using a variety of flavors and all the marketing genius that Stanford Design School (where JUUL was hatched, originally as a class project) could offer, the company was initially shunned by Silicon Valley VCs who believed it would be perceived as too exploitative and unhealthy to be a net positive for their portfolios, though venture capital nonetheless later sank nine figures into other similar nicotine-vaping companies.[43] After 35 percent of JUUL was acquired for $12.8 billion by Altria—formerly the tobacco company Philip Morris, which changed its name to seem less a death-dealing tumor manufacturer and more a "high performance" tech company—the company also brought in more than $700 million in additional investments.[44] These are the kinds of decisions that get made in hierarchical worlds, in which the exceptional geniuses in power can, tautologically, do no wrong.

I worry aloud about what kind of inner life the people making such decisions must have—where are their values? What motivates the mostly white, male group of leaders who are responsible for establishing this sort of culture? What are these individual tech leaders *feeling*, and how might their mental health be suffering?

"I'm going to go back to the systemic problem," says Pao. "Nobody is asked what their values are. Nobody is asked about the ethics of what they're doing. Their focus is on the bottom line, the financial outcome, at all costs. . . . The assumption is you want to get rich and you want to build some huge monolith. Nobody cares about you, your development, your emotional maturity, or whether you can manage a team or not. [Investors are], in this old school way, focused on growth and on what are you doing and what levers are you using to make your product more addictive and more engaging?"

"Are you saying that they almost literally have no inner life?" I ask.

"I think they're not valued for their inner life," Pao responds, "so they don't work on it. I don't think that it's that they don't have an inner life, but it's pushed down, because in the end it's like, I need to build this cult."

It's an elegant explanation that could also be applied toward understanding why many religious clergy who violate ethical codes do so. You'd think they'd have some sort of prayerful revelation that what they're doing is wrong. Maybe they do, from time to time. But they become adept at ignoring such impulses, because, like too many business, tech, and political "leaders," they've been indoctrinated into a faith in success, in followers, in a "bottom line" that seemingly requires a race to the bottom.

"I remember one CEO I worked for," Pao continued. "He would ask: 'On a scale of one to ten, how happy are you?' And I would ask him back: 'On a scale of one to ten, how happy are *you*?' At one point I'm like, you're always at a nine, and he's like, 'I have to be at a nine. I have to be. I can't *not* be at a nine, because everybody's watching me.'"

But as I blurted out as soon as Pao had shared the anecdote, forcing yourself to be "at a nine" of happiness because you think that is what other people expect or require is the very definition of *not* being at a nine. Because your own personal happiness isn't something you can program like an algorithm or acquire as if it were a competing startup. It is not a product but an emotion, a temporary state of being that one experiences but does not decide.

The idea of being at a nine not because one is *actually* at a nine but because one *has to be* at a nine might simply be dishonest—one might understandably claim to be happy when one is not—but in the case of Pao's former employer, I see a different dynamic at play. In my view, his statement is a stunningly clear example of what theologian Jonathan Z. Smith. saw as part of the essence of religion. To restate Smith's insight: religion is characterized in large part by its "capacity for rationalization, especially as it concerns that ideological issue of relating what we do to what we say or think we do."[45] It is how we justify and explain, to ourselves and others, the contradiction between who we are inside and who we want to be, "performing the way things ought to be in conscious tension to the way things are."[46]

In other words, when Pao's boss, the CEO of a very successful tech company, convinced himself that in order to be successful he must *believe* that "not happy" is actually "happy," he was not just lying to her, or even to himself, about his happiness level. By performing himself to be one thing when he was in fact another, on a daily basis, he was engaging in a kind of religious and spiritual ritual. Once such a practice becomes an ingrained pattern across an industry, where does it stop? Down can be made into up, and up into down. Hierarchies can be simultaneously justified as healthy and denied as if nonexistent. And everyone involved is made miserable.

Consider this anecdote about Ellen Pao's former boss in the context of Mark Zuckerberg's early inspiration for building Facebook. Zuckerberg spent a lovelorn evening in the fall of 2023 in his dorm room at Harvard's Kirkland House, staring at the Harvard College "facebooks," which were then official university publications meant to help students get to know their classmates. Zuckerberg famously decided to hack into them to create a "Hot or Not"-style website to rank their looks.

In a journal on the site, Zuckerberg wrote that some of the pictures were "pretty horrendous," saying, "I almost want to put some of these faces next to pictures of farm animals and have people vote on which is more attractive." A decade later, the entire Harvard men's soccer team was suspended from play for a year after being found to have conducted a similar ranking experiment on the Harvard women's team, creating and maintaining a spreadsheet, over multiple years, that ranked the women according to their sexual appeal on a scale of one to ten, "including explicit descriptions of their physical traits and musings about the women's preferred sexual positions," according to the New York Times.[47] "Doggy style," team members said of one woman. Viewing another woman on the team as "manly," they wrote, "Not much needs to be said on this one, folks."

Maybe the reason Harvard men and men from similar backgrounds spend so much energy on ranking women—and on creating and maintaining hierarchies of all kinds—is that subconsciously, we are constantly ranking ourselves. We rank our own happiness and desirability and find ourselves unacceptably lacking, so we convince ourselves that by ranking others, we can elevate our own standing. But ultimately we have trapped ourselves not only in what I discussed in chapter 3 as a cult of the founder, but in a cult of the self. The self, rather than representing a human being with an emotional life who deserves acceptance simply for being alive, is treated like an external product to be marketed, invested in, and sold.

THE BODY KEEPS SCORE

The day before I met with Ellen Pao in San Francisco, I had a chance to sit down with one of her close colleagues, Laura Gómez. Gómez, a founding member of and current advisor to Pao's Project Include, is a VC-backed startup founder and was an early employee of YouTube (starting in 2007) and Twitter (2009). I met her in Redwood City, California, her adopted hometown after her family immigrated to the United States from Mexico. Redwood City is, today, a rapidly gentrifying small city on the San

Francisco Peninsula, a straight shot eight miles northwest of Stanford
University's Palo Alto on Route 101, or fourteen miles from Menlo Park
(home of the Googleplex) in the same direction.[48] When Laura came of
age in the 1990s and went to Berkeley, when she told people in and around
the local tech industry where she was from, they often responded that yes,
they knew Redwood City; that's where their house cleaners and gardeners
lived. Indeed, Gómez's mother worked as a nanny for Silicon Valley families, driving that stretch every day, as did so many of their neighbors. As
a haven for Mexican immigrants, Redwood City paralleled East Palo Alto
for Black communities and the East Bay for Samoans, Filipinos, and Pacific
Islanders. All of these communities are currently struggling to cope with
skyrocketing rents and relatively stagnant pay as the behemoth that is tech
eats the entire region.

Gómez and I met on a hot, sunny Saturday afternoon under an umbrella
in a new-looking outdoor mall with a movie theater and several restaurant options. Our table faced the headquarters of a major cloud-service
company. As you might guess, that wasn't what Redwood City looked like
during Laura's childhood. When I was a kid growing up in New York City,
I remember thinking such changes were good because of the new luxuries
and amenities they brought. I hadn't yet learned to see the displacements of
whole communities that often came from such "upgrades."[49]

As the daughter of a nanny, Laura had a unique window into Silicon
Valley, and she was good at math and science, so after high school she got an
internship with nearby computing giant Hewlett-Packard. At HP she was
the only female intern in the software department, out of around a hundred. There was one other Latina in the entire department—an administrative assistant, or secretary at the time, who was more like a departmental
COO and might well have been just that had her identity and background
been different. That woman took Laura under her wing and helped her
get an early sense of how to climb the rungs of the industry's ladder—or,
rather, the religion's hierarchy. (How could there not be a hierarchy in tech
if people are constantly referencing climbing the ladder?) But the impressive successes Gómez would go on to experience came at a high cost, psy-
chologically and physically.

"Founders are two times more likely to suffer from depression," she told
me, as well as "three times more likely to suffer from some sort of addiction
or substance abuse, five times more likely to suffer from anxiety."[50] And, as

she pointed out, those numbers are for founders in general—a professional class overwhelmingly likely to come from more privileged backgrounds than someone like her, and thus have more resources to fall back on in coping with intense stress.

These statistics became part of her reality after Gómez founded Atipica. A startup that built the first platform focused on diversity, equity, and inclusion for predictive analytics in talent acquisition, Gómez's endeavor raised the largest financing seed round for a solo Latinx founder in the history of Silicon Valley. The problem? That was only $4 million in total. And based on stories I've heard many times from women of color founders over the course of more than four years researching this book, Laura's experience was all too common. Brilliant, talented, and exceptionally accomplished women leaders found new tech companies and initiatives that attract significant investments, but the dollar figures involved are dwarfed by the support that goes to projects founded by members of the dominant group. And leading an organization funded at high levels relative to other industries, but underfunded compared to industry peers, can be an enormous burden on founders like Gómez, who deal with racism, sexism, and other malignant expressions of tech's hierarchies while seeing more "typical" founders benefit from nepotism and other invisible advantages.[51]

Indeed, through her time in different tech roles, Gómez told me, she experienced every kind of micro- (and macro-) aggression she could imagine, from sexual harassment to investors doubting that she could be capable of developing her own code. So perhaps it is no wonder that, as her tumor grew, she didn't notice it. As it began pressing on the parts of her brain that controlled her capacity for spoken language, causing her to pause in mid-speech and fail to pronounce things she would normally be able to, she figured it was just stress. Even when it had grown to two inches in diameter, as her doctors later informed her, causing symptoms that were impossible to ignore, she thought it was just a headache. She asked for an aspirin, telling herself it must be the travel. Vomiting, lightheadedness—she tried to sleep them off. Until she was surprised to find herself in the emergency room and ultimately at the neurology department at UCSF Medical Center. When she awoke, she was already well along the process of multiple MRIs to determine the tumor's size and whether it was malignant.

"I do believe that the body keeps score," Laura told me as we sat in Redwood City a couple of years after her surgery. "I believe that trauma is inherited into our bodies and that it can display itself in either our mental or physical health." I'm no expert on the physiology of trauma, but that makes sense to me.

Fortunately, Gómez's tumor was benign. She has a titanium plate inside her skull now, which she told me was a metaphor for her relationship with tech. I wasn't quite sure what she meant by that, but I wondered if maybe she was glad to always have a piece of technology in her head, Iron Man–style, protecting her brain from the circumstances that endangered her. Maybe, on the other hand, she imagined the plate protected her from the hostility, sometimes villainy, of Silicon Valley. We didn't get into it in our conversation, but I wanted to imagine she meant both, and she later told me that was "a beautiful way of framing" what she'd told me.

Laura has what you could call a love-hate relationship with tech, which treats drivers, content moderators, and other service workers; those displaced by gentrification; and even exceptional "success stories" like her so poorly. But even she acknowledges that she has also benefited from her associations with tech. Rather than leave tech altogether for a simpler life of some kind, to criticize it from afar, women like her and Ellen Pao work to change the system, remaining optimistic enough about the possibility of change to keep trying. It's complicated. And it's why every time I hear a story like hers, I return to the idea of tech agnosticism.

Every thoughtful person alive today will be at least somewhat ambivalent about their relationship with the dominant religion of our time. Each of us must decide how much we can afford to participate in an endeavor that oppresses and divides at least as much as it uplifts and heals. And yet people like Ijeoma Oluo, Ellen Pao, and Laura Gómez not only use technology every day but work within—or at least near—its systems to make them fairer.

"WHAT'S YOUR WORLD CHANGING IDEA?"

When we met at her Shoreline home in the summer of 2021, Ijeoma Oluo generously spoke with me for nearly three hours to help me prepare to write this chapter, but during that conversation she cautioned me not to regurgitate work done by people like her or like Safiya Noble. I could and

Ijeoma Oluo.

Ellen Pao. *Source:* Charlie Grosso.

should use my unique social position, she advised, to gain and share insight into, for example, more people like Miles Lasater, with whom we began this chapter. Ellen Pao encouraged me to share stories that go beyond tech's current mainstream—the ideas and lives of tech's usual suspects don't need to be shared much more at this point, Pao argued. They've already been shared too much as it is. I struggled with this contradiction as well, and how to balance between the two conflicting impulses in my own analysis of tech hierarchies. Ultimately, however, I thought back to the afternoon of my meeting with Gómez.

After she left our café table to attend her friend's wedding, I was feeling restless in the hot summer sun of Redwood City, wondering what to make of tech's frustrating hierarchies and of efforts, like hers, to reform them from within. To distract myself, I wandered down the block to the San Mateo County History Museum, an attractive stone building with Corinthian columns and a Roman dome. I found a gleaming courthouse under the dome's stained-glass ceiling. In the adjoining room were displays commemorating San Mateo as having once been "The Most Corrupt County in America," a bastion of political misdealing and illegal dueling. The museum's layout guides guests through an exhibit called "Land of OPPORTU-NITY," documenting the immigrant experience in the county. African American, Filipino, Chinese, Japanese, Mexican, Italian, Irish, Portuguese, and other groups arrived in waves from the post-gold-rush years through World War II.

The museum culminates in a large exhibition on entrepreneurship and technology, in a seeming and understandable bid to ensure all visitors will remember that this is the birthplace of Intel, which became the largest semiconductor company in the world. Taking it in, I passed by a purple board adorned with messages in children's handwriting, on large yellow sticky notes, underneath the question: What's Your World Changing Idea? Answers included "global instantaneous food sharing to end world hunger," "stop killing the forest," "make prisons more of a rehabilitation or self-evaluation center rather than more of a place to make inmates feel trapped and helpless," and, "I haven't thought of it yet but it will probably be really world changing." (Who can argue with that last one?)

Finally, I sat down, peacefully alone, in the exhibit's Theater of Innovation, a darkened corner of the museum in which a menu allows visitors to select from among eight videos on "San Mateo County History Makers."

All were focused on white male entrepreneurs. Later, I'd think of this place when Ellen Pao advised me against talking to more white male tech leaders for this chapter, or for this book as a whole—because, as the exhibition on tech entrepreneurship made clear, such people's stories have been told already. Not just told; they've been recorded and placed in displays like this one as a canon. "The Intel company culture, from the very beginning, was one of, 'Talk to me with facts. Get me real information,'" says Dave House, an early Intel employee, in a video about Intel cofounder Gordon Moore. In the same video, Moore describes the origins of "Moore's Law," the famous maxim (it's not actually any kind of proven scientific law) that the number of transistors on microchips doubles about every two years, with ever-declining costs per transistor, resulting in exponential growth. House recalls his bewilderment at Moore's genius when the latter at one point accurately brainstormed that cosmic rays might be interfering with their new inventions.

The ambition—the chutzpah—of such theories is truly amazing, and not unworthy of preservation in a history museum. But it should perhaps equally amaze us that such brilliant leaders never figured out or pursued equally aggressive plans for tackling the massive systemic inequalities and inequities baked into their own systems.

The agnostic approach to tech leaves room for the idea that any given technology might hold potential to be of real benefit to humanity. It also, however, remains fiercely skeptical of any technology or technologist claiming to benefit all humans when they cannot demonstrate, in their hiring, firing, coding, and wiring, that they value different kinds and castes of humans equally.

It won't be easy to find an appropriate balance whereby we can enjoy and appreciate the best tech has to offer while fighting effectively against the worst and most harmful religious aspects of its culture. But as an ancient rabbi and teacher named Tarfon said in the *Mishnah*, a two-thousand-year-old work of Talmudic literature, "It is not your duty to finish the work, but neither are you free to desist from it."[52]

In fall 2020, more than a year after her surgery, Gómez founded Proyecto Solace (now Proyecto Sol). A nonprofit rather than a tech venture, the

project's LinkedIn page invokes how "the intersection of racism, xenopho-
bia, sexism, homophobia, and other forms of bigotry leaves many in the
Latinx community searching for harassment-free, safe places as part of their
mental health journey and to . . . socialize without judgment."

Solace means comfort, consolation, and relief in both English and
Spanish (*solaz*, in the latter). She is proud to lead "a community of Latinx peoples, investing in our emotional and mental safe spaces for themselves, communities, and in collective healing." I love the choice of the
word *investing* here, flipping its meaning from something associated with
financial profits alone.

These days, Laura is mostly grateful. Not only in the general sense of
being thankful to the universe for continued success but specifically for fellow activists fighting for a less hierarchical Silicon Valley. Especially for
Ellen Pao. When Laura awoke at the hospital, after her surgery, Ellen was
at her bedside. And when Laura realized that the operation had repaired her
brain's speech center and she was now able to speak normally for the first
time in over a year, she longed to talk. Ellen sat with her for over two hours,
late into the night.

When Laura is transported back to those times, or when she is prompted
to speculate on the future of technology, she often thinks of one of her
heroes: the Italian sociologist and political theorist Antonio Gramsci, who
was imprisoned for thirty years during the Mussolini era. When asked how
he maintained hope, Gramsci said, "I'm a pessimist because of intelligence,
and an optimist because of will."[53] Laura's pessimism is that we haven't
done enough to address the hierarchies of tech. Her optimism is that she
and countless peers are doing more, every day.

4
RITUAL

In the summer and fall of 2020, families like ours went hiking, an activity that, at least some of the time, distracted small children from the absence of school, or camp, or anywhere else they could see and play with other children. At least it was something to focus on that wasn't a screen. But as the seasons turned once more, the rocky tedium of indoor isolation would meet the hard place of a New England winter. The still uncontrolled coronavirus threatened to grow deadlier, and the summer's racial unrest unfurled into a presidential election many of us feared could finally end the nation's experiment in democracy. We all buckled under the stress, each of us in our own way. But somehow everyone's way involved technology. We turned to screens and devices and apps, not only to connect with one another but as psychological life rafts, things to do in the frequent moments in which we couldn't bear to do one more thing.

One evening, to break my own compulsive streak of pandemic nights spent emailing and tweeting from my phone while half-watching HBO on my laptop, I uploaded some PDFs to my reMarkable 2 tablet: comic book scripts, actually, for the apocalyptic graphic novel passion project I'd procrastinated on for years.

The reMarkable, an e-ink-based tablet created by a Norwegian startup by the same name, was designed for reading, writing, and drawing—*only*—and advertises its purpose as "helping you think."[1] Its implicit purpose, beyond that, is to be an antidote to digital technology's maniacal pursuit of our time and attention. While others were understandably obsessed with another kind of vaccine, a tech antidote was just what I was in the market for.

In bed, I spent hours reading the book-like gadget—it had the satisfying shape and heft of a graphic novel, and the woven polymer fabric of the folio cover I purchased felt like a clothbound journal. Within minutes I was digitizing and uploading handwritten notes (with or without the doodles

that felt, well, remarkably natural to create), as well as reading and mark-
ing up various documents. With all my other electronics in another room,
I experienced something like peace. Instead of blasting my eyeballs with
UV lights, I switched on a forgotten invention called a table lamp. Flipping paperlike pages with the flick of an index finger, I used the "marker"
to scratch ideas in the margins. Of course, you might reasonably object,
I could simply have read . . . a book. I could have picked up a real, paper
notebook to write notes in. Except then I'd have to confront my usual dis-
organization and anxiety: if I took important notes on paper, would I lose
the paper? I felt reassured knowing that any ideas I might scratch out on the
reMarkable were instantly uploaded to my drive for access later.

Ditching the smartphone the next morning, I swapped its SIM card into
my Light Phone II, which was loaded only with some music podcasts I was
eager to try and the ability for close contacts to reach me if the sky, which I
suddenly had some mental space to notice, were to fall. I took a long walk. I
saw trees; I heard the bass lines under guitars and vocals. I wondered where
I even wanted my thoughts to wander. Absent the phantom limb of endless
connectivity, I alternated between blissful awareness and a twitchy impulse
to check . . . something. *Anything.*

"That's the question," said Light Phone cofounder and CEO Kaiwei
Tang about how his customers, like me, struggle with this uniquely modern ritual. "Can we handle boredom? I mean, these are timeless questions,
right? Like, what are we going to think about [if we're not online]? The
people we love. Our pain, our hopes."

TECH EFFERVESCENCE

This second part of *Tech Agnostic* is about tech as religious *practice*, if not
necessarily as religious belief. In the previous chapter, we looked at how
hierarchies in tech can function like hierarchies in religion. Both serve to
bring communities closer together by defining and solidifying in-groups
and out-groups; both concentrate power, convenience, and charisma in
the hands of some at the expense of others; and both are strengthened by
mythologies that convince individuals on all sides to believe in the importance and justification of the hierarchy.

Now we will turn to another cornerstone of religious practice: rituals,
how all religions thrive. Doctrines may get more attention as the primary

element of religion. In no major religion, however, have leaders success-fully imposed uniformity of belief. If common faith alone were the mea-sure of a religion's potency—in other words, if religion amounted solely to the shared understanding that a given God exists or that a specific vision of the afterlife is real—no religious sect in history would or could have risen to great influence. Human minds are simply too stubborn, not to men-tion too fickle. But as the great sociologist of religion Émile Durkheim recognized, by bringing people together to engage in common activities, rituals create the potential for people to share meaningful experiences and coalesce into a community. The more effective the rituals in each tradition are at bringing people together—the more they promote what Durkheim called "collective effervescence"—the more power that tradition can gain to influence the lives of its adherents.

The tech religion's rituals are effervescent indeed.

Formal, organized religions might prescribe prayer rituals once a week, once a day, three times a day, or five times a day. The average American now opens their phone some 344 times a day, or once every four minutes, according to a survey by Reviews.org published in early 2022.[2] Many of the approximately 97 percent of Americans who are smartphone users touch, tap, or swipe their devices thousands of times each day.[3] In guidelines pub-lished in February 2020—just before a pandemic overturned norms and drove a dramatic increase in screen time—the American Academy of Child and Adolescent Psychiatry noted that children aged eight to twelve in the United States spent four to six hours a day watching or using screens, and teens spent up to nine hours.[4] During the COVID-19 pandemic, according to a study in *JAMA Pediatrics*, US teens spent an average of eight hours daily on screens—not including school or homework, much of which was also completed using screens.[5]

We'll come back to the question of what makes a ritual, and why it can be important to think of tech use as such, but let's set beliefs aside for a moment. Which makes the more forceful statement about our values, priorities, inner state: mouthing words of devotion to an unseen God, or offering intense devotion to a force so irresistible that 74 percent of Amer-icans "feel uneasy leaving their phone at home," that 70 percent check their phones within five minutes of receiving a notification and 71 percent within the first ten minutes of waking up? In this chapter, we will examine tech addiction (mine, certainly, and perhaps yours, or a loved one's, as well) as the defining ritual of the tech religion, or at least as the behavioral and social equivalent of religious ritual.

SOME SORT OF WORSHIP

The comparison between obsessive tech use and religious ritual is not entirely my own. Author Joyce Carol Oates, for example, noted in a tweet the resemblance between our digital lives and spiritual worship.

As in the picture she posted, of a New York City subway car filled with mesmerized supplicants, we too-frequently freeze our gaze downward and

Joyce Carol Oates @
@JoyceCarolOates

Entire subway car mesmerized by small gizmo clutched in hand. Some sort of worship?

11:44 AM · Oct 8, 2019 · Twitter for iPhone

Tweet by Joyce Carol Oates.

lift our hands upward in a novel prayer position. The object of our vener-
ation is a miniature altar fashioned from aluminum, copper, iron, and no
fewer than seven rare-earth minerals. Electricity, flowing through lithium
cobalt oxide and carbon graphite, lights its eternal flame. Its single pane of
glass, a chemically strengthened alkali-aluminosilicate, undergoes an ion-
exchange process and is dipped in a bath of molten salt. When combined
with a magical repository of sacred texts and bawdy tales called the Internet, it shines forth with a light more mesmerizing and numinous than that
of any stained-glass window.

I'm being lighthearted, as was Oates, in this comparison of the physical
artifacts of worship with tech addiction. A more serious question, however,
might be: What is the difference between the devoted practice of religious
ritual and addiction? As in, if someone prays throughout each day, or for
many hours each weekend, are they addicted to prayer? The answers to
such questions lie largely in the eye of the beholder. An advocate for religion might call such a comparison ridiculous, and on one hand I'm inclined
to agree. We all know that many religions are socially sanctioned, well-
respected, time-tested activities in which people feel they have good reason
to participate. On the other hand, some religions are abusive and harmful,
and these more manipulative sects benefit from blurring the lines between
healthy devotion and addictive worship. For certain faith communities, in
other words, ritual practices that mimic addiction are a sign of their success.
And there is an obvious parallel here to the makers of smartphones and similar devices, who can talk all they like about helping us preserve our "digital
well-being" but make money hand over fist when we put that well-being
aside to devote ourselves to their products.

A number of scholars and authors have spoken of inventing rituals for
digital life, including Stanford Design School lecturers Kürşat Özenç and
Glenn Fajardo, who note that many of the norms and conventions of in-
person meetings don't work well in the virtual lives we've all been thrown
into in recent years.[6] Özenç and Fajardo's suggested alternatives include
a number of exercises that sound like spiritual rituals, including "guided
breathing," "team positivity contagion," and "spread the warmth," as ways
to improve productivity. But what could be more diametrically opposed to
spiritual ritual than rituals to increase production under capitalism? When
I spoke with Fajardo via Zoom in early 2022, he acknowledged the validity
of such critiques with forthrightness (and perhaps resignation).

Implicit in any call for better digital rituals is the notion that digital communications themselves are already full of rituals. Like signing in. Checking for notifications. Reading our feed. Opening a new browser tab. Clicking on a viral video. Allowing the algorithm to guide us to the next video. And so on. These are the experiences I associate most with my time, these past decades, around computers. As an ordained rabbi and longtime chaplain who has spent a lifetime around religion, I can say for myself that few religious experiences of candle lighting, kneeling, chanting, or walking through a labyrinth have as consistently altered my consciousness or created out-of-body experiences as surfing the World Wide Web. Even if such rituals do not best serve my personal goals, or the goals of any given person or civic institution, they certainly do serve the interests of the various houses of tech worship: Meta, Twitter, Zoom, Google.

Communications scholar James Carey, notably, compared communication in general, and technological communication in particular, to ritual. Carey's influential "ritual view of communication" is a theory by which human communication is seen as an increasingly sophisticated effort not simply to send messages but to maintain society through "the representation of shared beliefs."[7] Communication is not just about advancing ideas, in other words, but maintaining common values. Carey's theory "derives from a view of religion that downplays the role of the sermon, the instruction and admonition, in order to highlight the role of the prayer, the chant, and the ceremony."[8] Highlighting the common ancient roots of words such as *commonness*, *communion*, *community*, and *communication*, Carey invokes Samuel Morse's quip that the telegraph was not invented to share the price of pork, but to ask, "What Hath God Wrought?" If that was true of such a primitive communications technology, then today's ever-evolving tech and its ability to galvanize the populace, for better and for worse, is like drinking holy water from a fire hose.

THE CROWDS USED TO BE ON 100

I don't know exactly when or how I decided I had an addictive personality, but it was before I hit puberty. At a Greek restaurant the night I came home from summer camp, my parents agreed I was now old enough for coffee after dinner. I declined, which in my family was like telling Santa you didn't want any presents. I thought they were going to flip the table when I finally explained it was because I didn't want to be addicted, like them.

We were a small family of recent immigrants on both sides, all battling complex, intergenerational traumas, and I was terrified to wind up alone, impoverished, as worthless in the eyes of others as I felt to myself. Abstention helped me maintain a (false) sense of control over my life and future. We tabled the subject after that night and continued to avoid the topic even while my dad battled terminal lung cancer as he struggled to give up smoking.

I never did touch a cigarette or a drug, or even let myself drink much. So it was odd to find myself in recovery meetings in my early thirties, hooked on the Internet.

In May 2008 I found myself in a classroom at McLean, Harvard Medical School's psychiatric hospital, and not for academic purposes. After a forty-minute bus ride from my office in Harvard Yard, I arrived for my weekly session of SMART Recovery, a wonderful recovery method pioneered by a Harvard Med professor and mentor of mine as a secular, science-based alternative to the "higher power" and other religious concepts central to Alcoholics Anonymous and Narcotics Anonymous.

Slouching into one of those high school chairs with the tablet desks, I try to explain myself. Hands upturned in preemptive defense, I told the other attendees how I couldn't log off. How I would just . . . refresh: email, Facebook, NYTimes.com, ESPN, again and again, searching for anything to break the tension and escape the pressure as I attempted to fulfill a book contract. I described how the longtime girlfriend who had just (very deservedly) dumped me used to complain—way back in 2006, when I had a PalmPilot Treo and the iPhone was still just a glimmer in Steve Jobs's eye—that maybe I should date the phone instead of her, seeing as how it was the last thing I wanted to see at night and the first thing I wanted to see in the morning. I revealed my fear that I wouldn't stop before I destroyed more relationships and torched my writing career.

I was thinking they'd laugh me out of the recovery room. And there was some bleary-eyed bewilderment at my story. But several attendees kindly acknowledged that my digital predicament bore similarities to their substance issues. Like them, I was "using" to escape painful emotions. Like them, I craved both a state of dissociation and a dopamine high, because both helped me avoid what I didn't want to face about my life.

Missy Elliott ✔   •••
@MissyElliott

I LOVE we have phones to record shows for memories BUT....@ the
same time it take away from the ENERGY 💫 It's hard to dance &
enjoy yaself when you trying to hold the phone steady to record ya
fav artist performance 🧎 😩 the crowds used to be on 100! 🙌

7:22 PM · Jun 19, 2022

Missy Elliot Tweet.

I put myself forward as an unusual example to demonstrate the importance of this comparison. But I don't think the problems I describe here—
which I continue to manage to this day—are so different from what many
(surely millions, if not more) are currently experiencing. The inability to
log off is now a widespread malady. As the rapper Missy Elliott, one of
the best-selling hip-hop artists of all time, recently put it, about one of the
most numinous forms of secular ritual (at least, to me), the summer con-
cert: "I LOVE we have phones to record shows for memories, BUT [at the
same time it takes] away from the ENERGY . . . it's hard to dance & enjoy
yaself when you trying to hold the phone steady to record ya fav artist . . .
the crowds used to be on 100!"[9]

And yet: If I had finally found the ethical tech holy grail I was seeking back in 2020, when I began to experiment with alternative tech like
the reMarkable 2 and the Light Phone II, why did I so often ignore, even
avoid, these simpler devices? And why, even after I published a review of
the devices as part of a *Boston Globe* story exploring tech addiction, did my
struggles with smartphones and other forms of tech continue well into the
process of writing this book?[10]

When I asked Tang, at Light Phone, how his customers tended to react
after purchasing his product, most often as what Tang calls a "secondary
phone," he told me that many are reluctant to settle into a "lighter" routine.
After all, as Tang discovered in his prior lives as a designer for Motorola,
Nokia, and BlackBerry and in a prestigious Google incubator for design-
ers, algorithms in social media and other smartphone apps of mass distraction are genuinely addictive. They are built with the express intention of

monopolizing users' time and becoming artificially indispensable to our lives. They are "digital crack cocaine," to paraphrase Dr. Julie Albright, a digital sociologist and expert on how we "hyper-attach" to tech at the expense of in-person human relationships.

Still, is all of this really best understood as a ritual? Many tech leaders, like the secularists and other intellectuals with whom I've spent so much time with over the course of my career, don't like to think of themselves as being inclined towards ritual at all. Ritual is too "tribalistic," as one philanthropist I know says. We like to think we're above such things. Maybe we'd even say that we've overtaken ritual, replacing it with science, logic, *thinking*. But have we?

If I were to compare the behavior of most people I know around their tech with their behavior around religion, their tech use would seem much less rational and much more ritualistic—especially when you consider data suggesting that our tech behavior can actively harm us. Increased social media use, for example, can lead to increases in depressive symptoms and suicide rates among adolescents.[11] Excessive screen time can lead to inattention problems in preschoolers, to language development delays among eighteen-month-olds, and to a range of other developmental difficulties in children of various ages.[12] It has been connected to strained bonds among families around the world.[13] Screen time was even associated, in one UK study, with less-healthy levels of insulin resistance among children.[14] This level of usage, in the face of such consequences, is not undertaken by "rational choice." It is not something we think through. It is a ritual.

While social scientists have not demonstrated such clear-cut connections between screen time and negative outcomes for adults, neither have they ruled them out. In the face of unclear formal data, I would ask: How do *you* think the adults around you are doing in dealing with our digital age? Do people seem healthy and well-adjusted to you? Does democracy seem to be functioning well, for example, or does it seem negatively impacted by systemic disinformation and constant social media skirmishes? How do we appear to be doing, in this ostensibly well-informed age, at addressing climate change, creating an equitable society, and promoting emotional and physical health and well-being for all? If you are feeling optimistic about our current status with regard to these and other major political and social developments . . . congratulations.

THE ADVERSARIAL PERSUASION MACHINE

"Digital technologies are making all forms of politics worth having impossible," James Williams argues, "as they privilege our impulses over our intentions and are designed to exploit our psychological vulnerabilities in order to direct us toward goals that may or may not align with our own."

Williams may not be as much of a household name in tech circles as some of his ex-Google peers, but his ideas are important. In 2010, Williams won the Founders' Award, Google's highest honor for its employees. Then in 2017 he won an even rarer award for his scorching criticism of the entire digital technology industry in which he had worked so successfully. The inaugural winner of Cambridge University's $100,000 Nine Dots Prize for original thinking, Williams was recognized for his book *Stand out of Our Light*, the fruit of his doctoral research at Oxford University. As he describes it, he was at Google, earnestly working to organize the world's information in useful ways, when he had "an epiphany: there was more technology in my life than ever before, but it felt harder than ever to do the things I wanted to do. . . . This was some new mode of deep distraction I didn't have words for."[15]

In his book Williams details how the algorithms that beat the world champion at the game Go are now aimed at beating us, dominating our ever-dwindling attention spans, and "standing in the light" that might otherwise allow us to see the way to a better human future. They are at least convincing us to watch more YouTube videos and stay on our phones a lot longer than we otherwise would.

When I met Williams, first on Skype in early 2019 and then when he came to visit Harvard from his erstwhile home in Russia, he was kinder and friendlier than one might expect from a man known for staying off the grid while intensely criticizing the entire project of Big Tech companies like his former employer. This was back when I was beginning my own transition from leading humanist communities to thinking and writing actively about tech, and though Williams ultimately expressed some optimism about humanity's long-term prospects, his dark outlook on the present and near future influenced my own. "If it happens that we are in multiple crises," I asked him, "where in the next ten or twenty years we need to get our act together on emissions and on carbon in the atmosphere and all of that, or we're going to face a severe degradation of our way of life . . . it's inconvenient if we lack the ability to have a coherent conversation with one another. Isn't it?"

"Absolutely," he said. "I mean, I think living in a society together requires some amount of minimum viable trust, and . . . if you wanted to design a set of media platforms to undermine that kind of trust, that kind of common story, you could do a lot worse than what we've got now."

Or as Varun Soni, dean of religious life at the University of Southern California, pointed out in a lecture for my colleagues the Harvard chaplains, addressing our concern for our students: since the 2008 release of the iPhone, rates of self-harm and anxiety have doubled on college campuses; anxiety and depression have increased exponentially among high-school students; and almost every study shows that the more time we spend online, the worse we feel about ourselves.[16]

Indeed, joining Soni in such concerns are an increasing number of psychology researchers, including Dr. Albright and Jean Twenge, a professor of psychology at San Diego State and prominent author on the digitally addled young people she calls "iGen."[17] In a landmark 2017 essay in the *Atlantic*, "Have Smartphones Destroyed a Generation?," Twenge writes, "It's not an exaggeration to describe iGen as being on the brink of the worst mental health crisis in decades. Much of this deterioration can be traced to their phones."[18] Twenge points out, among other startling figures, that twelfth graders in 2015 were going out less than eighth graders did as recently as 2009.

Maybe all of this will turn out just fine. Maybe I'm being alarmist. Some studies have shown that screen time itself isn't bad for kids, albeit with caveats and uncertainties.[19] And reviews of recent studies on teen social media use have shown that it isn't demonstratively negative for the health of all the young people who use it . . . just some. Mostly girls. But it may be more than that, and it may affect even some of the teens who *don't* use social media, because of the so-called network effect of being surrounded by peers who are immersed in it. I won't attempt a review of the reviews of the reviews here because others have done so, and the debate rages on. The short version is: it's all a moving target and will continue to be so for some time. And I am trying to remain—appropriately, I hope—agnostic.

I just know that, as Steven Hassan, the cult recruitment and deprogramming expert I spoke to (see chapter 2), put it, picking up his iPhone and waving it around in a podcast episode I discussed with him, "This device is the perfect mind control device." Hassan, who advised Democratic politicians like Maryland representative Jamie Raskin on how to think about the controlled minds of the rioters who stormed the Capitol on January 6, referred to our devices' propensity to heighten both the political instabilities that worry James Williams and the intrapersonal psychological disruptions that concern people like my USC colleague Varun Soni. As the saying goes, "the personal is political"—a principle that very much applies to the downsides of mass addiction to our phones.

Under quarantine, when I started experimenting with the reMarkable and the Light Phone, our devices were our lives. The dark predictions Williams had written about years earlier, that the "adversarial persuasion machine" Google and others had unleashed upon us all would distract us all so profoundly as to make it difficult for civil society to function, looked scarier by the minute.

"I don't know where I see any media dynamics working in the direction of mutual understanding, in the direction of reflection, restraint, charitability, autonomy, or any of the qualities that the maintenance of a free society requires," Williams told me in 2019. "Where exactly are our forms of media shoring up civilization? They all seem to be working in the other direction." Where indeed? Perhaps seeing profound worry on my face and sensing that I wasn't used to or prepared for existing in such a state—in humanist circles, I've long been known as an irrepressible optimist!—Williams quickly added that he would love to be proven wrong about what lay ahead. Thus far, he hasn't been.

Once again: Am I saying that any of this is literally a religion in the exact same sense that, say, Catholicism has a pope and a catechism and encyclicals and puffs of smoke emerging from the Vatican? No. It's not. Because in many ways, that would be giving tech too much credit, as its rituals are probably better compared to what Durkheim called "magic." "Magic also consists of beliefs and rites," Durkheim wrote, "but they are more rudimentary, probably because . . . magic does not waste time in pure speculation."[20] In other words, magic, in the Durkheimian view, is the deployment of the same kinds of miraculous beliefs and practices as one finds in religion, but without the communal purposes or higher ideals. It is easy to picture tech filling this kind of role in our modern society. Yes, there are many communal elements in social media and tech of various kinds. Possibilities for connection remain plentiful even now, even if a company like Meta has

of late downgraded the vaunted "social graph" to compete with TikTok
and YouTube, becoming more like an algorithmic broadcast network than
a place for friends to bring the world closer together. But do the benefits of
contemporary tech use for the communal needs of the users exceed those
for advancing the profit motives of the corporations that deploy the technologies? The burden of proof, on this point, lies with the corporations.

I've always felt one of the most obnoxious traits of the Old Testament
God, even if seen simply as a literary figure, is His obsequious need for
attention. He demands not only devotion but constant and almost manic
praise. Compare that temperament, just as a thought experiment, with
the idea of a powerful being like the ones tech companies are building: an
entity that inspires obsession but doesn't even need us to acknowledge its
existence. If the tech of our present and near future were a deity, its self-
effacing stealth would be as admirable as it was terrifying.

ALONE IN THE CROWD

Is it really enough, however, to think that in the face of such forces, we
can simply slap down a few hundred bucks for new "alternative" devices,
swap SIMs, and become more evolved human beings? Of course not, says
Dr. Emily Anhalt, a San Francisco–based clinical psychologist and the
cofounder of the world's first "mental health gym."[21] Growing up in Silicon Valley, Anhalt developed an interest in tech culture and the psychology of the entrepreneur, finding that people who take on masochistic tasks
like starting companies frequently struggle inwardly, waiting for a crisis to
seek help.

"If it's not our phone distracting us, it'll be something else," Anhalt told
me. "We need practice sitting still with the uncomfortable feelings and
thoughts that being human brings up. . . . You can't numb yourself to bad
feelings without also numbing yourself to the good ones. To feel joy, ela-
tion, and peace, you also have to be willing to feel sadness, loneliness, grief,
and anger."

This is the point. And the paradox. Ancient religions were not exactly
the cause of—or the solution to—human suffering. They were the medium
through which we processed our struggle to be human on an unfamiliar
and uncaring planet. Likewise, today, tech itself is neither the perfect savior
nor the absolute villain of our daily experience. It is the medium through

which we process our humanity, which makes our tech rituals like religious rites of the past.

The Bronze and Iron Ages—the period between approximately 3300 BCE and 600 BCE, in which the basic beliefs and practices of what became the world's great religions were forged—were a time of extraordinary uncertainty, turmoil, and striving. As we built some of the first sophisticated tools ever known and pioneered agricultural techniques that were vulnerable to disastrous collapse, of course we asked, "why are we here?," "what happens when we die?," and "how should we live?" Temples, priests, myths, and sacrificial rites were understandable responses to such a world. Sometimes those who invented such responses preyed on their flock. Sometimes they cared for it. Often their motivation was mixed, because they, too, were uncertain. Most important, at all times, was the struggle to cope, to live with the pain of death and the fear of the unknown and uncontrollable. When we look back at the worst moral failures of ancient religion—say, child sacrifice, the subjugation of women, or the division of society into hierarchical in- and out-groups—it can help to keep context in mind. But the context doesn't excuse or erase oppression, just contextualizes it. We can lament and condemn ancient examples of dehumanization *and* remember that all ancient peoples were, like us, human. Gripped by grief while striving for joy, they built our complex world, where we would do best to keep in mind our own inner complexity even as we look for outward oppression. Not because the oppression isn't there; it is. But in order to replace it with anything worthwhile, we will need to do our own inner work as well.

I'm not interested in removing all digital technology from society any more than I want to eliminate cars, planes, air conditioners, televisions, or any other technical innovation that might have negative side effects. Clearly there have been positive reasons for inventing and spreading some of our most ubiquitous recent inventions. I can get lost for hours without a GPS system like Google Maps. It's simply easier to carry a phone than to carry a bulky Polaroid, a Walkman, cassette tapes, a Game Boy, a portable TV and radio, and . . . a post office? And each of those was better, in at least some meaningful way, than what came before it. Sure, I'd vote to eliminate all guns or bombs, but I don't think such a vote will be taken any time soon.

But does it necessarily follow that every invention is a good one? Or does the market's tautological wisdom always win out—must we simply accept the presence of anything that sells *because* it sells? With remarkable speed, we've rearranged humanity's moment-by-moment existence around the rhythm and pace of our smart devices. Yes, the devices do serve some of our needs. However, I am most interested in what is good for us as human beings. What kind of society do we want? What kind of lives are we trying to live? Are we *sure* these exact devices, only in existence for the miniscule blip of history that a decade or two represents, have been or will be a net benefit to humanity? And if we aren't sure of that, even as we invest trillions of dollars into a global infrastructure to support and be supported by them, then what on earth are we doing?

As I was drafting this chapter, in mid-August 2022, my wife and I took our kids on a five-day family camping trip in New Hampshire. Joining a dozen or so families with children ages zero to twelve, we drove to the grounds of Camp Glen Brook, a 250-acre forest and farm on land inhabited by the Pennacook people of the Abenaki Nation until settled by white Europeans around the time of the American Revolution. I attended the camp from ages eight to thirteen and worked as a counselor there in my teens. On our trip, which ran after the sleepaway campers had gone home but while some of the counselors were still available to act as paid staff and groundskeepers, we lived in a simple A-frame wooden cabin and slept on metal cots. Waking around sunrise and ending our days shortly after sunset, we passed from activity to activity—chores and other responsibilities, like cleaning our own cabins, were mandatory, while swimming, paddleboarding, wood shop, hiking nearby mountains, archery, and the like were optional. After the evening meal, communal singing was followed by a simple game like kickball. Nights wrapped up with bedtime stories told in circles: tall tales for the older children, Dr. Seuss or Encyclopedia Brown for the younger ones.

Notably absent from all of this? Smartphones.

At Glen Brook, important times are communicated to campers by the ringing of a large bell. Technology is gently forbidden for children—it's not a sin if they somehow catch a glimpse of an iPhone, but they aren't allowed access to their personal devices. For adults, taking occasional pictures is fine, but scrolling or streaming is expected to be done in private. Knowing the camp's approach well after many years of visiting, on this trip I didn't even bring a phone charger, and I spent perhaps half an hour online. If that sounds Spartan to you, consider that I'm the one writing a chapter about my tech addiction, and I found it easy. Why? Community.

In environments in which I feel genuinely connected with other people, my tech is not such a big deal. I can take it or leave it. The problem is that these environments can be few and far between. This is a problem when it comes to tech addiction, because lonely rats, as studies have repeatedly shown, are much more likely to become addicted to heroin.[22] Humans (like rats) evolved to be social creatures, and when we don't have access to needed psychological rewards like interhuman closeness, we seek those rewards elsewhere. And the makers of apps and devices are more than happy to oblige. Or at least they oblige, perhaps sometimes feeling "tremendous guilt," as Chamath Palihapitiya, former vice president of user growth at Facebook, described it.[23] "The short-term, dopamine-driven feedback loops that we have created are destroying how society works," Palihapitiya told an audience at Stanford in 2018, referring to research that is well-known in social media circles.[24] Essentially, the dopamine reward systems in our brains, which regulate pleasure and pain, are extremely susceptible to random, intermittent positive feedback. When something pleasurable suddenly happens, we go in search of it again, and we're willing to endure quite a bit of pain and loss to get it, which is why gambling is also so addictive. The makers of our most commercially successful devices and apps are experts at this, as Palihapitiya admits, crafting more and more tiny rewards in the form of likes, follows, retweets, messages, points, and other incremental updates that feel, in the deep recesses of our evolutionary minds, like drugs.[25]

In a communal camping environment like Glen Brook's end-of-summer family retreat, however artificially constructed, there is constant human interaction, which provides rewards of its own. Social connection disrupts the feedback loop of artificial reward systems like those in slot machines or Twitter feeds. Just as lonely rats get more powerfully addicted when offered drugs, rats prefer social interaction to those drugs, forgoing heroin and methamphetamine in favor of spending time with another rat.[26]

Now, you may have plenty of social interaction at your office or on your campus. I do, too, and at its best, this interchange is enough to keep me from spiraling into doomscrolling. The problem, however, is that at work or in school, many of our interactions are devoted to the pursuit of the capitalist faith from which the tech religion emerged as a dominant strain. Everything is a competition. We strive for more sales and revenue, promotions and recognition, better grades, better recommendations, better publications.

Homer Simpson lesbian bar meme. *Source:* IMGflip.com.

Even hobbies are too often conducted in the pursuit of the perfect CV or résumé, whether golfing to bring in new customers or getting elected president of a club to help your college applications. The underlying message of all this striving is that we are what we do. Our value as human beings is defined by what we accomplish, and there is a finite amount of value available, so while we *can* make friends amid our efforts to influence people, too many of our would-be friendships are colored by this never-ending search for more. Even when we have time to spend around others, we know that time is limited, and we can rush to spend it with the *right* people, engaging in the *right* activities. It's hard to connect that way, and even harder to *feel* connected. So we end up feeling lonely, even in a crowd.

Enter: tech.

At the family dinner table after my post-camp day of writing, my wife asked if I'd told anyone about the trip. "No," I said, "I didn't talk to anyone today." Indeed, I hadn't had an in-person conversation or even a phone call with another human the entire day, though I did engage with other humans via Twitter, three different email accounts, reading texts in a group chat about movies, and even lurking for a few minutes on the Facebook account on which I refuse to post. The difference in my own happiness level in the camp setting (not elated all the time, but nearly always content) versus elsewhere (often miserable) was extraordinary. But much of our tech use is no longer about whether it will make us happy. We do it for the ritual. I check my phone now because I checked it earlier, and I want to see if anything has happened since. I'll check it again later because I'm posting something now, and I'll want to know if there's been a response. I do these things because I've been taught that doing them is important, whether by the companies whose content I am consuming and providing, by my peers, or by the constantly updating status that is my dopamine and serotonin level. And yet my own tech "choices," if they can even be called that, have degraded the quality of my everyday experience, in a reversal of collective effervescence.

What is one to do? How does the tech agnostic solve this paradox, where we're damned if we do worship at our tiny altars, and we're also damned if we don't?

If there is a new and important solution to be found to this dilemma, perhaps it can be found in a common thread I heard in the comments offered to me by both Dr. Emily Anhalt, the entrepreneurial Silicon Valley psychotherapist, and Kaiwei Tang, the philosophical smartphone engineer turned CEO. Both likened tech like smartphones to fast food. "It's better than going hungry," said Anhalt, "but it's not nourishing."

Connections among nutrition, eating, and tech came up throughout my research on tech addiction. Tech use has become almost like eating for many young people (and for the not-so-young person whose work you are reading right now). How? First, in the sense that we eat just about every single day of our lives, usually eating multiple meals and snacks over the course of the day, trying our best to balance nutritious content that is good for our growth and health with the fact that we all crave a sugary treat (or a deep dive into some random YouTube influencer's Instagram cat videos) from time to time. But also in the sense that if we develop a pattern of addictive behavior around either food or tech, it's not possible (in the case of food) or realistic (in the case of tech, for most young people) to recover from that problem by abstaining completely, the way one might try to do with alcohol or narcotics.

I could try to tell myself, or students or others with whom I work, that I'm going to stay completely off of email or social media for a day. And that attempt might work, though in my now many years of experience I can say it will fail more often than not. But what if I were to try to stay off the Internet completely for a day, let alone a longer period? In the 2020s? While not attending an expensive retreat? When seemingly every aspect of my life is now online somehow? Trying to retreat from tech rituals entirely would go about as well as going cold turkey from eating—not just from fast food, but from all food.

Which is how I found myself, for the conclusion of this chapter, talking with experts on eating disorders.

THE END OF AVOIDANCE

First I reached out to Dr. Tara Deliberto, a clinical psychologist and author of the medical textbook *Treating Eating Disorders in Adolescents*. A former professor at Cornell University's Weill Medical College, Deliberto served as the director of New York-Presbyterian Hospital's partial hospitalization eating-disorder program before moving to California during the COVID-19 pandemic to work with a startup attempting to leverage tech to treat eating disorders. I've known of her since her days as a graduate student in Boston in 2007 through my work as a humanist chaplain. Science was her religion back then, she told me when we Zoomed for me to ask whether her field might offer insights for addressing tech addiction, though she's mellowed into a more spiritual place now.

"I agree there is an overlap between eating pathology and tech addiction," Deliberto said at the outset of our dialogue. Not only do both involve compulsive issues with experiences we can't realistically just avoid, she explained, but both serve the function of what therapists call "experiential avoidance," avoiding aversive internal states at the expense of connection with ourselves. Both are widespread. We've already covered the prevalence of tech addiction–like behaviors in this chapter, and as Deliberto points out, perhaps 70 million people worldwide suffer from eating disorders such as anorexia nervosa, binge eating disorder, bulimia nervosa, food obsessionality, or "mindless eating," the last of which she notes is not an eating disorder per se but can be problematic in that it "fosters disconnection with the body." This especially reminded me of many tech rituals, which may not rise to the level of needing clinical intervention but aren't good for us, either.

Another expert who affirmed my comparison between addictive tech rituals and eating disorders is Bahia El Oddi, the founder of an organization called Human Sustainability Inside Out, a Moroccan-born Harvard Business School graduate, and a professional advocate for young people's mental health.[27] I discovered El Oddi's work through a dialogue she organized called "I Instagram, Therefore I Am: Your Child's Brain on Social Media," sponsored by the group Harvard Alumni for Mental Health (HAMH).[28] At the virtual event, El Oddi led HAMH in conversation with Julie Jargon, a *Wall Street Journal* columnist on family and technology. Jargon described how social media can have so much power over young people that it affects their neurology in externally visible ways, such as the recent trend of otherwise healthy teen girls watching videos of teens with Tourette's syndrome and seemingly developing Tourette's symptoms, like facial tics.[29]

El Oddi, who was raised amid intense poverty in the teeming city of Casablanca, had her life transformed when her mother's Spanish citizenship afforded her the opportunity to attend European schools on scholarship while her peers could not afford an education. Seeing their suffering, she wanted to give up what she had to help them, but her father responded with advice that stuck with her: As a child in her circumstances, she simply could not do anything for other children. The best she could hope for, he explained, would be to achieve excellence abroad, attain a better life, and return one day to tend to them. El Oddi took her father's advice, pursuing academic achievement with near obsession and making it to Harvard. Along the way she developed her interest in and concerns about youth social media use because she personally experienced a link between social media and mental and physical vulnerability, as one of the world's 70 million individuals struggling with eating disorders.

When I spoke with El Oddi, she detailed her intense concern about young people's mental health in a technological world and her efforts to combat this phenomenon, ranging from giving talks at Harvard to leading art therapy workshops for street children in Tétouan, her birth city in Morocco. She was, however, reluctant to blame such problems entirely on tech. Instead, she blamed perfectionism. In striving to be an exception who could change the rule, she had internalized that only by being perfect or exceptional could she find self-worth. It was this belief that led her to seek out images, online and elsewhere, of "perfect" Western beauty standards— ideals she wanted so badly to live up to that they ultimately endangered her life, as they do the lives of many others like her.

When Tara Deliberto works with patients, her emphasis is often less on the act of eating itself, or on gaining or losing weight, and more on helping patients understand why and how much they are afraid of eating and weight. We are hardwired to fear many things, she points out, like a lion hunting us down, but we are not hardwired to fear gaining weight.[30] So she invites patients to inventory the fears and anxieties they've developed along their way: perhaps of being socially rejected, having people say or think disparaging things about their body, losing a competitive edge, or not finding a romantic partner. Such fears are nontrivial, despite the popular stereotype of anorexia nervosa as primarily the ailment of thin, young, well-educated white women of higher socioeconomic status. In fact, Deliberto says, "We see anorexia nervosa in all populations, and quite frankly, the more trauma and colonialism that you see in a population, the more mental health issues you're going to get, and that includes eating disorders."

After patients inventory their fears and anxieties, Deliberto points out that it's not weight gain per se that they fear but their own many and varied associations with it. A chain reaction emerges in which the presence of food or the sensation of fullness can trigger a fear of fatness, which comes from associating fatness with some other fear, such as of dying young or of romantic rejection. Those fears then cause internal suffering, which then leads to vomiting or dieting or binge-eating to avoid the experience of suffering. Avoidance then creates a pattern called "negative reinforcement," in which an association is formed between the disordered eating behavior and the absence of suffering, which makes it more likely that one will return to the disordered eating to cope with future suffering.

Reflecting on my own experiences of tech addiction considering Deliberto's insightful analysis, I began to see how I've created my own tech rituals in a similar pattern of experiential avoidance and negative reinforcement. Early in life, for reasons beyond my or anyone's full control, I internalized the message that my value as a human being was contingent on what I did. I could only be worthy as a human being, I came to believe, if I not only achieved excellence but demonstrated that I was extraordinary. Even the slightest bit of mediocrity was unacceptable, the thought followed, because it meant I was unworthy of love, admiration, or self-esteem. Thus began a vicious cycle: the relentless pursuit of some mythical greatness on the one hand, and a frantic avoidance of the terror I would experience whenever situations inevitably emerged in which it was impossible for me to be "great."

My father grew up believing that the only way he could ever feel okay was to achieve an unassailable victory in a life he was taught to see as an endless competition. By the time I was born, my dad had come to see himself as a failure, and thus he struggled with the fear that he was unworthy of love—receiving or even giving it. Unintentionally, and even with genuine affection, he passed these fears on to me. "Champ," he called me when I did well. But I worried about what would happen when I wasn't a champion. Which, as with any kid, no matter how talented, was often.

My dad's greatest failed dream was to become a writer. I learned early on to bask in his rare praise when he liked my writing, and then to fear the absence of that praise. My fears led to experiential avoidance; I would turn papers in late or incomplete rather than finish them and face my terror. Then, as personal computing became ubiquitous, writing itself became synonymous with technology, and then technology became a place to be distracted: by the Internet, by email, by social media, and on and on. I got stuck in a subconscious pattern: any writing assignment would trigger thoughts of unworthiness, which led to fears of abandonment, which made me miserable. *Makes* me miserable—I still sometimes get stuck this way even after many years of therapy. My breathing becomes shallow. My body tenses. As my thoughts race, attention and focus evaporate. I search, like prey in the presence of a predator, for escape. And there to welcome and comfort me are my constant companions. For the binge eater it would be food. For the alcoholic, a drink of choice. For me it's another tab to open. Another notification to follow like a white rabbit down a hole into blissful unawareness. Until the next moment of awareness triggers the next chain reaction.

It's embarrassing to share all of this. But I know it's a story many have in some form, because it's normal and natural and understandable if you feel incomplete and unlovable and like you are running from something you barely even understand. And just as I know for a fact that others share this cycle of avoidance and shame in their relationship with food or drugs, I know many of us are repeating this unhealthy pattern with our tech. I know it because of former tech executives, like Chamath Palihapitiya, who speak of their "tremendous guilt" for having taken advantage of our human frailties for profit.

Camp Glen Brook, the seemingly idyllic little retreat I described above, is affiliated with the Waldorf educational movement—a century-old attempt to live holistically and to develop artistic imagination and creativity over rote memorization of dry facts, or, most recently, over the use of computers in and around the classroom. Waldorf is an imperfect movement based on an obscure early nineteenth-century mystical philosophy.[31] But that hasn't stopped it from gaining enormous popularity in Silicon Valley, specifically because of the antitech, screen-restrictive stance taken by Waldorf School of the Peninsula, a Bay Area branch of the movement with campuses in Mountain View and Los Altos, California. "A Silicon Valley School That Doesn't Compute," read one of many headlines about this "most sought-after private school in Silicon Valley," a place that can cost as much as $30,000 per year for kindergarten and $50,000 for high school.[32] Why so popular? Because just as Steve Jobs famously did not let his own children use iPads or iPhones, and other iconic tech executives like Bill Gates and Sundar Pichai are extremely restrictive of their children's screen time, even rank-and-file technology leaders are increasingly concerned about the damage their "disruptive innovations" might do by preying on our all-too-common psychological vulnerabilities.

All of us feel weak and inadequate at times. When we click on a notification as a comforting, addictive ritual to momentarily ease our anxiety at the expense of emotional awareness, that is a feature, not a bug. And when our operating systems get buggy, maybe the way forward can best be encapsulated by an insight Bahia El Oddi shared with me, one from her own humanistic Sufi spirituality. Whereas the religious worldview of places like Silicon Valley and Harvard Business School is essentially to be a bright star, unique among all others, Sufism offers an alternative. Instead, El Oddi told me, we can each strive to be stars that come together to form brilliant constellations. When we all shine on one another, everyone's way is brighter.

THE ONLY WAY OUT

Once Tara Deliberto has helped her eating-disorder patients to become aware of their own chains of anxiety, she coaches them to interrogate those fears in therapy. When possible, she encourages a break in the chain of behavior, avoiding unhealthy eating even when unhelpful thoughts remain, because this change in one's actions can free up psychological energy for examining one's thoughts and feelings. We, too, might begin to create new rituals around tech use, learning to keep our attention on one another or on the painful task at hand (in my case, on my writing) when we know how and why we're tempted to click.

After Deliberto explained to me all the basics of eating-disorder treatment, she mentioned a way in which her approach is different. Modern psychological theory around eating disorders, she said, encourages individuals to look at their own values in search of alternatives to their associative chain of anxiety, fear, and disorder. The problem, she told me, is that the things we value are themselves often not healthy for us. Instead, Deliberto recommends thinking about our ideals and whether they are based on fear or on compassion and justice. It comes back to the question of what kind of life we want to live. Do we want to spend our days in fearful pursuit of beauty and perfection, knowing we can't ever fully attain such things? Or do we want to orient our lives toward something else, a different kind of ideal that we may never completely achieve but that places our day-to-day struggles in the context of collective liberation and shared flourishing?

As Dr. William Shunkamolah, an Indigenous clinical psychologist with whom Deliberto has worked on treating eating disorders among Native American communities, has said, even institutions like the American Psychological Association can err when framing their treatment around individual choice.[33] Individual choice is fine, Shunkamolah says, but what about the wisdom of elders? Collective intention is an ideal that the tech religion and its individualistic rites and practices have placed at the bottom of its hierarchies. By returning some of our attention to collective values—by asking, How are my actions and inactions contributing to my community? Am I helping to build a healthy, compassionate, equitable world for all?—we can cast a new light not only on eating disorders but on tech use, too.

Your elders, in other words, don't care if you use tech a bunch. What they want, I invite you to imagine, is for you to have a positive life goal that affirms the community justly and equitably while also giving you personal satisfaction, meaning, and strength. They want you to let your compassion overtake your fear. They want you to practice sitting through discomfort to get what is more lastingly worthwhile for everyone. And nobody worthwhile cares if you have an occasional cookie, digital or otherwise. "Enjoy yaself," as Missy Elliott said.

Anhalt, the psychotherapist, offered advice my addicted and anxious younger self could have used in trying to create healthier rituals, technological and otherwise. Maybe you can use it, too, regardless of what electronic purchases or screen-time goals you may have in mind for this year.

Psychologists William Shunkamolah and Tara Deliberto.

"Almost everything in our society is a shortcut, a cheat code. . . . Just take the straight path through, and it won't feel good all the time," but by feeling our feelings without panic or judgment, we can fold them into who we are in such a way that we don't need to constantly shoo them away. "The only way out," Anhalt told me, "is through."

For a tech agnostic, the way through means accepting that checking our devices is, sometimes for better and certainly for worse, the dominant religious ritual of our time. As with many rituals, it is almost impossible to opt out of a practice to which almost everyone alive is devoted. So our ritual life will continue, for at least the generation to come if not many more, bringing with it some pleasure and gratification along with pain and frustration. But also: in this new religious world, the work to focus, to be mindful enough to endure discomfort, and to pursue ideals of compassion and justice, too, can and must become a ritual.

Here on earth, there is no exalted place where you will ever be perfect, in nirvana, or one with the universal consciousness. At least, that's not how I see it. Instead, you evolve every day. You build new neural connections, new relationships. You deepen your sense of who, beyond yourself, you are fighting for—like both eating-disorder experts Bahia El Oddi and Tara Deliberto have done, not to mention Light Phone founder Kaiwei Tang and, yes, even me.

Whether you see it metaphorically or literally, we are all radiant stars. The key is not to shine the brightest or the most perfectly but to enjoy and appreciate your own light and that of all others.

APOCALYPSE(S)

The concept of apocalypse is so discussed in today's popular imagination that it can seem to have lost nearly all the original meaning of the word: a revelation, unveiling, or uncovering, in the famous New Testament book of the same name. The first word of that Book of Revelation—apocalypse, or *apokalypsis* in the Hellenistic or Biblical Greek dialect—refers to more than just the sort of epically disastrous event with which the word has become associated in the contemporary popular imagination.

In the ancient Mediterranean world there was an entire apocalyptic literary genre, including several prominent Jewish and Christian books dating back to at least 200 BCE, and even Islamic apocalypticism. Each of these ancient apocalyptic texts shared common characteristics, such as narratives, esoteric language, and a dour view of present times, to describe "supernaturally inspired cataclysmic events that will transpire at the (imminent) end of the world."[1] Some of this ancient apocalyptic literature, as in the Book of Daniel, was included in the Old Testament canon. Other very similar texts would have been lost to history had they not been translated by early Christians from their original Hebrew and Aramaic into languages like Greek, Latin, and Ethiopic, or discovered in fragments amid the Dead Sea Scrolls, a collection of ancient documents that sat for two millennia in pots in the dry caves of Qumran, Palestine, until an Arab boy discovered them in 1946.

By far the most influential ancient apocalypse, however, is the Book of Revelation. And what makes that book so famous—beyond the fact that it, unlike other books of its genre and time, happened to be canonized in the New Testament—is the vividness of its fantasy. From the disappearance of millions due to "premillennial dispensationalism" (the Rapture), to cataclysmic war between angels and a dragon-like satanic beast, the book spills over with archetypal themes. Plentiful and memorable supporting details, like four horsemen, seven-headed monsters, and locusts with human faces and hair and lion's teeth wearing iron breastplates, give Revelation some of the drama that made it a key to the evolution of modern, Western civilization. But beyond what amounts to the ancient version of movie special effects, the real power of the text came through its message.

According to the biblical author of Revelation—a first-century prophet, John, who was both Jewish and Christian (as most early Christians were)— the destruction will make way for the New Jerusalem and the revelation of God's true intentions for the world and for humanity. John was most likely a survivor of the 70 CE war in which the grand Jewish second temple in Jerusalem was destroyed and Jews were largely expelled from the Holy Land by the Roman Empire. Whether or not he had participated in the conflict himself, his life was marked by these calamitous events, turning the greatest symbol of his people's history and identity into barely recognizable wreckage that Roman historian Flavius Josephus described thus:

> The holy house burnt down. . . . Nor can one imagine anything greater or more terrible than this noise; for there was at once a shout of the Roman Legions, who were marching all together, and a sad clamor of the seditious, who were now surrounded with fire and sword. . . . The people under a great consternation, made sad moans at the calamity they were under. . . . Yet was the misery itself more terrible than the disorder; for one would have thought that the hill itself, on which the Temple stood, was seething hot, as full of fire on every part of it.[2]

John's message, some two to three decades later, was one of both vengeance and hope. The Roman emperor and his forces were mighty enough to overwhelm the Jewish earthly community, but they would one day receive divine comeuppance. There would, quite literally, be hell to pay. Eternal heavenly rewards, for those on the side of good, would trump not only anything they had seen and lost while mortal but also anything their enemy had ever possessed.

This, then, is the story of Revelation: ultimate reward for good people, along with final, overwhelming suffering for evil ones. The canonization of such a text had a profound psychological impact on the next two millennia of human history, especially when, just over two hundred years later, the conversion of the emperor Constantine elevated the New Testament as the defining ethical vision of the most powerful institution the world had known to that point. Since then, warring factions on many sides have seen themselves as the good, as depicted in this book, and seen their enemies as the evil.

Western society is a polarized, resentful, and competitive affair that loves nothing if not spectacle. The apocalyptic tradition's elaborate message of future glory for one side and torturous suffering for the other has been a compelling fantasy for aggrieved parties in conflicts of all kinds, from religious schisms to hip-hop battles. When the Confederate South fought the Union North in the American Civil War, for example, both forces claimed Revelation as an affirmation of their righteousness and a roadmap of what was to come.

HYPERVISIBLE

Chris Gilliard works as a tenured professor of English at Macomb Community College, a school founded in 1954, when eighty-four students from Michigan's Macomb County rented classrooms in a local high school.[3] Macomb County, about twenty minutes northeast of Detroit, is the third-most-populous county in Michigan. Most of its eight or nine hundred thousand residents are white and working class; Donald Trump visited the campus no fewer than three times during his two presidential campaigns. In his spare time, Gilliard has become one of the world's leading critics of surveillance capitalism.

@Hypervisible, the handle for Gilliard's Twitter account (which in my opinion was one of the most tragic losses caused by what Gilliard and I both view as the death of Twitter upon its takeover by Elon Musk) is partly a reference to "the feeling of being overly visible because of an individual's race or ethnicity, sometimes to the point of overshadowing their unique skills and personality."[4] But the notion of hypervisibility, a decades-old term of art in academic discussions of racial justice, takes on added power when considering the new reality of police surveillance technology.[5] Once, Black, Indigenous, and other people of color may have sensed or intuited that eyes were on them because of their skin color and identity in a society dominated by whiteness. Now, however, it is increasingly common that they're also in the crosshairs of cameras intended to surveil them in a society dominated by structurally racist tech.

Early in his academic career as a scholar of rhetoric, Gilliard had experiences in which his mere presence was seen as both notable and questionable.

Chris "Hypervisible" Gilliard. In his profile pic, Chris Gilliard obscures his face
behind a ski helmet and goggles. *Source:* "Dr. Chris Gilliard," Our Team, Surveillance
Technology Oversight Project, accessed August 8, 2022, https://www.stopspying.org
/chris-gilliard-bio.

At one of the schools where he served as a new professor, for example,
he once taught the English portion of a summer outreach program that
was housed in the university's engineering building. When he would spend
time in that building making copies and such, engineering faculty repeatedly suggested he did not belong, as if he were an interloper who should
leave for everyone's safety. These academics failed to imagine that a relatively young Black man with dreadlocks might be no more or less than a
fellow scholar, with much knowledge of technology and related matters. In
such environments Gilliard was, in a word, hypervisible.

Chris Gilliard was born in Detroit in the early 1970s and came of age in
the wake of violent uprisings (they have often been called "riots," but that
term has been challenged as promoting prejudice by historians of the events;[6]

I will use "uprising," or "rebellion") that famously drove "white flight" from his birth city. His large Catholic family had been part of the Great Migration, in which thousands of African Americans settled in Detroit as part of a wave of 1.5 million Black people leaving the South in the first half of the twentieth century.[7] Gilliard's father was a repairman and his mother worked for the University of Detroit Mercy, a Jesuit Catholic institution with a campus stretching across the city. On their combined working-class salaries, Gilliard's parents put him and his several siblings through excellent private Catholic schools in a city that became, in the years after the uprising, best known to outsiders as a place not of success stories but of poverty and racial division—a characterization Gilliard resists. A place of violence and poverty is certainly how I thought of the city when I entered the nearby University of Michigan–Ann Arbor, as an eighteen-year-old freshman in the 1990s. Though I eventually grew more familiar with the city and its rich history, I confess that when I've thought of the city over the years, I've been much less likely to reflect on good things that have happened there than to see it as a focal point of struggle and strife. There is certainly data to support such a negative view, if one is inclined to look for it: more than 40 percent of Detroit's residents, including 57 percent of children under the age of eighteen, live below the federal poverty line, according to 2017 census data—the highest such rates among the twenty largest US cities.[8] Within city limits, 80 percent of the population is Black, compared with the population of the entire Detroit metro area, which is 22 percent Black and 67 percent white.[9] But while such facts have some relevance, one can't truly understand Gilliard's Detroit upbringing without an awareness of the goodness, decency, and inherent human wisdom of countless families like his, even as they coped with sometimes-difficult situations. In his eulogy for his father some years ago, Gilliard recalled being driven to book conventions as a kid. Gilliard sat for hours at a time sorting through boxes, eventually settling on a stack about a foot thick to buy and bring home. His father did not read the books or know much about what they contained, but he knew his son loved them, so he waited, never complaining or rushing, his patience a lasting expression of love.

I was thrilled that Chris Gilliard gave me permission to come to Detroit to meet with him and learn about the effects of surveillance technologies on Detroit's citizens and his own related work. Gilliard has been a leading voice in writing about Detroit's Project Green Light, an initiative that began in 2016 as a technological partnership between the Detroit police department and eight gas stations. Each station agreed to have surveillance cameras installed, secured, and monitored by the Detroit police, accompanied by a bright, flashing green light: a sign that surveillance was underway, but also meant to connote the idea that it is safe to "go," as in to move around an inner city associated in the minds of many with crime and danger. The project has since expanded dramatically: hundreds if not thousands of green lights and cameras now dot the landscape of Detroit and several surrounding municipalities. In combination with concerns about the racist application of facial recognition technology and the proliferation of doorbell cameras, Detroit-based writers like Gilliard and the tech justice activist and poet Tawana Petty have noted that their city is being used as a kind of modern panopticon, in which a majority-Black population is tracked, intimidated, and in many cases abused for the benefit and convenience of mostly white outsiders who might occasionally want to pass within the city's lines without having to think of the social implications of the area's savage inequities.[10]

Project Green Light did not evolve as an isolated experiment. Since the 1967 rebellion just before Gilliard's birth, Detroit has been one of the most heavily surveilled cities in the world. The violence itself was sparked when police raided a Black speakeasy and then became a "hostile occupying force" as the city's Black institutions burned and its white citizens fled.[11] Even that, however, is far from the beginning of this story.

Black southern migrants of the Great Migration—people fleeing from the virulently racist aftermath of hundreds of years of kidnapping, murder, and enslavement, traveling at great difficulty and with very little, if anything, to their name—arrived in Detroit, where there had been fewer than 6,000 Black residents before 1910. Just two decades later, the Black population numbered over 120,000 in this metropolis of the American North, a place that, mythologically at least, was much better to and for Blacks than the South they had fled. But these migrants were not allowed to live in most parts of what was then, thanks to the automotive industry, a booming metropolis.

Their labor was needed, however, as the city and industry expanded, and they were allowed to settle mostly in a single neighborhood: Black Bottom, originally named, according to Jamon Jordan, a historian of Detroit's African American community, for its rich dark soil, but ultimately associated by name and otherwise with the Black community.[12] Black churches, schools, and businesses thrived in Black Bottom as it became a diverse area of more than a hundred thousand residents that included Jewish and other European immigrants in addition to its heavily Black population.[13]

Until it was destroyed.[14] Between 1953 and 1967, under a white mayor and city leadership, a plan cynically labeled "urban renewal" wiped out the entire Black Bottom area, including the Paradise Valley business and entertainment district that bustled, between the 1930s and 1950s, with brilliant performances from famous names: Duke Ellington, Dinah Washington, and other performers excluded from white hotels and venues. Every single structure in that entire area was demolished. These actions provide context for the uprising that unfolded in Detroit in 1967. Imagine people growing up around this period of state destruction, who had known little but a hostile and oppressive state their whole lives, joining others rising up violently during a tumultuous time in America's history.

Then, in subsequent years, a new kind of police unit called STRESS was created to deal with the violent aftermath of violent times. As detailed by journalist Mark Binelli in the *New Republic*, when faced with a dramatic increase in robberies—eighteen thousand in Detroit in 1970, several dozen of which ended in death—a new chief of police named John Nichols tasked his staff with constructing a study that would detail the "typical robbery."[15] Using computers to complete their assignment, Detroit police found that criminals were most likely to be "young, nonwhite, and armed," and that victims were also generally nonwhite and male, though not young,

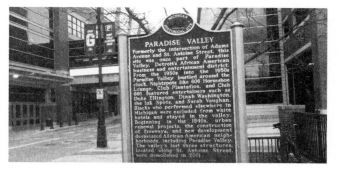

Paradise Valley historical marker.

suggesting a society at odds with itself in the wake of not only poverty and inequity but massive, multigenerational trauma. Instead of attempting in any systematic or ambitious way to address this trauma, Nichols and his colleagues designed a strategy of deception: undercover police officers posed as "drunks or derelicts," priests, hippies, or even a "helpless looking little old lady."[16] These officers in disguise would bait would-be muggers, with other officers lying in wait. The result? Along with some marginal reductions in crime rates, thirteen Detroit citizens, all but one Black, were killed by STRESS officers between April and December 1971—the highest number of civilian killings per capita of any American police department.[17] What could have been a season of righting wrongs and bringing people together instead became a data-driven incubator of mistrust and resentment.

Let's fast-forward to present-day Detroit, near which I comfortably enjoyed my education at the University of Michigan, and where Chris Gilliard has lived and developed his work on the technologies he calls "luxury surveillance." We set out to see the city, which is dotted with once-elegant buildings that have been either abandoned or allowed to fall into disrepair since their origins at the early peak of Detroit automaking. Religion is ubiquitous, and at one point we pass a hand-painted church sign reading "Souls for Christ: Deliverance Center"—in contrast with Amazon's fulfillment centers. You sacrifice your life in fulfillment for the future of others. But not in an altruistic way. In an oppressive way.

"We're coming up on Cass Corridor here," Gilliard points out as we drive. "Over a dozen people that STRESS killed, right in this area." The first stop on our driving tour of the city is Ford's Piquette Avenue and Highland Avenue plants, the first-ever factories for the mass manufacturing of the famous Ford Model T. In the first years of the twentieth century, these adjacent manufacturing facilities produced arguably the finest example of mass personal technology the world had ever seen.[18] "And here it sits," Gilliard says.

The 102-acre Highland Park plant, which was designated a National Historic Landmark in 1978, looks from the outside to be in ruins, with broken or obviously repaired windows, crumbling concrete, and missing bricks. Surveillance cameras were the only defining features of an expanse that otherwise looked like an abandoned building of little significance, abutted by a very modest strip mall called Model T Plaza.

The old manufacturing facility is so big it's almost impossible to photograph, though I try, as you can see below. I'm reminded, in doing so, of the old parable of the blind men who, having never seen an elephant, come upon one and try to identify it by feeling its different parts. "This is what Detroit does to its monuments," Gilliard sighs. What could be a monument to the most important piece of automotive tech in history is in utter, abject disrepair. "Nowhere else would they let something like this just sit there."

Gilliard then tells me about the Woodward Avenue Dream Cruise, where people come from all over the world to drive their classic cars down this street, which passes from Detroit through several neighboring suburbs. People camp out on the streets to see and celebrate the cars with barbecues and parties. But the cruise starts in Ferndale, just past 8 Mile, the border of the city, because the event's participants would hesitate to spend time inside Detroit proper. "I've always said"—he smiles ironically—"they should start here."

Meanwhile, he points out, "the rehabilitation project for Henry Ford has been significant." Not grasping his intention yet, I ask, "You mean the Henry Ford Museum? Or Henry Ford High School?"

"No, I mean Henry Ford the man," he says with a gentle, knowing sigh. "He was a pretty gross individual."

The point, in retrospect, is clear: the Model T factory, because it sits in Detroit, a Black city, hasn't been rehabilitated. But Henry Ford *the person* with his whiteness and the ability to exist in the space of imagination instead of here in real Detroit, can be posthumously fixed up and polished into better shape than he deserves. Ford's antisemitism is a known fact, but it's certainly not one that attracts much attention when one visits the Henry Ford Museum of American Innovation in Detroit-adjacent Dearborn, Michigan.[19] "The Henry Ford," as the museum calls itself, is an enormous campus—"250 acres of inspiration"—that features numerous historic trains and planes along with automobiles and attracts nearly 2 million annual visitors. Few of these visitors, based on my own family's visits to the museum in 2019 and 2023, will be confronted in any significant way with moral qualms about Ford's legacy while they take in attractions such as John F. Kennedy's limousine, Rosa Parks's bus, or hot dog–maker Oscar Mayer's twenty-two-foot-long, ten-foot-high Wienermobile. The museum and its website acknowledge that Ford's antisemitism has "tarnished his reputation," but unlike how, say, Planned Parenthood has recently disavowed its

Ford's Model T factory, 1913. *Source:* Collections of The Henry Ford.

historic connection to cofounder Margaret Sanger over concerns about her belief in eugenics, the Henry Ford stands today as a monument to the overall positive memory of a problematic historical figure.[20]

I thank Gilliard for thinking to bring me here to the Ford factory at the start of our trip, noting it as a fitting origin story for tech as a whole: this historic landmark abandoned, left for a total loss, with a Nazi-adjacent founder who is now admired by many as a symbol of American leadership, ingenuity, and strength.

"It's true," Gilliard says simply. Could such a fate, I muse, await the Googleplex or Apple's Spaceship headquarters a century from now? What cataclysmic events might have to occur for that to happen, I wonder, and how likely might such events be?

[*Note:* In January 2024—over a year and a half after I originally wrote the section above, and just as this book was being prepared for publication, a large site of a century-old auto manufacturing plant near the Ford factory was acquired for the purpose of electric car battery manufacturing. Fortescue, the Australian mining juggernaut receiving extensive tax breaks to develop the reported $210 million facility, is a major carbon emitter that has pledged, with uncertain results thus far, that its new mission is decarbonizing and "turning the world green."[21] The company's own billionaire CEO Andrew Forrest has admitted that modern day

Ford's Model T factory, 2022.

slavery is "at least present, if not fundamental," to his own supply chain
and supply chains like it, around the world.[22]]

SECULAR REVELATION

Having run through a short history of ancient apocalyptic literature and
mythology earlier in this chapter, let us now turn our attention to an
entirely different and arguably more influential form of apocalyptic literature. For shorthand, I'll call the biblical and apocryphal apocalypses we
discussed above type I, or religious apocalypses. Type II apocalypses, then,
would be apocalypses in popular culture, including postapocalyptic movies
and TV shows. Understanding how these works compare to their theological predecessors might be useful for understanding the mindset of—and
therefore the risks posed by—those who might place humanity at risk of
an actual tech apocalypse today.

In modern postapocalyptic art, you typically find a world devastated,
or at least a civilization severely threatened, by some enigmatic cataclysmic
event. All the better if it depicts a series, chain reaction, or "perfect storm"
of events, after which a bewildered remaining fragment of humanity slowly
and haltingly attempts to rebuild civilization, via time travel or superpowers or good old-fashioned human pluck. For examples of these, see *The
Terminator*. Or *Mad Max*. Or *The Walking Dead*. Or *The Road*, *Planet of the
Apes*, *Station Eleven*, *Snowpiercer*, *Interstellar*, *World War Z* . . . I'll refrain from
naming all 211 of the "200+ Best Apocalyptic Films"[23] listed on IMDb,
but the fact that such a list exists says a lot. A favorite of mine would be

the 1984 BBC film *Threads*. It starts with a close-up of a spider's distended abdomen as she weaves her web. The narrator says, over dignified but foreboding music, that "in an urban society, everything connects. . . . Our lives are woven together in a fabric, but the connections that make society strong also make it *vulnerable*."

Workaday characters then go about their lives in a middling industrial city in the most advanced and sophisticated society ever known. Then, suddenly, hundreds of megatons of nuclear warheads drop, and British society is so thoroughly destroyed that the next generation can barely speak English. They are raised in a world in which there is almost no community or infrastructure left for them.

Apocalyptic narratives can be literary works as well, whether fiction like Octavia E. Butler's *Parable of the Sower* or nonfiction like Rachel Carson's *Silent Spring*. They can be documentaries, like the HBO or BBC explorations of Hiroshima or Chernobyl. Some works bend genres, like the "future nonfiction" of award-winning historians of science Naomi Oreskes and Erik Conway in their novella *The Collapse of Western Civilization*. Informed by the authors' deep expertise in the social science and history of climate change, *Collapse* is a narrative account of humanity's history from the perspective of three hundred years from now, where data went unheeded until it was too late, causing hundreds of millions of excess deaths and leaving cities like New York completely abandoned for higher ground. China, where only a fifth of the entire country's population was killed, represents the best case for how well any nation on earth coped with what, in our time, is still to come.

Few of these type II apocalyptic and postapocalyptic stories are triumphal tales of victory and glory, at least not without significant undertones of ambivalence or satire. Unlike the Book of Revelation and similar ancient works, these modern stories tend to invoke feelings of fear, sadness, anger, or disgust. Some involve detailed character studies of flawed characters on multiple sides of a conflict, and even in works with clear "good guys" and "bad guys," there is rarely a clear sense of what theologians would call *teleology*—a vision of the purpose of history and human events, as in the biblical apocalyptic notion that all that came before the Revelation was God's will leading up to it. Instead, the purpose of type II secular apocalypses seems to me to be to leave us with a question: Are these cautionary tales helping us to understand what we need to change in order to avoid the fate they depict? Or are they instructional manuals, preparing our psyches for trials and tribulations to come?

One of the most frustrating problems with tech culture today is that it is more like a type I than a type II apocalypse. Tech's most publicly articulated visions offer the revelation of a dramatically better future for those who deserve it and damnation for those who don't. This is certainly the case for the mainstream visions of Big Tech—"bringing the world closer," not being "evil," "making history," "changing the world," and making it "a better place."[24] Such promises are demonstrably true . . . for the founders, investors, and other senior leaders of such companies. They live in a changed world thanks to their endeavors. They possess historic wealth and privilege. Their lives and communities are, materially at least, better for their success. So are those of many others. The problem is that such victories also come replete with losses. Machine learning can be great for automating manufacturing, phasing out tedious tasks, or analyzing masses of data. It can also be great for other things, like how one drug-developing algorithm needed just six hours to invent "40,000 potentially lethal molecules," otherwise known as biochemical weapons, in one study run by people who otherwise work on such benign endeavors as drug discovery.[25]

For every tech executive or highly educated Westerner reaping the benefits of AI and social media connection, how many traumatized content moderators are there in Manila, Philippines? That's where the documentary *The Cleaners* exposed conditions so stressful and unsafe as to endanger the health and even lives of thousands of outsourced or underpaid workers. How many of the estimated hundred thousand lithium miners in the Congo alone, using "hand tools to dig hundreds of feet underground with little oversight and few safety measures," live without reliable access to necessities while their communities are exposed to "levels of toxic metals that appear to be linked to ailments that include breathing problems and birth defects"?[26] How many Chinese factory workers must live under repressive leadership, losing fingers, as part of a brutal "996" work culture (9 a.m. to 9 p.m., 6 days a week, if not even more)?[27]

Even the proponents of the idealistic effective altruism and longtermist movements, while concerned about existential risks and the potential for future disaster, are not particularly inclined to reckoning with whether their own beliefs and actions might contribute to such risks of disaster. The vision tends to be us versus them—our way, toward tech utopia, or the highway, to tech hell.

No one I know is asking for tech leaders to disavow their every creation or shut down every factory. No one is blaming them for every problem that humanity faces. It does, however, seem that the quest for ultimate glory so multiplies ultimate risks that humility and willingness to move more slowly would be a better long-term bet for humanity. Tech needs fewer sweeping, certain narratives and many more close-up character studies of the actors it currently considers extras. Can those who reap the most benefits from tech's many successes take time to better understand the reality of those suffering here and now—before it's too late?

KIND OF A MOONSHOT

I discovered Chris Gilliard on Twitter, where he has compared Silicon Valley CEOs to supervillains. Before it was purchased by Elon Musk, I spent a lot of time on what was sometimes called a "hellsite" reading about various apocalyptic and dystopian possibilities: disinformation and misinformation facilitating the election of Donald Trump, news of gargantuan polar ice shelves crumbling or "holding on by their fingertips," mass shootings again (and again), race- and gender-based violence and prejudice spilling over into campaigns for awareness that required repeated viral viewings of traumatic events, and many more. I'd also spent my entire twenty-year career as a humanist and atheist resisting calls to label religion or religious leaders evil or delusional or poisonous. For every predatory pastor or priest or guru, there is a heartwarming story of a religious cleric defending social justice or comforting the bereaved. My instincts are to see tech executives similarly, but I must admit, as a comic-book lover, that certain resemblances between tech CEOs' behavior and that of comic supervillains seem more than coincidental.

In July 2018, telecommunications giant AT&T acquired HBO, which had turned nearly $6 billion in profit over the previous three years, as part of an $85 billion takeover of Time Warner. At a town hall meeting for employees of the new company, newly minted chief executive of Warner Media, John Stankey, was blunt. "It's going to be a tough year," Stankey said, explaining the company's pivot to a new business model. "We need hours a day," he explained about his plan to add billions of dollars and millions of subscribers to the highly regarded film company's bottom line. "Not hours a week . . . not hours a month. You are competing with devices that sit in people's hands that capture their attention every 15 minutes."[28]

If his language of capturing and conquering had not been spelled out clearly enough, Stankey continued, "More hours of engagement [are important] because you get more data and information about a customer that then allows you to . . . monetize . . . which I think is very important to play in tomorrow's world."

Gilliard's response? Quote-tweeting the *New York Times*'s Kashmir Hill comparing this sort of tech mentality to the tobacco industry, he wrote, "And they say villains spilling their whole plan is unrealistic. . . ."[29]

For Gilliard, this is not a one-off quip. When Amazon asked the Federal Communications Commission for permission to launch more than 3,200 satellites in orbit around the earth, he tweeted, "Bezos looking to up his super-villain game by blanketing the globe with satellites."[30] Or when Elon Musk sought to follow in Bezos's footsteps as a media mogul by purchasing, in Twitter, one of the world's largest media outlets: "Can't be a modern-day super villain without owning a media company."[31]

Such lines may make good comedy, but they are more than just a repeated punchline. They are an insightful analysis of the psychology of the Silicon Valley leadership landscape. Gilliard even sees this analysis as an eventual graphic novel: the *Illustrated Guide to Silicon Valley Super Villains*.

Why, it is often asked, do so many tech leaders behave in ways that not only seem dismissive of their fellow human beings but fit a familiar pattern in terms of *how* they relate to said human beings? What if that pattern is similar to the corporate leadership style depicted by the writers of comic books when they portray certain villains? After all, classic characters like the Joker, Lex Luthor, Dr. Doom, or Thanos do not simply emerge from nowhere.

When the US border patrol turned to weaponized robot dogs for its patrols, Gilliard focused on the story of Gavin Kenneally, who grew up wanting to be a veterinarian but instead became a robotics expert who built the "animals." In an essay for *Newsweek*, Kenneally painted his invention of robotic "progeny" as a feel-good story of creativity and service. But, one might reasonably ask, what about the potential consequences when these devices, deployed in sensitive areas and situations, might be given weapons and the power over life and death? "Weaponizing the robot dogs is something that we're not doing *ourselves*," Kenneally wrote (emphasis mine). "That's really a decision that the government is making . . . The robot is really a tool."[32]

Gilliard's pithy response: "Every single time these guys tell their story, it sounds like super villain shit. This guy wanted to be a veterinarian but instead winds up building robot dogs and writing about how he has no responsibility for how they are used."[33]

And when Silicon Valley elites and other tech billionaires funded Altos Labs, a biotech startup intent on using cellular regeneration to help achieve a version of what some called "eternal life":

> It's like the evil sci-fi villain checklist:
>
> Facilitate global destruction (check)
> Develop plan to escape earth (check)
> Invest in immortality schemes so you can be a colonizer forever (check)[34]

Once we finally met in Detroit, I discovered that Gilliard seems just as eager to discuss Bezos and his various schemes as he is to tweet about them. "You get to the point where you think you're the pharaoh. You think you shouldn't die." There's even a Teen Titans comics character, Brother Blood, who bathes in blood to get his powers, Gilliard points out. On the HBO satire *Silicon Valley*, fictional tech executive Gavin Belson receives transfusions from a "blood boy" who looks like a "Nazi propaganda poster."[35] Such transfusions, called "parabiosis," actually exist as a real thing; Peter Thiel is the most noteworthy of people who have publicly associated themselves with the notion. It's amazing how so many conversations in today's tech world involve both potential plotlines from superhero or sci-fi epics and prep for a looming apocalypse. "And why not just preserve the earth?" Gilliard added later in the conversation, on the topic of Bezos framing his space travel ambitions as a kind of disaster escape plan. "Because there's no way to amass extraordinary wealth *and* save the earth. That's the absurdist thing about Jeff Bezos," he continued, "At the stage we're at now, saving the earth is kind of a moonshot."

If you're not as much of a comics geek as Gilliard and I are, let me assure you, the resemblance to supervillains is real, from the lust for additional riches to the love for new frontiers and equipment, the steadfast belief that one is never responsible for the negative consequences of one's actions, the tendency to overexplain to one's enemies, and the boasts of potential for

Reverse Invigilation ✓
@hypervisible

···

It's like the evil sci-fi villain checklist:

Facilitate global destruction ☑

Develop plan to escape earth ☑

Invest in immortality schemes so you can be a colonizer forever ☑

technologyreview.com
Meet Altos Labs, Silicon Valley's latest wild bet on living forever
Funders of a deep-pocketed new "rejuvenation" startup are said to include Jeff Bezos and Yuri Milner.

9:42 AM · Sep 6, 2021 · Twitter for iPhone

Sci-fi villain checklist tweet. *Source:* Chris Gilliard.

positive world transformation leading to an uneasy relationship with the possible end of the world.[36] The home planet of one of the most prominent such characters in DC Comics, Darkseid, is named Apokolips, while one of the most powerful and fearsome beings in the Marvel universe, a demigod with a passion for science and social Darwinism, is also called Apocalypse.

From Gilliard's feed I learned about a seemingly endless parade of outlandish but all-too-real innovations and proposals in the field of surveillance capitalism: Roomba vacuum cleaners doubling as security cameras;[37] video game AI for telling if a player is depressed;[38] Zoom using AI to detect the "emotional state" of its users in real time;[39] social media camera drones[40] mounted with Tasers;[41] autonomous military vehicles;[42] and of course, lists of ways in which US law enforcement agencies can legally access your mobile data, including search history, location history, texts, and more.[43] Because what Gilliard calls "luxury surveillance" isn't just charismatic evil geniuses like Jeff Bezos using tech to amass unprecedented wealth, though that is absolutely happening. Its most worrisome characteristic is the dangerous alliances between Big Tech and the powerful government agencies we are supposed to trust to ensure our safety and protect our civil rights. This entanglement of overwhelming powers for which no American has ever directly cast a vote is so dangerous because it represents an era in which surveillance is an essentially ungovernable social norm, a kind of modern force of nature to which we can only submit. As Shoshana Zuboff, Harvard professor and author of *The Age of Surveillance Capitalism*, says, surveillance capitalism is "a logic in action and not a technology," a mentality that presupposes that some people's vision of "safety" or "security" justifies almost any incursion into the civil rights of others.[44]

Evidence of the dangers of surveillance is piling up like boxcars in a freight-train wreck. In a 2019 study, the National Institute of Standards and Technology analyzed large photo sets such as mug shots and visa photographs and found certain algorithms were more likely to misidentify African American or Asian individuals than white males "by factors of 10 to beyond 100 times."[45] And during Black Lives Matter demonstrations, AI company Dataminr, a $1.8 billion startup and Twitter "official partner," scanned the contents of millions of social media posts, forwarding information to police departments who used the data to track and surveil protests.[46]

Internationally, 80 percent of erroneous arrests in Rio de Janeiro over the past decade caused by the city's ubiquitous facial recognition systems were of Black people.[47] During the COVID-19 pandemic, many European

Union member states stepped up tech surveillance, with Europe's growing military-industrial complex increasingly using cell phones to spy on "politicians, journalists, activists, business leaders and ordinary citizens."[48] The Israeli cybersecurity firm Cellebrite, the self-proclaimed "undisputed leader in the investigative digital intelligence space," held, as of 2021, nearly seven thousand contracts with governments and private groups.[49] These clients included twenty-five of the twenty-seven European Union member states' national police forces. Then there is Russia, with new laws requiring banks and state agencies to enter clients' biometrics, including facial images and voice samples, into a central biometrics database to be used, among other things, to repress antiwar and antiregime protests.[50]

Biometric surveillance data or spyware is currently being used, or is at risk of being used, by the Taliban against Afghan citizens, by China against antiautocracy protestors in Hong Kong, and by over a dozen other countries against their own people.[51] Those under surveillance include journalists, who cannot safely research and report on key stories while being tracked, and human-rights workers, who can't help the vulnerable without potentially exposing them to additional danger.[52]

The US government is using surveillance to amass truly extraordinary and unprecedented powers. During the first fourteen months of President Donald Trump's administration, US Immigration and Customs Enforcement (ICE) made 58,010 arrests of people without criminal convictions, many of whom were longtime US residents.[53] This was a huge increase from previous periods, facilitated by tech surveillance tools from companies like Palantir. Such abuses of authority continue even as of this writing: between December 2020 and November 2021, the FBI "queried," or searched, the data of potentially more than three million people in the United States without a warrant.[54] While I was researching this chapter, the United States Supreme Court overturned *Roe v. Wade*. Women on social media advised one another to stop using period-tracker apps lest they track users in more sinister ways. And Nick Bostrom, the influential philosopher of technological superintelligence and "existential risk" whose work we considered in depth in chapter 2, has actually advocated for a preemptive surveillance police state in order to preserve the "heaven" he envisions for our galactic post-human descendants. In his 2019 paper "The Vulnerable World Hypothesis," Bostrom writes that in order to avoid the "apocalyptic" hell of total civilizational destruction in the future, humanity should consider developing an "*extremely* well-developed preventive policing capacity" (emphasis his), including, perhaps, everyone being required to wear an "advanced appliance," around their neck, "bedecked with multidirectional cameras and microphones."[55]

In the here and now, most large US tech companies have refused to articulate how they plan to respond to demands by law enforcement for information about, say, women seeking abortions. But as I learned from the antisurveillance activist Albert Fox Cahn, the tech needed to lock up every single such woman is already in place. In the summer of 2022, I sat in on a meeting Cahn held with New York legislators on the dangers of geofencing warrants, or search warrants that can allow police and other government agencies to use cell phone geolocation data to find out everyone who has been in the vicinity of a given target at a given time. Like, just for example, a known provider of illegal abortions in a post-*Roe* world. Yes, motivated cops, given legal permission, would be able to find out exactly who had been in that clandestine apartment with the proverbial wire hanger. Meaning, of course, that—thanks to an assist from a Supreme Court packed by Donald Trump's social media–driven presidential disinformation campaign—surveillance tech has taken one of the most dystopian aspects of the twentieth century and made it much, much worse.

"We have a pretty nuanced position [on geofencing surveillance]," Cahn said during that meeting, "which is just: ban it because it's a threat to humanity." But will such a ban ever happen?

While I was in the middle of writing one of these paragraphs, I checked my notifications on Twitter and saw a report, first out of *Vice*, that the US military had spent millions to access a new mass surveillance tool named Augury, invoking the ancient spiritual practice of attempting to interpret omens and signs of what will happen in the future. The tool, developed by cybersecurity firm Team Cymru, promises to provide government and military clients with a "super majority of all activity on the internet," including individual users' web browsing and email data.[56] "It's everything," one cybersecurity expert told *Vice*. "There's nothing else to capture except the smell of electricity."[57]

A THEOLOGICAL—AND PERSONAL—APOCALYPSE

Though my own grandparents and mother fled multiple regimes that had used spying and surveillance to oppress and murder their family members and friends (more on that in a bit), I spent my first decades of life in the digital world blissfully unconcerned that any such thing would ever affect me. "I have nothing to hide," I remember thinking to myself, using the exact phrase Harvard's surveillance scholar Shoshana Zuboff says represents the mentality that allowed such systems to flourish. In retrospect, I was convinced I could rely on my own privilege, thinking of myself as the kind of person—white, male, American—to whom nothing bad would happen. If I wasn't committing any crimes, online or elsewhere, my logic went, I had nothing to worry about, even if I ever were surveilled. For Zuboff, however, the problem is far larger, and indeed more apocalyptic, than whether any individual person might run afoul of the law of any given regime at any time:

> Just as industrial civilization flourished at the expense of nature and now threatens to cost us the Earth, an information civilization shaped by surveillance capitalism and its new instrumentarian power will thrive at the expense of human nature and will threaten to cost us our humanity.[58]

Zuboff's work is part of a tapestry of scholarship on and criticism of surveillance. In her 2015 book, *Dark Matters: on the Surveillance of Blackness*, Simone Browne highlights the critical relationship between surveillance and anti-Black racism. "Understanding of the ontological conditions of blackness is integral to developing a general theory of surveillance," Browne writes, arguing that race has been underestimated as a factor in why American and Western powers devote so much energy to surveilling people in the first place.[59] Her work places the field of surveillance studies in dialogue with historical studies of "transatlantic slavery and its afterlife."[60] She notes, for example, that biometric identification—a much-noted form of surveillance including facial recognition and other technologies that have been repeatedly deployed for racist purposes—did not arise, out of nowhere, with modern tech.[61] The original racist biometric identification, for Browne, took place as a key part of the slave trade: "by branding the slave's body with hot irons" to track their identity for the rest of their life.[62]

As we'll explore in the next chapter, a number of individuals and organizations are doing wonderful and necessary work to push back against modern surveillance tech. In addition to Browne, Gilliard, and Albert Fox Cahn's organization S.T.O.P., there are writers like Virginia Eubanks, a scholar and journalist who started out asking questions about how marginalized people could get better access to modern technology.[63] Eubanks became convinced this was the wrong question. in fact, she says, "technology is ubiquitous in their lives. But their interactions with it are pretty awful. It's exploitative and makes them feel more vulnerable. It's often dangerous to their families and our neighborhoods."[64] Other related contemporary discussions include the surveillance of Muslims, Arabs, and others in the name of a "war on terror"[65]; the roots of such work lie in studies of the FBI spying on Black leaders such as Martin Luther King, Jr., during the civil rights era.[66]

I recommend reading the above works, along with Zuboff's, as part of any in-depth exploration of surveillance beyond what I can offer in this chapter. I've returned to Zuboff here, however, because she acknowledges that surveillance itself is a theological topic. In *Surveillance Capitalism*, she phrases her concerns about surveillance tech in explicitly religious terms: Google's "narrow priesthood" of computational specialists; the attempt to produce "digital omniscience" through ubiquitous cameras, sensors, and data; "proselytizers" of ubiquitous computing; the belief that everything will one day be digitally connected as an "article of faith"; and "evangelists" of the inevitability of such connectedness.[67]

I find all this faith in surveillance as something that will solve more problems than it causes understandable from the perspective of a person raised to see themselves as deserving of special status and above-average means in an unequal world, but afraid to lose these things. If you have, or experience, or can imagine no intrinsic worth beyond what you own, then it makes sense to do what you can to protect your possessions. And I can relate to the idea that some possessions are not only valuable but vulnerable and worth protecting. I just prefer a more optimistic vision: that the best way to make people safe and moral is to satisfy their needs and educate them generously.

I'm not sure when I first became fascinated by apocalyptic stories. I've always loved falling asleep with postapocalyptic movies in the background, a habit my wife found odd when we moved in together. But it's probably because my feeling at home among such tales dates to before I was born.

Though my grandparents were never in Nazi concentration camps, their journey to escape that fate made its mark on multiple generations. By the time my maternal grandfather, Max, arrived in Cuba from Eastern Europe (we refer to their place of origin by region, rather than calling it Russia or Poland, because the land had been conquered so many times that its borders had lost much of their meaning), almost everyone in his large family had escaped the anti-Jewish pogroms of the nineteenth and early twentieth centuries only to perish in the influenza pandemic of 1918.

As a young man, alone in the world except for one brother, he met and married my grandmother Irene, whose large Polish Jewish family had made its way to Cuba more or less intact. They had two daughters: my aunt, Norma, born during World War II, and my mother, Judy, at the beginning of the postwar baby boom. They learned Spanish and settled into a modest family life in Matanzas, a small city in central Cuba, where they opened a shop in the small storefront beneath their second-floor flat, selling porcelain dolls and other children's playthings. They sent Norma and Judy to Catholic schools.

Not even twenty years after Max and Irene fled their language, their continent, and everything they had known, revolution came to Cuba. A Communist military takeover. Businesses like Max and Irene's shop could be seized. Castro's alignment with the USSR meant global conflict loomed once again. And so my mother, then thirteen years old, was given two days' notice that she was the only member of her family to qualify for a program called Operación Pedro Pan, named after the eternally young title character of a 1950s Disney fantasy. Along with fourteen thousand other children, my mom was forced into a clandestine exodus to the United States, where she arrived, speaking no English, with nothing to her name but two silk dresses that would prove completely useless. For the next two years she lived in the foster-care system among middle American families while her mother, father, and sister attempted to navigate US immigration policies to join her. As that waiting period came to an end, the Cuban missile crisis threatened to destroy the entire civilized world.

Eventually my grandparents arrived and reunited with my mother. My mother shrugged off her horrific teenage years, graduated from college, became a hippie, and moved to New York City. She met my father, a twice-married American Jewish man eleven years her senior. In his family's half-century journey from the shtetl to the Bronx, he'd somehow sustained even more childhood trauma than she. And I was born in 1977 as a white, blond, American boy and raised in a Queens, New York, neighborhood where such characteristics were rare. Most of my elementary-school friends were refugees, or people of color, or both. What I knew of this history felt, then, like it had been forever ago. But really, Nazi death camps and two atomic bombs were no farther in the rearview mirror then than the movie *Back to the Future* is now. I could not possibly have understood it then, but my entire childhood was spent amid the psychological rubble of dislocation and inherited fear. All I knew with confidence was that despite it all, I still had it better and easier than many of my peers.

As I told my wife when we moved in together, I find disaster narratives comforting. The gorier and more grittily realistic, the better. I've always been given to bouts of anxiety and depression. I worry I won't achieve enough, see enough, do enough, love enough, be loved enough. With much therapy and life experience, I've learned to put such fears in perspective and stay focused on the present, which is (I'm fortunate to be able to say) rarely as bad as whatever is in my imagination. But even now, if I'm not careful, I can go through periods where negative thoughts dominate. Apocalypse stories soothe me with a fundamental truth of human life that I have never (thus far) experienced but those who came before me knew by muscle memory.

Things could always be worse. Which means the status quo isn't nearly as bad as you think, and you should probably pause to appreciate it.

I grew up understanding, unconsciously at least, that things could get very bad for me, or others I care about, or a whole lot of people, quickly. My whole life, I've been bracing for tragedy, even though I have always known that it's extremely unlikely that such worst-case scenarios would ever unfold. And in the meanwhile, my tendency to anticipate disaster has caused considerable angst and anxiety for me and for my loved ones and would-be loved ones. Because it's no fun living with someone who refuses to look up and talk to you because he's too busy prepping for an apocalyptic disaster. Yet I've never quite been able—or willing—to give up this part of myself entirely.

UNEVENLY

As Chris Gilliard told me enthusiastically the first time he and I ever spoke, one of his biggest heroes and inspirations is Jim Starlin. Born into a Catholic family in Detroit in 1949, Starlin served as an aviation photographer in the US Navy in Vietnam. In moments of downtime during the war, he started writing and illustrating comics and submitting them for publication. In the 1970s he had a striking run of success at both DC Comics (*Batman, Legion of Super-Heroes*) and Marvel Comics, where he became famous as one of the great storytellers in the history of characters like the Avengers, inventing or developing many of the characters now featured in the enormously profitable Disney-acquired films and TV series based on Marvel's universe. Starlin's comics made a big impression on me, too, as a youth. So, realizing we had the topic in common, I sought out Starlin and arranged an interview with Gilliard and me during my visit to Detroit. After our initial tour to and around the Ford Model T factory, we sat down, sipping mochas at an outdoor café, to work on our interview questions.

At one point during this conversation, Gilliard related, from memory, the story of an arc in Jim Starlin's 1982 *Dreadstar* comic-book series. (In the 1980s, Starlin, seeking financial and creative independence, created Dreadstar, a character who could embody a universe of characters all Starlin's own, based on his own life and experiences in Detroit and beyond.) Dreadstar, the lone survivor of the Milky Way Galaxy, and his futuristic space-warrior crew are pursued by a villainous group known as the Church of the Instrumentality, or the Church, for short. At one point their ship malfunctions, and Dr. Delphi attempts to fix it, getting irradiated and lost between dimensions until the Church is defeated . . .

I should've listened more closely to what happened next, but instead I interjected with a more basic question: What do you make of the villains in this intergalactic story about a technological future being called the Church?

Gilliard laughed a distinctive laugh that I would hear several times over the course of our two-day visit and subsequent conversations, a cross between a child's giggle and a satirist's slow clap. He wasn't sure, he said, as it had been at least twenty-five years since he'd read the books in question. The times, and he himself, had changed since. He hadn't thought until recently, he told me, about how the Monarchy, the other major force battling the Church, represented tech—their cyborg characters, teleportation devices, and such. The heroes of the Dreadstar narrative find themselves in an uneasy alignment with this powerful group in its war against the Church, which uses some technology but primarily represents a kind of mysticism. It fascinated me to think that as early as forty years ago, a writer like Starlin was interrogating the moral culpability of technology as an institution relative to other comparable institutions—like mysticism and religion.

We then had a classically nerdy debate over the motivations of characters in colorful tights and capes: how the Disney/Marvel movie version of Starlin's most famous invention—the archvillain Thanos—misunderstands that character's basic motivation. Starlin's Thanos was in love with Death, a feminine version of the Grim Reaper meant to represent mortality throughout the Marvel Comics multiverse. In order to get her attention and persuade her to fall in love with him, Gilliard reminded me, Thanos "like, decides to go on a mass campaign of genocide." Complex antiheroes committing atrocities in the name of romantic love becomes a major theme of Starlin's work, intentionally raising a question that cuts to the heart of not just Gilliard's work on Silicon Valley supervillains, but my own. Namely, what *is* a villain? If even the most antisocial character can have a relatable backstory and some loving motivations, where do we draw the line in assigning responsibility and blame? How do we fight a kind of ambivalent evil that, however menacing, ultimately resembles ourselves?

We sat down to conduct our Zoom interview with Jim Starlin in Gilliard's modest home office, a room filled with long, white comic-book storage boxes: 7½ × 10¾ × 27¼ inches, made of unmarked, yellowing cardboard folded into place, they hold 250–300 comics each, often for decades. A pile of stickers that Gilliard had printed up sat on a shelf in the corner. Accompanying a skull and circuit board (instead of crossbones) insignia was the slogan "Every future imagined by a tech company is worse than the previous iteration."

"I'd be surprised," Gilliard said to me before the interview, "if there's a human who knows more about Starlin's work than I do. I mean, except for Starlin." When we repeated the line to Starlin a couple of minutes later, the comics auteur said, "Okay, you know more." This remark made Chris bust out laughing. It occurred to me at that moment that comic books are a lot like the Bible, for the obvious reason that the texts inside his boxes were our shared sacred scriptures when we were growing up—a few hundred miles, a decade or so, and much history apart. But more than that, the comics Gilliard and I liked best were, like biblical works, written by multiple authors over long periods of time, with moral messages dramatized through myriad characters, supernatural powers, and improbable events.

We asked Starlin to reflect on his famous stories of villains like Thanos and the Joker, as well as the mystical and technological antiheroes in his *Dreadstar* series, because we wanted to know what he thought of their sometimes striking resemblance to the tech leaders of today. As we discussed the connections between Starlin's stories and Gilliard's current work critiquing mass surveillance, and my interest in both, we each revealed something of ourselves. For Chris, it was his unprotected fandom—he doesn't get many moments where he can express sincere admiration. For Starlin, it was that his stories of ambivalent and morally compromised love echo his own personal life, including his own struggles to find and cultivate love when he felt less than admirable.

For my part, I shared some of my deepest professional struggles. Because over two decades as a chaplain at two of the most elite educational institutions in the world, I've come to worry that schools like Harvard and MIT are part of the problem. It's not just their enormous budgets, or even their penchant for creating weapons and other devices that facilitate global domination. It's the fact that so many of the world's most talented people—people like me, who got to be so talented less by being born better and more by being lucky enough to soak up disproportionate amounts of the resources, love, nurturing, and education in their communities (and indeed in the world)—will spend most of their energy and time *striving*. We say, and we are told, that all this striving is for the betterment of humankind, to "change the world." But sometimes the change we end up facilitating is in our own worlds, as we gain money, convenience, and esteem for work that again and again leaves more than half the globe's population undernourished, under-resourced, and afraid for the future. Fear, of course, makes a population angry and volatile, which motivates those in power to surveil it, which, under a tech-centric regime, leads to a constant back-and-forth struggle whose ending may not yet be determined but doesn't seem promising.

It's all a moral tragedy. And it can read like the collective origin story of a class of people who want, in our hearts, to do good, but who end up acting in a way that all too often advances the evils of injustice and inequity.

Should I leave such places, I ask both Starlin and Gilliard? How does one fight the sort of villain whose superpower is the granting of success and prestige, at the expense of others?

"You talk about tech as a villain," Starlin says to both of us. "It's also a great benefactor. Half of what I do now is so much easier because of computers. At the end of the day, when I come down to dinner, I don't have to scrub paint off my hands [because comic book art is now all digital]. . . . The buggy whips may have gone, and the horses may have appreciated that, but we get a lot more people killed by cars now, than by buggies. So, you know, you figure out the balance and hope it doesn't go one way or the other."

Metaphors of balance, when applied to great moral and spiritual questions, are rarely satisfying. When we cry out for answers and someone answers with, in effect, "it's a little of this, and a little of that," we tend to experience a revulsive disappointment, a primeval wish for something starker. But even arguably the greatest secular achievement in human history—the concept of democracies, governed by laws drafted and ratified according to principles of human judgment and compassion—has been symbolized, for centuries, by a cartoonish (in a good way) character, Lady Justice, whose signature act is balancing a scale. Since ancient Roman times, a robed and armed woman named Justitia, modeled after the Greek goddess Dike, has been depicted as weighing two sides of a story on scales as a way of determining truth and goodness. Since at least the 1500s, Justitia has worn a blindfold to accompany her swords, armor, and scales. These visual details would be at home in Jim Starlin's comic books, which often feature extraordinarily powerful and noble female characters with whom Starlin's male heroes have ambivalent relationships.

Existence is complex. In any attempt to craft a meaningful life or a good society, we must confront not only our uncertainty but our love and fear, in agonizing conflict with one another. And change never comes easily. Even Starlin's hero Dreadstar, who physically resembles Starlin himself, embodies these frustrating truths. Like Harrison Ford's character in the classic cyberpunk film *Blade Runner* (spoiler alert!), Dreadstar turns out to be, if not exactly the villain, then at the very least guilty to some degree. Part of the problem. That's me, too. And it's many of those reading this, now.

"As I get older," Starlin says after several minutes of pondering such topics, "I'm finding it harder and harder to label things as good or bad."

"Do you love your work?" Starlin then asks Gilliard, who seems to strain at the question.

"I'm more focused on hating the bad, the harm [that surveillance tech does]," my new friend replies to his hero. "I'm more motivated by that, than love."

This is where Gilliard's perspective, as a younger Black artist and intellectual, builds on and perhaps even transcends Starlin's vision of heroism. Maybe we will never end up loving our work. Maybe we'll never be so cape-wearingly good as to earn full *super* status. But none of that necessarily makes us the antihero, either. Rather than worrying about our own normal, personal moral failings, we can focus on the extraordinary suffering caused by out-of-control systems, and we can dedicate our art to pushing back against those harms in real and meaningful ways. This is the radical vision for research, writing, and storytelling that I see in the work of @Hypervisible, and it sets a standard to which I can only hope my own work will measure up.

Portrait of Chris Gilliard by Jebb Riley. *Source:* Chris Gilliard.

Ultimately, if I were to distill my conversations with Gilliard on these topics down to a single insight or maxim—Hypervisible's law, if you will—it would be this: Superheroes are rare at best, and complicated even if you find them, but villains with superpowers are all too real, because of the devastation their actions cause. Tech supervillains don't tend to be especially "great" or exceptionally ingenious, Gilliard argues; their super-power, if anything, is their "immeasurable greed."[68]

Mad Max: Fury Road is haunted by the question of who killed the world. The movie's answer is that guilt is widespread, as it depicts multiple lethal strains of toxic masculinity choking humanity to death. This message "grows more prescient by the day," as my friend Eric put it as we chatted about the movie.

Moral and legal responsibility for crimes against humanity will always be hard to fully determine, as such crimes almost by necessity involve a web of actions and reactions that stretch across time and around the globe: the manufacturers of the gun, the builders of the bomb, the elitist educational system that taught all of them complicity, the relatively privileged citizens who enjoyed a measure of seeming safety because others lived under threat.

Who is to blame? Or let's presume for a moment that the worst of the tech apocalypse still exists only as a mere possibility and has not yet been unleashed upon the world (a big assumption!). Then the question would be not only who would kill the world, but also, who is trying to kill the world? The answer may essentially be no one. At the very least, I see no evidence sufficient to try or convict any of the most powerful leaders in tech of willfully or deliberately trying to destroy the world as we know it with malice aforethought. Would that the world were simple enough that it was otherwise; then we could all arrest that person and our strife and suffering would all be over.

Nonetheless, the absence of lethal intent does not mean the presence of innocence. Even if none of the prophets and demigods of the tech reli-gion are or would be murderers, what if they are (or might yet be) guilty of destroying the lives of far too many people, but in a less culpable way? What if too many of tech's beliefs and practices cause extreme risk of spark-ing a slaughter that would not owe itself to malicious premeditation but to an absence of wisdom, humility, and consciousness?

If we ever experience a deadly tech apocalypse—or if one has already begun—then the crime involved will, most likely, be manslaughter. But whether voluntary or involuntary, manslaughter is a serious crime. In American jurisprudence, for example, it typically involves decades of incarceration. Manslaughter cases are, in fact, often double tragedies precisely because they involve not only the loss of innocent life but the loss of freedom for individuals who never meant for horrific things to happen. We recognize, with sobriety and sadness, that death is always with us. Tragedy is never more than a careless accident or a crime of momentary passion away. As a society, therefore, we invest heavily in caution, doing all we can, often at considerable inconvenience, to decrease the likelihood that such crimes will take place. We require seatbelts and airbags, have speed limits and highway police and laws against drunk driving (or driving while distracted by text messages and social media, for that matter). Surely bucket seats and the lawless open road would be more fun much of the time, but that is the point.

In the mythological world of our popular culture, which can be as close a representation of what Carl Jung called "the collective unconscious" as anything on earth, how do we refer to characters who commit mass manslaughter, or those who would do so if not for extreme measures to stop them? How do we regard such individuals, regardless of how tragic their backstories or how noble their grand aspirations? We call them supervillains, obviously. For good reason.

CLEANING UP, TOGETHER

After our conversation with Starlin, Gilliard and I both wanted to do something with our nervous energy. He'd told me he was fixing up a small home gym in his backyard garage, to help him recover from a stressful and somewhat sedentary pandemic, and he showed it to me. The space was dusty and a bit disorganized, so we set to cleaning it, together. We swept, dusted, reorganized, and then went shopping for supplies for further renovations, like strings of LED lights and comic-book stickers to decorate the insulation. They were a couple of the best hours I'd had at work in years.

As a child, if one is lucky, one learns to help with routine tasks around the home. Summer-camp bunks, college dorm rooms, or postgraduate apartments can also be times where people who care about one another pitch in on mundane labor for the collective good. But after a certain point, in highly urban and overeducated communities, manual labor is either delegated to service workers or done in relative isolation. Men, especially, too often leave domestic jobs like cleaning, childcare, and household planning to women; even when we take on these tasks ourselves, in my experience at least, it is extremely rare that we ask one another for help. Jerry Seinfeld famously documented this dynamic in the 1989 *Seinfeld* episode "The Boyfriend," in which Seinfeld meets professional baseball player Keith Hernandez, the star first baseman for the comedian's beloved New York Mets. Though initially starstruck, Jerry is unsettled when Hernandez asks him for help moving. "That's a big *step* in a male relationship," Seinfeld exclaims, wide-eyed, to his friend Elaine. "That's like going all the way!" Moments later, Jerry feels compelled to ask his neighbor, Kramer, not to mention Hernandez's request to anyone.

So much of the US tech culture that has emerged over the three decades since is encapsulated therein. We men can be so afraid of any form of intimacy whatsoever that even basic signs and gestures of friendship are off limits, terrifying. And what are we so ashamed of? Not even homoeroticism, which is perhaps an undertone in the *Seinfeld* episode but not intended as any real possibility. No: the shame, in *Seinfeld* and in the lives of so many Americans, is simply that we might need almost anything from almost anyone beyond our closest (and most neurotic) connections. Tech did not create this shame, but tech companies of all kinds profit by expanding it, offering so-called innovations that often amount to little more than monetizable ways of incentivizing us to choose isolation over connection, technological transaction over organic human interchange. Because human beings are not comic-book characters, and thus we are not invulnerable. We do need help, often. And so we pay for that help, and if it can't be paid for in the form of a person showing up at our doorstep—to teach us when we are ignorant, to feed us when we are too tired or frail to feed ourselves, to drive us when we can't muster independence—then help must be automated, mechanized. We need an app for that, or one day a robot.

Maybe the process of repeatedly attempting to prove how powerful we are, by demonstrating that we don't ever need any help, is in fact the very thing that brings a possible tech apocalypse closer by the minute. Because the less we allow ourselves to rely on one another, the less we trust one another. Without mutual trust, we overemphasize power and the need to possess it. Weapons and tech give us power. Surveilling others, in order to control them or dominate them or even defend against them, gives us power. Asking for face-to-face help with a daily task does not. But unlike the act of asking for and receiving such help, with tech and weapons and spying and the power they provide, enough is never enough. There is a simple beauty and satisfaction, like eating a healthy, savory family meal and knowing there will be another such meal tomorrow, to basic acts like assisting a friend or community member with a daily burden. But no one in the history of the world has ever felt sustainably satisfied by buying a new piece of luxury tech or accessing another trove of surveillance data. There must always be the next piece of tech. There must always be more awareness of what others are doing, saying, thinking. No matter what the cost.

TOO MUCH HUMANITY

After cleaning up and shopping, Gilliard and I finally ventured around Detroit to see Project Green Light together, which was the original purpose of my visit. But it was pouring rain that afternoon, so all we could muster was a drive around the city, during which we viewed many of the infamous lights—bright green bulbs the color and almost the size of a traffic light, mounted onto buildings and other structures under protective plastic domes. We saw them at gas stations and overlooking strip clubs, protecting abandoned churches and formerly luxurious mansions. Ironically, it didn't feel like there was much to see, because it quickly became obvious that we would see the same thing again and again. Detroit, after all, is a sprawling city: "six hundred thousand people in a footprint of two million," as Gilliard put it to me over delicious, spicy fried-chicken sandwiches that we ate in his car while watching the rain fall.

What kind of future should we want for a city like Detroit? Gilliard, for his part, doesn't want any kind of tech gentrification, with tech workers and companies moving into Detroit en masse to seize upon cheap real estate and other incentives local government might conceivably offer. He'd rather keep the spaciousness that the city's relative underpopulation affords, as it leaves plenty of room for aromatic cafes, ambitious murals, and sprawling art markets, not to mention the sorts of spacious living accommodations (a backyard gym for a community college professor!) that are unheard of in places like New York City, Boston, or San Francisco. Of course, he'd prefer that safety emerge not from surveillance cameras and lights but from equitable economic opportunity, perhaps via reparations-style reinvestment meant to make right the injustices of urban renewal, race-baited violence, white flight, and predatory policing—not to mention addressing the legacy of slavery in a mostly Black city.

Sadly, no such changes seem to be coming to Detroit any time soon, because without the carrot of gentrification (and maybe even with it) there are no hockey-stick profits to be had in a socioeconomic model based on restorative justice. Too much "opportunity cost." Too much humanity, in other words.

The next morning, just before my flight home, Gilliard and I sat at Anthology Coffee, an almost cavernous gourmet café whose generous white walls display large paintings by local artists along with a sign behind the counter: "White Silence is Violence." An elaborate nearby mural, with fluorescent colors and pictures of animals covering the entire brick wall of a building near an abandoned factory, reads "THERE IS NO BEAUTY WITHOUT STRANGENESS."

As we sipped coffee and tea, I finally asked him the big, overarching question that had ostensibly brought me here in the first place: "How close are we, really, to a tech apocalypse?"

"A tech apocalypse?" he asked. Think about the genocide in Myanmar facilitated by Facebook, he reminded me. Migrant refugees terrorized by ICE agents at the southern border. COVID disinformation online, contributing to hundreds of thousands of preventable deaths in the United States alone, including thousands who'd died of the virus in Detroit. "[Tech] already does all those things to a whole bunch of people. . . . [The tech apocalypse] is already here," he told me, echoing a comment attributed to famous science fiction writer William Gibson about the future. "It's just unevenly distributed."

III

BELOVED COMMUNITY, OR, THE REFORMATION

APOSTATES AND HERETICS

More than five hundred years ago, a soft-spoken monk posted a list of theological complaints. The man had grown up in a small town and watched his father do well in the local copper-refining trade; he'd thought of becoming a lawyer until a terrifying thunderstorm may or may not have persuaded him to pursue clerical robes. In his early twenties, he studied in a heatless room containing little more than a table and chair until moving on to more advanced theological education. In his late twenties, he became a professor and began to teach. But he wasn't teaching a revolutionary new doctrine; he wasn't undergoing a spiritual struggle.

When Martin Luther posted his *Ninety-Five Theses* in the autumn of 1517, they were a reaction to a fellow Catholic cleric who had preached a message Luther found objectionable—that one could pay for a letter of indulgence and have one's sins forgiven. But Luther was not yet so scandalized or outraged as to completely cut himself off from the Roman Catholic Church. In fact, the tone of his famous theses was "searching rather than doctrinaire," and its text gave no indication of any formal break from papal authority.[1] Still, Luther's critique drew attention and sparked debate. Over the next few years, his critiques of authority grew fiercer and his stances bolder.

In 1520 Pope Leo X issued a papal bull charging that forty-one sentences from Luther's prolific writings were "heretical, scandalous, offensive to pious ears" and issuing an ultimatum that Luther recant his blasphemous teachings, though Leo had not specified which ones. Luther responded with a new piece of writing he called *Against the Execrable Bull of the Antichrist*. (The word *bull* did not have the slang meaning then that it does now, but I like to think he delivered it with the same tone.) In early 1521, the Pope officially declared Luther a heretic.

Not only was this monk from Wittenberg, Germany, willing to denounce the most powerful institution ever known to humankind at that point, but he did so under the blinding light and intense heat of what was at that time the greatest revolution in the history of Western media technology. Though others in China, Korea, and elsewhere had used woodblock and metal printing long before, in Europe, the book-printing process was less than a century old, and Luther played a major role in expanding its reach. As German audiences demanded more and more printed manuscripts of his works of criticism, the Catholic Church found it could stamp out neither Luther himself nor the message reaching an unprecedented number of Europeans through the means of this innovation in communications tech. So they did what religious authorities have essentially always done when faced with criticism too strong to defeat: they condemned Luther for heresy, declared him an outlaw, and called for him to be rounded up for punishment.

It is no coincidence, after all, that the Ten Commandments—arguably the most famous summary of moral principles in the world—begins with the exhortation of a jealous God. The faithful, commandment number one says, must not commit the ultimate betrayal of following or believing in another God. "If your brother, the son of your mother, your son or your daughter, the wife of your bosom, or your friend who is as your own soul, secretly entices you, saying, 'Let us go and serve other gods,' which you have not known," goes the text of Deuteronomy 13, following the commandments by a few chapters, "you shall not consent to him or listen to him, nor shall your eye pity him, nor shall you spare him or conceal him; but you shall surely kill him; your hand shall be first against him to put him to death, and afterward the hand of all the people. And you shall stone him with stones until he dies, because he sought to entice you away from the Lord your God, who brought you out of the land of Egypt, from the house of bondage."

That biblically heavy-handed demand for obedience—you might think of it as an "unfunny valentine," as the writer Salman Rushdie called the infamous fatwah that the Ayatollah of Iran issued in 1989, demanding Rushdie's death—is more or less what Luther encountered in the sixteenth century and is just the sort of message heretics and apostates have faced since time immemorial. Power does not like to be challenged. And yet, when Luther was publicly called to account by the Holy Roman Empire in 1521, in an assembly, or a kind of primitive trial, in the city of Worms, he did not give in. "I cannot and will not recant anything, for it is dangerous and a threat to salvation to act against one's conscience," he offered defiantly. "Here I stand, I can do no other, God help me. Amen."[2]

We might call them infidels, heathens, unbelievers, or any number of similar names, depending on the context. But every religion ever created has had its vocal doubters, because the human will to think critically and question authority follows closely behind our instinct toward faith.

In this chapter I focus on one particular tech apostate, or heretic, though we will also encounter a number of her contemporaries. Read from one perspective, her story, and theirs, can seem to be one of great victory against difficult odds. It is the story of a loose network of courageous nonconformists reshaping the tech religion by the force of their unwillingness to accept its theologies, doctrines, and practices without question. From another angle, however, the story of tech's dissenting class can read as a tragedy. The truth is probably somewhere in between, and we will try to find it here.

THE TECH APOSTATE AS RISING STAR?

"People fix things," Veena Dubal told me when I met her in 2019, "Tech doesn't fix things."

At the time, Dubal was making an unlikely star turn in the tech world. A scholar of labor law who studies the relationship between technology and work and now a professor of law at the University of California, Irvine, Dubal and her work on the ethics of the gig economy were covered that year by the *New York Times*, NBC News, *New York Magazine*, and more. She was in public dialogue with world-famous authors like Naomi Klein and stretched out of her scholarly comfort zone by penning a prominent op-ed on facial recognition tech in San Francisco, all while winning prestigious, if more obscure, awards for significant contributions to scholarship in labor and employment law.

When I met Dubal in October 2019, she was about to step on stage at the annual symposium of the AI Now Institute, an NYU-affiliated research group examining the social implications of artificial intelligence. At NYU's largest theater, alongside two other labor rights activists and moderator Meredith Whittaker, Dubal said that that the rise of ridesharing had brought with it "not just the dehumanization of thousands, millions of workers . . . but actually an exploitative business model that is independent contracting met with . . . app based technology . . . [spreading] all over the world from the streets of San Francisco, to the streets of Bombay."[3]

"Rather than just tell drivers what to do and how to do it, which would trigger employment status," she wrote in one paper about the history of gig companies' efforts to avoid treating drivers as full employees, "Uber and Lyft use 'psychological inducements' derived from social science and deployed remotely through algorithms to influence when, where, and how long their drivers work."[4] Uber, she pointed out, had even gone so far as to hire hundreds of social scientists and data scientists to shape driver behavior. By studying social psychology and video game technologies, "Uber notoriously tricks drivers into working at undesirable hours and locations."[5]

In other words, in Dubal's telling, these multibillion-dollar venture-backed enterprises were not merely behaving inappropriately, committing minor policy violations, or making ethical missteps. They were behaving in bad faith, using the tools of modern science, data, and technology to trick, exploit, and dehumanize their workers. All because "direct employment," or taking responsibility for the fair treatment of drivers as workers, increases corporate costs—by roughly one-third, Dubal and others have estimated.[6]

The gig companies' manipulation had consequences. "Almost overnight," Dubal wrote of the period in which companies "weaseled their way out of century-old regulations on fares, vehicle caps, and licenses," full-time drivers were making 65 to 80 percent of taxi drivers' earnings in 2012.[7]

It would be difficult to levy such dramatic accusations without working toward equally dramatic counteractions. As Dubal rose to prominence, she did not shy away from demanding that her work help change the political and economic reality of the gig industry, aspiring that her legal, historical, and ethnographic research might "destabilize, however slightly, the dualism of legal worker identities and compel new visions and approaches to achieving fairness and equality for workers in the modern labor movement."[8]

For a while at least it seemed Dubal had succeeded. At one point in her AI Now symposium remarks, Dubal received resounding applause after speaking about the labor movement's victory around AB 5, a statute passed a month earlier by the California State Legislature, that required companies hiring independent contractors to reclassify them as employees.[9]

"It's *so* major," replied the whistleblower and scholar-activist Meredith Whittaker in her role as moderator.

Veena Dubal.

The terms *heretic* and *apostate* have slightly different meanings. A heretic is technically a proponent of any heresy—any "opinion or doctrine at variance with the orthodox or accepted doctrine, especially of a church or religious system."[10] An apostate, by contrast, is someone who commits apostasy—"an act of refusing to follow, obey, or recognize a religious faith."[11] The difference here is, in theory, that apostasy requires a formal orthodoxy to renounce, whereas heresy need only be against established beliefs, perhaps a lower bar to clear.

The real difference between the two concepts is often fuzzy at best. There are a range of stances one can take against religious authority, ranging from heretics or blasphemers who deny one or more forms of agreed-on belief but who continue to consider themselves and be considered more or less fully Muslim, to, mostly in modern times, the notion that groups of people can formally renounce and deny both Islamic belief and their own Muslim identity itself. In my role as the humanist chaplain at Harvard, I have even helped sponsor or host meetings of both kinds of people: a group of people who call themselves "Muslimish," diplomatic heretics who recognize that they are Muslim by culture and heritage, and their friendly rival group, calling themselves "Ex-Muslims," who proclaim themselves fully separate from anything to do with Islam. What I have observed with these two constituencies, however, is like the relationship between heresy and apostasy more generally: they have overlapping memberships and values. Ultimately, one can seek to be a reformer from within or a reformer from without, and the question of which is which can come down to what type of meeting one happens to attend.

What interests me about the warnings against heretics and apostates over history is that they confirm that forms of iconoclasm have long existed. Imagine the efforts of religious history's winners, over countless generations, to wipe out any memory of their most irritating critics. *Historiography*—the study of how history is recorded by historians and others—teaches that when ancient sacred texts repeatedly warn readers of what *not* to be, people who disagreed not only existed but were probably important.

One person's custom can be another's dogma, so a heretic against one tradition or view might be an apostate about another, and vice versa. In this chapter, therefore, I will use the terms interchangeably. One of the key features of the tech religion that makes it more powerful than any previous world religion is that it has avoided any formalism from which one might officially distance oneself. As we've explored in previous chapters, tech has its theology, prophecies, prophets, priests, rituals, and apocalyptic nightmares. But it has no official church and no divinities that demand worship—even though some tech leaders have proposed we worship latent tech Gods anyway.[12] This is a winning proposition for tech in the sense of the famous Harry Truman quote that "it is amazing what you can accomplish if you do not care who gets the credit."[13] Traditional religions want credit for their dominance and influence, so the nuances of how you deny their purported truth matter more to them than the nuances of how you might deny the truth of a tech gospel that has never been formally proven to even exist. Yet tech heretics and apostates do exist. Just ask Molly White, the millennial software engineer, Wikipedia editor, and crypto skeptic who has relentlessly pursued the excesses and untruths of the cryptocurrency community. In a manifesto called "Letter in Support of Responsible Fintech Policy," White and more than 1500 technologists nailed theses to the door of the cathedral of crypto and Web 3.0:

Not all innovation is unqualifiedly good. Not everything that we can build should be built. The history of technology is full of dead ends, false starts, and wrong turns. . . . We need to act now to protect investors and the global financial marketplace from the severe risks posed by crypto-assets and must not be distracted by technical obfuscations which mask an abject lack of technological utility.[14]

In order to reform the religion of technology, we need a movement of gutsy skeptics who refuse to be charmed or purchased by Big Tech. We need people who will not simply go along with bad ideas, worship false gospels, or swallow the propaganda put out by gilded empires from antiquity to Amazon. In the tech religion, thankfully, apostasy and heresy are not rare. There is an enormous and growing pushback against Big Tech's claims—and now, Big AI's as well—from scholars and journalists to activists, artists, and everyday people who start out simply trying to live until they are provoked by unjust and inhuman systems.

TAXI DRIVERS AND SILICON VALLEY HERESY

By the late 2010s, the misdeeds of companies like Uber and Lyft had generated new interest in their marginalized workers. Because they were now considered tech workers who labored through venture-backed tech platforms, they were suddenly considered relevant to mainstream conversations about who mattered and what was important in contemporary capitalism. "Because they are atomized and dispersed," Veena Dubal said to me, "and they have this multibillion-dollar organization that's hemorrhaging money and still not giving them any money and just such a messed up situation, people are interested." This was a turn of events that struck Dubal as not only unexpected, but even funny: just a few years ago, her husband had told her, "You have to find a way to make people interested in your work. No one is interested in taxi drivers."

Her husband's comments came around 2013 or 2014, when Dubal was positioning herself as a "workers' rights person" on the academic job market. She felt pessimistic about her prospects because, as she told me in our interview at NYU, "Taxi drivers were sort of viewed as these hairy immigrant men who carry your body from point A to point B. People had a tenuous relationship with them, even though they were also precarious workers." The subsequent years, however, had made a huge difference, not only in terms of media and other outside interest in drivers and other gig laborers but in the workers' internal sense of themselves as a constituency worth organizing, and fighting for. As she told me:

> Because you have these massive organizations that are using similar business models, like contracting temp workers at Google, and contract workers with Uber. They're creating such a mass space of workers who are exploited, that these blue-collar workers and these white-collar workers who, otherwise, wouldn't see themselves as having any kind of common interests, are forming alliances. In my twenty years of doing advocacy, I have never seen that.

A few months earlier, I had personally experienced tech-affiliated drivers and gig workers pushing back heretically against the Silicon Valley gospel of "innovation" disrupting their lives and destabilizing their communities. It was the summer of 2019, and I had just finished the first year after trading in a stressful full-time job as a congregational leader slash nonprofit CEO for a new life as a gig worker, working as a chaplain on two different campuses while also cranking out articles on ethics as a journalist at *Tech-Crunch*. Each of my roles came with its own form of modest compensation, and none provided any benefits or security, but in exchange each of them offered a theoretical "flexibility" to take on ever more such roles, until my schedule and perhaps my physical and mental health would collapse into a steaming, exhausted pile of "freedom" and "self-determination." Most people see my title as chaplain at Harvard and MIT and imagine me as, at minimum, a salaried employee with benefits, but that just isn't the case. (In 2007, the *Boston Globe Magazine* did a cover story on me and my work and listed my salary: $20,000 at the time. I lost track of how many people thought it was a misprint of $200,000.)

In 2019, I also used Uber and Lyft more than ever, shuttling back and forth across the two traffic-choked miles separating the Harvard and MIT campuses. Finally, I took a sabbatical from my chaplaincy roles (I was grateful my nonprofit's board was willing to pay $2,000 a month to help me experience the concept of work-life balance for perhaps the first time ever) and focused on tech writing.

So I was primed for a story when I heard that a group of Uber and Lyft drivers were planning a then-unprecedented protest—a caravan from Southern California through San Francisco to the state capital in Sacramento. Over two hundred drivers in more than seventy-five cars were preparing to drive south to north, with more drivers joining along the way, to advocate for California State Legislature Assembly Bill 5 (which would restrict California businesses from classifying their workers as independent contractors instead of employees) and for a drivers' union.

With AB 5 almost certain to pass the California Senate, the looming protests presented a crucial moment in the history of tech: an opportunity for drivers to demonstrate twenty-first-century modes of organizing labor even as Uber and Lyft ramped up efforts to kill AB 5, drop pay rates, and generally mistreat drivers. For the first time, drivers would use their sole work tool, their cars, to demonstrate publicly and disruptively at key locations like outside Uber's headquarters in downtown San Francisco and at the Capitol steps in Sacramento—a true David versus Goliath moment. I got in touch with one of the drivers leading the movement, and when she agreed to an in-depth interview in advance of the protest, my *Tech-Crunch* editors greenlit an extensive profile that would be my first breaking national news story as a tech ethics journalist. A canonical Silicon Valley publication that had risen to prominence and gained millions of readers as an unofficial mouthpiece for the venture capitalists and startup founders who built contemporary tech culture devoted four-thousand-plus words to the story and words of a gig worker in the rideshare industry.

Annette Rivero, the driver I spoke with, was thirty-seven years old at the time, driving eight to nine hours a day, seven days a week, while putting her five children through school. The San Jose native had gotten into driving over two years earlier after leaving an administrative job at Stanford with an annual salary of $80,000 to earn a degree in business management to better her family's life. The rideshare companies attracted her with the promise of what were initially decent wages, subsidized by the VC-backed industry's attempts to maximize the availability of cars and drivers in order to crush the taxi industry. Initially, she could support her family and put herself through school (earning straight As, she told me) driving five or six hours a day, most but not all days of the week. It wasn't easy, but she and her partner—a warehouse manager for a plumbing company who worked eight to twelve hours a day—faced the grind and made it all work, together.

Rivero even recruited her father into driving, excited to earn a $500 bonus for doing so. But soon after she noticed the algorithmic manipulation involved in both companies' bonus systems: initially high bonuses for continuing to drive when one might otherwise want to stop and rest, then decreasing such rewards so gradually as to go undetected for a while, and then dropping more steeply, as a higher number of drivers allowed the elimination of surge pricing for the busiest times. Not only did Rivero end up earning less than a living wage, but worse, she had to watch her father suffer at the hands of the job she had recommended. He lived in Los Banos, about ninety minutes away from the wealthier, tech-centric areas where he worked long shifts, and he was often too tired to drive home after work, reluctant to spend the gas money to travel back and forth and too proud to sleep on his daughter's couch. Veena Dubal also wrote about an Iranian refugee who had emigrated to California, where the gig work he found as a driver in San Francisco required him to sleep in his car for up to several days in a row rather than commute a hundred miles home. "We have no freedom," he told Dubal. "I sleep in my car. I eat in my car. I work in my car. That is not freedom. That is not flexibility."[15]

Sleeping in the car isn't the worst of what Annette has seen, however. As she proudly told me that she'd never caused an accident, that she always knew when to go home after hours on the road, even if she hadn't made her goal of $150 per day, she described the other drivers she knew who were worse off. "Without a doubt there are drivers out there who are driving beyond . . . when they realize they shouldn't be," she told me. "They're driving anywhere from ten, maybe fourteen, maybe even sixteen hours a day. . . . One of them has high blood pressure, and I could lose a friend because he can't afford his high blood pressure medicine, because he doesn't make enough [to pay for it out of pocket], but he doesn't qualify for Medi-Cal because we have to file all that money we're paying Uber and Lyft on our taxes. It looks like we're making all this money, but we're not."

Another acquaintance of Rivero had recently told her about how the loneliness of sleeping in his car—"because drivers don't talk to each other really," Rivero said—had led him to develop an addiction to cocaine. He initially took it to stay awake for long shifts out on the road. It soon became self-medication for depression, then an addiction that cost him his money, and, when he got pulled over for a speeding ticket, his license.

Frustrated and feeling alone, Rivero joined a Facebook group for the organization Gig Workers Rising, attended a new members' conference call, and then took a survey exploring the negative impacts driving was having on her health. She learned how she truly wasn't alone in her concerns. From there, she'd been at nearly every action and meeting, working passionately to get the word out about drivers' struggles and unfair treatment.

The Gig Workers Rising message that Annette learned to articulate was nothing if not a Silicon Valley heresy. They set out in direct opposition to what was at the time (and has since remained) a sustained media blitz of Uber commercials featuring a wide range of celebrities, from Kim Kardashian to Super Bowl–winning quarterback Russell Wilson to legendary actors like Mark Hamill and Sir Patrick Stewart. The obvious message of such commercials was that beloved personalities would not risk their reputations unless Uber was a force for good, an affordable luxury, and a win for society. When Annette and her new friends and colleagues showed up on the streets of some of California's biggest cities, blocking traffic and demanding attention from state legislators and Governor Newsom, they were doing more than just organizing against a large company or even against a wealthy industry. They were refusing to take part in a collective fiction about how to "reimagine the way the world moves for the better" (an Uber slogan) or whether certain people are, in fact, "improving people's lives" and "changing the way the world works" (from Lyft's mission, vision, and values statement).[16] They were exposing the truth that direct employment increases corporate costs, and given that the global gig economy boasts between hundreds of billions and over a trillion in total revenues, the maintenance of a collective fiction about the industry's virtues would be a valuable proposition indeed.[17]

In pushing for AB 5, a bill that would recognize them as the employees they were, Annette Rivero and her fellow drivers fought for people who were "barely making it, barely surviving, can't even put food on the table, can't even afford healthcare," as Rivero told me. And the gig companies, she said, should be "held responsible to give back to the community, not just take from the community." Which, in a world of prosperity gospels and grand, gilded temples of every creed and denomination, seemed to me both a profoundly defiant apostasy and a deeply moral statement of faith in humanity.

APOSTATES = ETHICISTS?

"Tech ethics," or #Tethics, a hashtag that appeared in the only-somewhat-fictional world of HBO's memorable series *Silicon Valley*, is a term that has

been bandied about to refer to an exceptionally wide range of conversations about issues in technology. It can be used to refer to the activism of
workers like Annette Rivero, or to the scholarship of academics like Veena
Dubal, or to provocative conversations hosted by institutions like the AI
Now institute. But it can also refer to work that is safe enough to take place
under the umbrella of major corporations run by tech billionaires, such as
Salesforce, which has an Office of Ethical and Humane Use of Technology,
or Microsoft's AI ethics and society team (which the company laid off in
March 2023). Will the real tech ethics please stand up?

Sara Watson, who has worked as a critic and ethicist of tech both independently (as an affiliate of Harvard University's venerable Berkman Klein
Center for Internet and Society) and as a tech analyst at consulting firms,
told me that it makes practical and financial sense for many companies to
enter the world of "responsible tech" (a term, she pointed out, that has been
tested and workshopped to sound less judgy and finger-wagging than "tech
ethics"). With tech having taken over every industry to the point where it
is now the biggest force in our lives, and with optimization and data now
the common faith of endeavors from oil and gas to chemical engineering,
from real estate to virtual reality, Watson told me, "There's no way you can
separate [attempts to define ethics] from the economic model." Tech would
rather self-regulate, she said, than face US Federal Trade Commission chair
Lina Khan's preferred model of aggressive antitrust regulation. Billions
upon billions of dollars in revenues, profits, and investment gains are thus
at stake in conversations about whether a given tech company, policy, or
endeavor can be classified as ethical.

For example, when Apple dealt Facebook a "stunning blow" in 2022,
costing the social media giant billions under "app tracking transparency,"
was that policy the result of ethical qualms with Facebook parent Meta's
lucrative practice of selling targeted advertising or Apple's desires to build
its own advertising business?[18] As hard as it may seem to know where such
ethical lines should be drawn, it is even more difficult when one considers
that many if not most of the growing number of ethics professionals in tech
are employed by corporations. Even academics and university-affiliated
researchers face potential conflicts of interest, such as MIT's cultivation of
a $500 million gift to name its new Stephen A. Schwarzman College of
Computing, which will require students to explore the "social and ethical responsibilities of computing."[19] The naming gift was provided by a

donor with close personal and professional ties to Donald Trump.[20] The same donor is also connected to Saudi crown prince Mohammed bin Salman, who has faced heavy criticism for wreaking havoc (per attorney Leilani Farha, in her former capacity as the United Nations special rapporteur on adequate housing) on global affordable housing efforts.[21]

None of this is intended to deny that institutionally affiliated tech ethics leaders are ethical, or to suggest that donations from center-right funders to support ethical initiatives are inappropriate. Most individuals who take the time and energy to pursue ethics as a field are motivated by a strong desire to do good and contribute positively to the world. At the risk of repeating myself, I am confident that much good is currently being done by all sorts of individual and collective efforts to make technology that is trustworthy, safe, and ethical. But with tremendous financial stakes, and with so much work being done by individuals who report to powerhouses of economic, political, and social influence, an agnostic stance seems necessary for evaluating each individual such effort. I don't rule out the possibility that efforts labeled as promoting ethical tech will produce more ethical outcomes, but I need to be persuaded by evidence in favor of belief.

Watson, the technology critic, has thought about the moral challenges in being close to technological power since she was a Harvard College student from 2003 to 2007, and thus one of the first thousand or so Facebook users. She has struggled with her own career transition, preferring to remain an independent voice for ethics but realistic about the fact that such a business model is not necessarily sustainable during economic downturns. In a world in which tech is growing more dominant—and potentially dangerous—by the day, what exactly should an expert like her do? Cry out against any innovation, product, policy, or person that exudes even the slightest hint of wrongdoing? That would surely require her to take aim at almost all tech. Or pick her battles conservatively, seeking incremental improvements where possible while conceding most moral battles? Middle ground between these two approaches exists, but it can be difficult to identify and even harder to tread.

By the time I finished my stint as *TechCrunch*'s ethicist in residence in mid-2020, I had concluded that while tech does need ethicists to weigh the ramifications of specific policies, proposals, inventions, and applications, it also needs much more. Internal tech ethics work often accepts the paradigms put forth by tech companies, calling for only modest changes.

When institutions acquire nearly divine powers, in order to reform them, we need what in divinity school circles we call "prophetic voices": people who speak truth to powerful leaders of powerful industries, often at great personal risk, about massive, big-picture problems with how tech functions in our society today. The tech reformation must begin with heretics and apostates.

JEDI, WHISTLEBLOWERS, AND LUDDITES

"After completing this course, you'll be equipped to orient your products towards human thriving," says an expensive-looking, besweatered cartoon avatar of Randima Fernando, cofounder of the Center for Humane Technology (CHT), a nonprofit helping to "realign technology with humanity."[22] Fernando—or at least, his voice and likeness—is narrating the introduction video to a course called Foundations of Humane Technology, offered by the CHT, a large and well-funded nonprofit whose other cofounder, ex-Googler Tristan Harris, argues for ideas such as making screen time "well spent" on Netflix documentaries and in conversation with celebrities such as Prince Harry and Meghan Markle.[23]

I discovered the class in March 2022 via a *Wired* article whose headline blared a question that piqued my interest: "Can an Online Course Help Big Tech Find Its Soul?" Initial course videos review statistics about climate change and increases in cyberattacks; prompt students to reflect on the United Nation's Sustainable Development Goals; and place the harms caused by modern tech on a ledger. The course promises to "respect human nature." But what is human nature, anyway? After an hour or so, I wondered, about the question raised by the *Wired* headline, can you really find something that never existed in the first place?

Another online course I took, meanwhile, doesn't need to wade into thorny debates about human thriving, because it is based on more clearly defined messages, like "stop killing us."[24]

The course, called TECH WARS: A #NoTechForICE Saga, was organized, starting in early 2022, by Mijente, an organizing and political community for Latinx and Chicanx people and allies seeking racial, economic, gender, and climate justice. Mijente formed in 2015 in the wake of Obama-era protests over immigration laws that targeted the Latinx community and quickly became intertwined with broader tech policy debates as large tech companies like Palantir were awarded major government contracts for surveillance technology that would facilitate deportations. I'd come across them before on the tech-justice protest circuit but was surprised to see them facilitating an entire course based around the *Star Wars* movies, promising to "build the force" against surveillance and policing in the digital world— complete with *Star Wars* fonts, references, and jokes galore.

Why *Star Wars*? Well, besides being a fan of Baby Yoda, jokes Mijente field director Jacinta González during introductory remarks, the George Lucas movies teach us a lot about rebellion against an empire. The course explored such relevant themes in today's tech as drone wars; Mijente's No Tech for ICE campaign against the use of surveillance software by agencies like US Immigration and Customs Enforcement, a galactic, or at least global menace; and perhaps most importantly, the Jedi movement-building training needed to resist the powerful empire and ultimately bring it down. "Policy is a lightsaber," said Julie Mao, an officer at the digital civil liberties organization the Electronic Frontier Foundation and Tech Wars course speaker. But "it's the organizing power . . . that can pass the next law."[25]

Ulises Mejias, the scholar of tech colonialism featured in chapter 2, looked genuinely happy to be a guest instructor for the course, smiling brightly while sharing his Día de los Muertos–themed *Star Wars* art. "That weapon," Mejias said, referring to a real-world equivalent of the nebulous but formidable Force that Jedi like Luke Skywalker and Princess (later General) Leia use in the films, "is imagination. Because when colonialism couldn't be resisted with the body, it could always be resisted by the mind."[26]

Mijente's forceful challenge to American legal authorities can be hard to take in for people like me, who grew up seeing American police and law-enforcement agencies as the ultimate good guys. But when an activist and course instructor like Cat Brooks explains that "policing in this community has never been about safety," the image of Darth Vader—a state-appointed official who was trained and empowered to think of himself as on the right side of history—drives the point home.

The course is summed up well by guest instructor and Princeton professor of African American studies Ruha Benjamin, who spoke about her book *Viral Justice* in a concluding session. "We're not just questioning external structures, we're reimagining internal structures," said Benjamin. "We're reimagining who we are to one another, and we are questioning structures

Santos Medrano, left, of La Puente, and Julia Diaz and Owen Diaz, 6, both of
Pomona, pose as Han Cholo, Princess Loca, and Artudito outside the Marriott
during the Comic-Con International in San Diego, 2016. *Source:* Photo by Jennifer
Cappuccio Maher, SCNG.

that . . . rely on us thinking of one another in a [hierarchical, deathly way]. We're thinking in a life-giving way."[27]

Still, just how far must tech criticism go to be the reformational apostasy I am calling for? Do we need a Luddite revolution to fix our tech culture?

If that sounds like an extreme question to you, it did to me, too. You may be familiar with a December 2022 *New York Times* story about a Luddite Club, in which teens support one another in using old tech like flip phones and pushing back against the excesses of Big Tech's demands for our constant attention. But before that story's publication, safe to say the concept of Luddism had not gained mainstream popularity. So I was intrigued when my editors at *TechCrunch* gave me the green light, in late 2019, to discuss it with Ben Tarnoff, a tech worker, writer, and magazine cofounder who wrote a column for the *Guardian* in September 2019 titled "To Decarbonize We Must Decomputerize: Why We Need a Luddite Revolution."

I've casually called myself a Luddite, over the years, any time I got frustrated with tech and wished there were less of it in my life. But I always imagined myself to be kidding. No one wants to live in a world in which "reinventing the wheel" isn't a cutesy expression a venture capitalist would use to describe a founder's pitch-deck but an actual thing we had to do. But Luddism, it turns out, is not what most of us think it is.

That's what I learned from Tarnoff, an unassuming thirtysomething given to stubble and plaid shirts, and also a prominent voice on the intersection of tech and socialism, who, with his wife and business partner Moira Weigel, created the insightful *Logic Magazine*, a paper-and-ink journal founded in 2017 to interrogate the politics of technology and ask big questions about what I call the tech religion.

As Tarnoff explained to me when we met at Harvard University's new Creative Commons—a stylish coworking space in the model of WeWork, but for university affiliates only—the historical Luddites were workers in early nineteenth-century England who smashed newly introduced machinery because it threatened their livelihoods. Capitalism emerged in England in the fifteenth and sixteenth centuries but found its industrialized form in the eighteenth and nineteenth centuries, a time characterized by hockey-stick growth. This huge jump in wealth was not distributed equitably, however, and eventually some of the workers whose labor powered it said, essentially, "not on our backs."

Because machinery was "threatening to deskill their labor or displace it entirely," Tarnoff told me, Luddism was "a very rational response to management coming in and trying to cut costs by pushing out workers." This dynamic of predatory managers using technology to destabilize the lives of workers or eliminate their jobs entirely is hardly just a nineteenth-century phenomenon. It is still how tech operates today, and it has been a foundational aspect of capitalism since the original Luddites. Which makes Luddism, Tarnoff argues, not really about the rejection of technology at all—it's about the rejection of a certain kind of political and economic deployment of tech.

Technology, as an abstraction, isn't useful to talk about, because at its most basic level technology is so ubiquitous as to not even exist. After all, cultivating fire to cook food is, technically, technology—as is a pen, a computer, a table, a chair, and so on. The ethical question that swirls around each of these things, Tarnoff said, is, Whose interest do they serve?

As one participant put it, during Mijente's Tech Wars course, "May the Force be with the new face of resistance."

CALLING BULLSHIT

Maybe harsh tech critics like Tarnoff, Veena Dubal, Annette Rivero, and their many thousands of like-minded colleagues are simply "calling bullshit," as academics Carl Bergstrom and Jevin West titled their recent book, a manifesto for skepticism in a world in which data—including, notably, data about the efficacy and risks of COVID-19 vaccines—is easily manipulated and weaponized against the public. Bergstrom and West cite *On Bullshit*, an elegant little gold-embossed tract from 2005, by now-retired Princeton University philosopher Harry Frankfurt, that I used to buy students as a graduation present when they finished at Harvard. Bullshit, Frankfurt argues, has become "one of the most salient features of our culture" and pointing it out can be an act of civic virtue. In *Calling Bullshit*, Bergstrom and West update Frankfurt, making a brilliant distinction between old-school and new-school bullshit.[28] Old-school BS, or the relatively meaningless use of rhetoric or fancy language, may be easier to detect and repel in a wired, interconnected world in which talk is cheap and tech is perceived as the more truly valuable commodity. New-school BS, however, "uses the language of math and science and statistics to create the impression of rigor and accuracy" and is ever more effective in a tech- and data-driven world, because no one knows everything, and therefore it is hard to feel qualified to refute suspect assertions.[29] Tech BS—the sorts of theology and prophecy we encountered throughout the first two parts of this book—is definitely new school. It is important to have knowledgeable voices to explain how to avoid being snookered by bad ideas.

All this calling BS can cause us to mistrust one another, as Ethan Zuckerman points out in his 2021 book, *Mistrust: Why Losing Faith in Institutions Provides the Tools to Transform Them*. Mistrust, Zuckerman argues, "is the single, critical factor that led to the election of Donald Trump in the United States and that may be empowering ethnonationalist, populist autocrats around the world."[30] Amid rising mistrust, Zuckerman points out, many ironically tend to place more trust in authoritarian leaders who refuse to be questioned, and the resulting dynamics can endanger not only millions of people but the very fabric of the societies in which we live.

After all, the institutions of liberal democracy—government and civil society, education, journalism, and private industries both free enough to act creatively and regulated enough to be trustworthy—were not simply created out of Adam's rib in the Garden of Eden. They are the product of countless decisions, over thousands of years, to create a more trusting, peaceful, and equitable society. Such decisions must compound on one another in order to positively impact our lives, and the fact that this will (continue to) happen is not in any way guaranteed. So we should be concerned when an entire enormous industry (or, as it were, a religion, replete with massive conferences and other congregations) declares disruption *for the sake of disruption* its goal. Yes, disrupting an individual industry such as taxis may provide enormous short-term benefits for some—especially the executives, investors, and shareholders behind the decision to do so. And plenty of others may experience some greater measure of convenience along the way. But such disruption is most smoothly and profitably achieved with the help of a widespread narrative that it is good: that its intentions are beneficent and that it will benefit society as a whole, not just the lucky few. The more that we, the public, believe these things, the more we will trust tech companies to organize themselves as they see fit.

Of course, sometimes the narrative isn't true: people get hurt, abused, or discriminated against; the public good is neglected in favor of private gain. In such cases, there is a tremendous amount of money and goodwill to be lost on the part of the disruptors. In order to suppress public discussion of such harms, disruptive companies have often undermined civic institutions: injecting billions of dollars of lobbying money into our government in order to sway legislation and attacking journalists and academics to discredit or stomp out criticism. The result? A truly vicious cycle of accusations of dishonesty and bad faith—against the powerful interests who disrupt, and then, in turn, against the disrupted. Rinse, repeat.

Another factor contributing to the cyclical nature of this conflict is that public protest, one of the strategies most frequently deployed to improve society and bolster democracy in this disruptive era, has itself become increasingly intertwined with the religious institutions of Big Tech. Powerful actors do their antidemocratic thing ("plutes gonna plute," as critic Anand Giridharadas says), and then, the story goes, the people take to the streets in massive waves facilitated by and organized on platforms like Twitter and Meta.[31] (Gig-economy platforms have sought to reap this good will as well, with, for example, Lyft offering discounts on rides to vote on election day.)[32]

The implication here is clear: How bad could such tech platforms be if they are integral to the (new and improved) functioning of the "good guys" in the comic-book narrative? Anyone calling out tech itself as part of the problem must be dangerously misled, a malicious crank, or worse. But it is long past time to stop the presses on the idea that tech itself facilitates social change, or at least that it facilitates movements for justice. As Turkish scholar and protest participant Zeynep Tufekci suggests, because of technology, protests are much easier to organize, and therefore more fragile and less impactful than they used to be.

We can now mobilize hundreds of thousands of people into the streets via tech platforms, with the promise that their collective voice is powerful enough to shake the world. If that promise is a lie—if such protests can be safely ignored by deeply entrenched rulers with powerful tech tools of their own, including those that enable spying on protestors—then mistrust, learned helplessness, and nihilism are even more likely to result. "For protests to achieve change," writes Ethan Zuckerman, referencing the work of Tufekci and others, "they require that institutions can be influenced and have the power to meet protesters' demands."[33] Our modern political (and technological) institutions, sadly, often fail to meet this basic standard.

What, then, are we to do? The answer cannot be to give up, become entirely coopted, or (continue to) develop some Silicon Valley strain of

Stockholm syndrome where we tell ourselves, despite evidence to the contrary, that if it's tech it must be good for us, or at least morally neutral. We also can't keep relying on "clicktivism" or the above-discussed tech-organized protest movements while hoping for different results—that being, to cite the much-abused cliché popularly attributed to Albert Einstein, the definition of insanity.

We can and must continue to educate and train smart, capable, and justice-minded leaders in fields like tech ethics. Such people should be placed in the halls of power, like tech companies big and small, academic departments, policy institutes, congressional offices, and more. These individuals will often help make better decisions and wiser investments. And yet—when ethical knowledge meets power, power often wins. Reforming the tech religion cannot be left entirely to the hands of in-house technology ethicists and "trust and safety" professionals, no matter how capable. Not because such people are guaranteed to act in the interest of their institution rather than on behalf of humanity as a whole, but because that is a possibility worth considering.

To build greater trust in our institutions and in one another, we must, ironically, draw on the passion and energy that come from righteous skepticism. In a world full of new-school bullshit, merely placing truth adjacent to such bullshit will rarely solve our biggest problems. We must first learn how to more vigorously, regularly, and collectively call BS. We need a reformation that builds a culture of doubt in the service of humanity. We need a revolution of agnosticism. Thankfully, there are a host of brilliant and brave leaders, traditionally religious and not, from all backgrounds (though women and people of color can be frequently found taking the lead), building just that.

REFUGEES OF DISRUPTION

In the early months of the COVID-19 pandemic—less than a year after AB 5 went into effect—Veena Dubal found herself at the center of a dramatic controversy.

Starting around February 2020, a stream of articles and posts in right-wing publications, blogs, and social media called her names like the "unelected puppet master behind california's AB5 law" and "a woman of privilege dictating how Californians can work" and accused her of writing the law herself, among many other imagined transgressions, opening flood-gates of vitriol towards her.[34] There were letters to her chancellor at the University of California, Public Records Act requests for all her emails, a Wikipedia war over her entry, and hashtag campaigns like #VileVeena. After her home address, salary, and husband's name were posted online, she slept on the floor of her two eldest children's room, she told one journalist, with a baby monitor close by.[35] She was scared, burnt out. "This is like a psyop," she remembers thinking—a targeted attack on her mental health carried out to make her advocacy less effective, as happens to a lot of scientists who study topics where huge profits are on the line, like climate change or the tobacco industry.[36]

Then journalist Sarah Lacy reached out to Dubal in mid-2020. Years before engineer Susan Fowler's famous 2017 callout of Uber's culture of sexual harassment and general hostility toward her and other women, in a scandal that ultimately led to the downfall of CEO Travis Kalanick, Lacy was the editor of tech website PandoDaily, where she called attention to misogyny and poor judgement at Uber. Lacy's writeups noted inconvenient details such as that a driver (who happened to be accused of sexual assault) had passed the company's "zero tolerance" background check despite a criminal history; that Uber was tracking passengers' one-night stands and doing business with escort services; and that then-CEO Travis Kalanick referred to the company as "Boob-er" in a cringey interview response about his own sex appeal. Kalanick and other Uber leaders spoke of spending "a million dollars" to hire researchers that would dig up dirt on Lacy and other journalists, looking into "your personal lives, your families," and giving the media "a taste of its own medicine."[37]

"They're going to find out everything about your personal life," Dubal remembers Lacy telling her, and when they can't find what they're looking for, "they'll start making things up. So . . . warn your husband." Lacy told Dubal she ought to talk to Fowler, who then called and told Dubal more about what to expect, including white vans outside of her house.

In October 2020, Dubal's beloved younger brother, Dr. Sam Dubal, a kindred spirit and source of inspiration all her life, went missing. Sam, an anthropologist and medical doctor, had nurtured a deep personal and professional passion for difficult questions of morality and human good-ness, even writing a book criticizing language framing ethics in terms of "humanity" as rooted in colonialism and potentially leading to unintended violence against marginalized and oppressed people seen as less human than the Western scientists and doctors applying that paradigm.[38] He was, in other words, a true heretic in his field. Sam Dubal also happened to be an expert hiker, but when he left for a solo hike of Washington's 14,411-foot-tall summit Mount Rainier, he never came home. After weeks of exhaustive searches that were even aided by the Trump administration, his body was never found, and he was pronounced dead.

"It was just fucking hell. It. Was. Hell," Dubal told me, two years later. "I respected him so much . . . he just helped me put things in perspective. When the third or fourth article [maligning me] in *Red State* came out, Sam said, 'good thing no one's ever heard of *Red State.*'"

And so Dubal was already drained in November 2020, when Proposition 22, a ballot initiative that classified drivers as independent contractors, essentially reversing the previous year's AB 5 victory, passed with 59 percent of the vote. Companies like Lyft, Uber, DoorDash, and Instacart had—successfully—spent more than $200 million to persuade the people who had voted to force gig companies to treat drivers as employees to take the opposite stance.[39] As part of that campaign, Lyft even went so far as to promote its efforts to "uplift" communities of color, with a one-minute YouTube ad set to the voice of the great African American poet Maya Angelou (a former public transportation worker!) reciting her poem for Bill Clinton's 1993 presidential inauguration, "On the Pulse of Morning." As the unicorn rideshare company used it, Angelou's poem about hope for reconciliation and justice after a legacy of brutal oppression served, as Dubal later explained it, "as the backdrop to scenes of workers of color, masked and happy" despite a pandemic that was disproportionately killing working people of color.[40] The cause of their smiles? The fact that Lyft provided occasional free rides to, as Dubal put it, "communities who lack access to food, jobs, and essential services." Even worse than such cynical posturing, for worker advocates like Dubal, was the fact that California law would now require a seven-eighths majority in the state legislature to change the new law.[41]

By the time I spoke again with Annette Rivero, the Uber driver and AB 5 campaigner, this time in person near her home in the San Jose area, the protests she had helped lead were a distant memory. She had moved on from protesting, she told me; she had come to believe most drivers and gig workers actually want to be independent contractors, however that might fly in the face of their own interests.

Rivero (still) thought the ridesharing companies "took advantage of those people, and . . . sold them a dream of having freedom to work . . . for the price of control." She still noted that a lot of drivers "have fallen to depression or anxiety because they have to stay on the road, they don't have any other options." She saw their desperation, their suffering, and thought with gratitude about how many of them opened up to her about these emotions that they rarely discussed with anyone. Still, she felt she no longer had the luxury of remaining in solidarity with those people. Instead, she'd taken a different message from her experience: that Uber and Lyft were just companies doing what companies do under capitalism, and so if she couldn't beat them or join them, she should try to emulate them. "For me, walking away from that scene, I've just been trying to teach myself how to start my own business because that's what I really want to do now after trying Uber and realizing this isn't really my own business."

So she'd started driving for Uber Eats while watching YouTube videos in her spare time to teach herself about how to thrift, DIY, and buy clearance items to resell—what is sometimes called "#fliplife," though she didn't use the term—seeing the small measure of independence in this digitally enhanced hustle lifestyle as something she might be able to pass on to her children one day, with a greater measure of success and self-determination than her driver peers, who have accepted Prop 22 because they "want the freedom to go and come as they please."

After my conversation with Rivero, I was depressed to find another tech apostate broken by the onslaught of Big Tech's lobbying and marketing efforts. With my calendar clear for the afternoon, I drove past a massive encampment of unhoused people near the San Jose airport, on and around a large empty lot owned by Apple, and adjacent to a large office park that includes the headquarters of companies like PayPal. In a sprawling area the size of at least several square city blocks, dozens if not hundreds of RVs were camped, along with countless pickup trucks and sedans, filled trash bins and dumpsters, pieces of bent fencing, tarps and tents, old mattresses, and portable potties next to portable sinks for washing up. There were even entire makeshift houses, three stories high, fashioned out of what appeared to be wooden shipping crates hammered together into complex structures. Such crates can cost hundreds of dollars each and are not the sorts of equipment I'd ever known people experiencing homelessness to possess when I was growing up in the heart of New York in the city's housing crisis of the '80s and '90s. But in the vacant lots of the tech Jerusalem, one can find people who have resources, but not enough to afford a stable place to live in Silicon Valley.

Witnessing Annette's change of heart and the nearby San Jose encampment in the same day, I saw just how little distance separated the world's most prominent disruptors, forging a new world through tech solutionism, from people whose lives had been completely disrupted. It would be one thing if tech produced disruption only at the scale of lives like Annette's: people who could see and experience tech's injustices and inhumanities but at least bounce back to try again in new ways. But there are worse scenarios that we can easily see increasing in the coming years: people living in utter destitution literally around the greatest riches in the history of the world.

Stapling shipping containers together suggests semipermanence, although we tend to picture homelessness as a transient state. I, at least, never pictured someone having the time to put together a home for themselves while squatting on or next to land owned by a trillion-dollar company. Conservative critics of the "welfare state" love to rail against the supposed laziness of their most destitute constituents, but that day in San Jose, I saw an analog technological solution that required both desperation and plucky initiative. these people, in one of the richest areas of the richest country in the history of the world, were so individually and collectively disenfranchised, over such a long period of time, that they effectively set up their own refugee camp. These people were not refugees from war, earthquakes, or even climate change—at least not directly. They were refugees of disruption. They were, and are, refugees of technology.

TIPS FOR TECH HERETICS, OR, HOW TO COPE WITH GREAT FRUSTRATION

When I asked Sara Watson, the Harvard grad and tech ethicist, to explain how she operated, day to day, as a self-identified Cassandra, she answered, "With great frustration." I was trying to figure out what I or my readers should make of the fact that the life of a tech heretic or apostate seems to be one of constantly crying out in pain and anguish—not to mention anger—and being ignored.

It might seem that Watson's reply was giving up, a shrug emoji of resignation. But I didn't take it that way, nor did she intend it as such. In fact, you might consider that the message behind her "great frustration" is to accept that you, too, will experience much discouragement, and that is okay. Great frustration is a necessary byproduct of doing anything challenging and worthwhile. To borrow from one of my favorite lines from the great and often-irreverent cognitive psychologist Albert Ellis, we shouldn't expect ourselves to be the only people in the history of the universe to never experience stress or discouragement . . . particularly so if we want to speak truth to power, and most especially if we intend to do so in connection with how we earn a living.

My conversation with Watson yielded several helpful tips for tech heretics." Such as "not giving a shit if people believe you," which perhaps is easier said than done, but we have to start somewhere, and the acknowledgement that there are more important things in life than being immediately praised or understood is a good start.

Another tip: "Letting go of the desire to say 'I told you so,'" because "that won't get you anywhere, either." A better saying than many Zen koans, which I can say as someone who has studied quite a few koans.

Finally, Watson shared that part of her experience as a heretic or Cassandra is what she called a "struggle with voicelessness" and a feeling of "screaming into the void." Who among us has ever worked to make the world better, whether from inside or outside of a powerhouse like the tech religion, without similar feelings?

Hers is, Watson admitted to me with humility and self-awareness, a "bleak vision" of a professional identity. But then again, she said, as if forcing herself to weigh the other side, unlike in the myth of Cassandra, in her case—and quite possibly in yours, if you choose to speak truth to the powerful interests in your life and career—"some people *will* listen." Watson's message is, therefore, deeply resonant when considered in the context of tech apostates like Veena Dubal, or whistleblowers like Timnit Gebru and Sarah Lacy, or social movements like Mijente. For such real-world Luddites and Sisyphuses-in-community, what matters is not just that they are fighting an uphill battle, but that they are fighting an uphill battle *together*.[42]

Over much of my own life and career, I saw myself as a professional optimist. As a "social entrepreneur," I sought to make the most of limited resources or difficult situations rather than complaining. The secular sermon I delivered at the ceremony celebrating my ordination as humanist/atheist clergy was entitled "To Sing and to Build" and argued that my often cantankerous and critical community of humanists and nonreligious people should look more to express positivity, creativity, and inclusion than to tear down traditional religious people or ideas.[43]

As I've stepped back from congregational organizing and into the world of tech, however, I've become more inclined to see the hole in the bucket, the fly in the ointment. I've come to admire the beauty of tearing down dangerously flawed ideas and assumptions, even as I recognize that it must be done carefully and that we must still consider what to build in their place (more to come on that later). Yet I still don't feel as natural or comfortable in the critic's chair as I did in the entrepreneur's, so I wanted to understand how Dubal developed her own apostate's mindset.

Growing up in the American South, going to high school in Kentucky as the daughter of Indian immigrants as the new millennium approached, Dubal witnessed a lot of racism and gendered racism, she explained. This continued as she went on to study at Stanford and at the University of California, Berkeley, where she was called a "homegrown terrorist" on the street and witnessed her partner coming home from a violent beating at the hands of racist thugs in what is generally considered one of the most progressive cities in the world. Dubal knew early on, just by experiencing such environments, that she would never have the social power of a teacher who said something terrible about her identity, let alone of the state that maintained a political and civic status quo in which that teacher, a public employee, could say what they did. What did give her a sense of her own strength, however, was that she understood why her teacher was wrong to say such a thing. It was the only power she felt she had: "to be able to artic-ulate why they were wrong." "Blind faith in any leader," Dubal told me in 2022, reflecting on that time, "is always bad, no matter what."

But I didn't know how to broach the subject of one of the last things I wanted to ask Veena about her experience. "You . . ." I said, "I . . ." Finally I spit it out. "How do you see the relationship between what you were going through at the time—the attacks on your character, your livelihood, your psychological and spiritual wellbeing—and what happened to your brother just as all of that was peaking? Is there any relationship?"

"No," she told me, "I didn't think there was any connection." Her father's mind apparently went to the same dark place as mine—he was "very concerned," she told me, about the possibility of Sam's disappearance being connected to targeted violence from someone in the orbit of Prop 22.

There was an uptick in racist violence generally at the time. But, Dubal told me, it wouldn't have made much sense for someone like that to go after Sam rather than her or a member of her immediate family, and there was such horrible weather on the day Sam attempted to climb Mount Rainier. It was, she believes, all just a devastating coincidence—if the kind that would make most of us desperate for an easier, more comfortable life.

Veena Dubal didn't give up, however. Not even after Prop 22, or after Sam's death. Instead, she focused on an insight she had actively cultivated since 2009, when, in writing about her fellow liberals' excitement over the Obama administration, she cautioned readers to "maintain a critical eye." "We must remember how to be critics," she said, calling on her peers "to hold this President and this Empire accountable to the people of this country and the world."[44]

And so, in 2021, Dubal picked herself up and finished a major scholarly article called "The New Racial Wage Code," in which she argues that Silicon Valley rests on a "tiered system" (evoking the caste system of her own Hindu heritage), of which Prop 22 is part, of "systemic racial inequalities."[45] Citing that in the United States, "app-deployed, in-person service work" is conducted primarily by immigrants and subordinated minorities, she writes that "Uber, Lyft, DoorDash, Instacart, and Postmates, like early twentieth century industrialists, used race as a resource to eliminate access to minimum wage and overtime protections (among other employment rights) and justified their actions through the mirage of racial benevolence."[46] Dubal even presented her essay as a repudiation of Lyft's efforts to use Maya Angelou's poetry in the service of "racial domination . . . as a centrifugal force in the legalization of partitioned, substandard protections for workforces of color."[47]

As we neared the end of our conversation in October 2022, I told her that this chapter was about the idea of the apostate.

"I love it," she said, "I've always been an apostate. I was the obnoxious person who wasn't willing to nod my head and say, 'it's great.'" She seemed to hearken all at once to her childhood, her views of India and America, and all of her work on Uber and beyond.

As her friend Meredith Whittaker told her during the fight for AB 5, "People like us—critics, apostates—we're never going to be rewarded. You make people uncomfortable, uneasy. And people want to be comfortable.

They want things to be easy." But, Whittaker continued to Dubal, "In your heart, there's going to be no festering."

To conclude our conversation, I asked Dubal a follow-up on a question I'd broached in our first conversation, back in fall 2019, as the world, unbeknownst to us, was poised for major change. As she'd written in 2017 and reaffirmed to me in the wake of the initial AB 5 victory in 2019, proworker activists were "winning the battle, and losing the war." What did she think, these few dramatic years later?

Her answer was the opposite of what she'd told me three years earlier. "We've lost the battles," Dubal said, "but maybe we're winning the war. There's so much public awareness now, of how damaging a lot of tech companies are to any sense of equality."

Reflecting on the influx of activist allies she had seen, in fights around not only her own area of expertise but against climate change, war, and election misinformation, she admitted she and her allies are losing many battles.

The years 2023 and 2024 have brought a fresh round of losses for tech apostates and allies battling dogmatic and hierarchical powers in Big Tech, AI, and beyond. But Dubal still believes that people, not technologies, corporations, or lobbying, are what fixes things: "I don't feel lonely in this fight anymore, and that's really exciting."[48]

HUMANISTS

He was the middle child of a construction foreman father and a box factory worker mother in Haverhill, Massachusetts, a Puritan-founded manufacturing town thirty-five miles north of Boston. When the Great Depression started just as Tom was born in 1929, Haverhill's economy sputtered. His father died of tuberculosis when he was three, or was he four? It's hard to know, because his mother then caught TB as well, even as the dreaded disease, also known as the Great White Plague for its victims' pale skin, was finally receding—after killing one out of every seven people who ever lived before the nineteenth century.[1]

Tom's mother carried on parenting for a few years, but eventually she had to go to a sanatorium, as they talked about it then, which meant the plague had consumed her, too. No one in the extended family was in much better position. The children became wards of the state.

His sister was taken in for a time by a family that came to love her. The social workers assigned to care for orphans did all they could to keep siblings together, so he and his brother ended up with her family too, though the boys felt unwanted.

It was a childhood, in other words, of almost constant existential insecurity—except for the security he found in the church.

Sitting on the back steps and staring up at the starry sky around age seven, he filled up with awe at the thought that a God had created it all. He relished the notion that God's home was their own Catholic Church. For the rest of his childhood, he dreamt that one day, he would overthrow the smallness and meanness of his circumstances as part of the Roman imperium—the rising of the church to the status of supreme world authority. Fantasizing himself a future cardinal, he preached to his elementary school classmates. He was "more Catholic than the pope."

When adolescence brought impure stirrings, he held them close like a beach ball underwater, making himself a regular presence at the confessional.

He chose a Catholic college and a direct path to priesthood. At seminary, he was the sacristan: the top student in his class. His family attended with pride as he delivered mass at ordination. The honor—the sense of finally *being someone*—was so seductive it made him dizzy.

The dam began to crack during Tom's early years as a priest. He was too independent-minded and upset many of his peers with liberal political ideals and a fastidious work ethic that irked those who preferred fine clothes and drinking and gossiping at night. But things looked up when the Catholic chaplain at Dartmouth College, a fabulous fundraiser, opened a new Catholic student center on the Ivy League campus and chose Tom as his new assistant. Boston's Cardinal Cushing himself approved loaning Tom to the archdiocese of Manchester, New Hampshire. It was a highfalutin assignment. With spirits already high—this was right in the middle of the Kennedy administration; there was a Catholic in the Oval Office!—Tom got a ski suit and all kinds of warm clothes as gifts, and suddenly he could afford a sleek new black Bonneville Pontiac coupe with white-rimmed tires and a blue interior.

At the college he was assigned to talk with young Greenies about their theological questions. He was good at it, as he didn't mind saying many years later, despite what those of us who knew him recognized as an almost debilitating modesty. The problem was that he was too good.

The more he talked to the young men (women didn't matriculate at Dartmouth until 1972) and discussed their doubts, the more he learned about how theology worked in practice, as a way of policing sexuality and buttressing authority. The more he learned, the better he got at questioning his own faith. He didn't appreciate his supervisor's unthinking devotion to the Virgin Mary as a purifier of real or perceived sins, nor how Father Nolan pushed rote prayer in a way that made their young charges, Tom later said, act "very much like sheep."[2] After Tom defied his superiors to help organize a march in honor of Martin Luther King, he was "reassigned," or politely fired.

Ending up at a blue-collar parish in Boston, he fell in with members who believed in Dr. King's message. Though not much of a public speaker, he learned to give earnest sermons about racial justice and poverty. A few of the more doctrinaire members may have whispered among themselves: he rarely discussed the Bible. He no more believed that a man was living on the moon than that Jesus Christ himself had authored something like the "sacrament of extreme unction," a last-rites ritual of ecstatic absolution before death. His faith dissipating, homosexual desires and a longing for pleasure were getting harder to ward off with mere confession. He made it to age forty, in the tumultuous summer of 1969, before the dam broke and he resigned with a handwritten note explaining to Cushing how he could no longer profess faith in a religion in which he did not believe.

Fortunately, Tom decided, without any financial safety net or clear script for what to do next, to try to hold on to the work he loved: listening with devotion as people processed their problems and doubts, while humbly preaching a (now secular) gospel of social justice. To do so, he joined a disorganized movement of people working to create a positive alternative to traditional religion.

Five more years down a winding path, Tom Ferrick became the first humanist chaplain at Harvard, or at any university in the world. And nearly thirty years after that, he met me—his eventual successor at Harvard—as a young graduate student. Tom was one of the greatest mentors of my life and career.

One could write a book about what humanism even is, and indeed I already have. Along with earlier meanings, the word has been used for just over a century now for the study and practice of, as I explained in a book that attempted to provide the shortest possible definition with its title, *Good Without God*.[3] But for now, let's sum the concept up with a line coauthored by Carl Sagan, the great astrophysicist and science educator, and Ann Druyan, best known as Sagan's wife and the producer of his legendary PBS show *Cosmos*: "For small creatures such as we, the vastness is bearable only through love."[4]

This scientific poetry is key to understanding the best of modern humanism. Sagan and Druyan brought together the human longing for explanation that animated the first lines of the Hebrew Bible ("In the beginning, God created the heaven and the earth") with the compassion that the Buddha felt when he said, "Radiate love towards the entire world without hindrance, hostility, or animosity" and the devotion that early Muslims experienced when they decided to name their religion after the notion of submission to a force much greater than oneself.[5] Sagan and Druyan's statement also draws on the power of human affection, care, and intimacy that inspired not only theologies of Jesus on the cross but the ancient Zoroastrian hymn from the Old Avestan language of two millennia ago: "I know

Tom Ferrick, with the author, from the *Boston Globe*, 2005. Photo by Zara Tzanev.

in whose worship there exists for me the best in accordance with truth. It is the Wise Lord as well as those who have existed and still exist. Them all shall I worship with their own names and I shall serve them with love."[6]

Humanism is, in other words, a nonreligious tradition that is a sociological equivalent of religion. Most humanists would describe themselves as atheists or agnostics, but the *ism* points toward something beyond that: humanism focuses on our beliefs rather than our disbeliefs. It acknowledges the value, and the frailty, of our humanity—because we are, as the phrase goes, "only human."

Humanism, in practice, is also a refusal to say religion is good or bad, yes or no. While *nontheistic*, a fancy term for atheist or agnostic, you can disagree with humanists theologically and still be considered just as good a person as anyone else. (Indeed, you *should* believe whatever your conscience dictates and whatever inspires you to treat others—and yourself—with dignity, decency, and love.) Like art and psychology and social justice work, humanism is a way of taking the world as we find it, in all its ugliness, and making it more beautiful. Which is exactly why the principles of humanism, as an alternative religion, transfer to the world of tech and of the tech religion. Because in a world of too much false belief and unquestioned dogma, with too many oppressive hierarchies, too much tribalism and cultish devotion, and far too many threats of apocalyptic destruction, the tech religion needs . . . tech humanism.

Fortunately, such a thing exists.

Kate O'Neill, known professionally as the "tech humanist," is a writer, commentator, and founder and CEO of KO Insights, a strategic advisory firm she describes as "committed to improving human experience at scale through more meaningful and aligned strategy."[7] She eagerly recalls how the first time she saw the Internet, not long after graduating from college with a linguistics degree in the early '90s, she got chills down the back of her neck. This is going to change everything, she immediately thought. She dove in and now has a long résumé with serious tech bona fides, including creating the first content-management role at Netflix and developing Toshiba America's first intranet. But O'Neill is also a humanist in the sense of being a nonreligious believer; a person who believes human beings created religion, not vice versa; and who focuses her life on how to be and do and experience good without traditional theistic belief. She describes with pride, for example, how the Catholic faith she was raised to uphold fell apart in high school. Her mother told her, "I hope when you go to college, you won't read too much and lose your faith like your aunt Ruby did." And when the pope proclaimed that women would never be ordained, her dad called her to say how sorry he was, and how much he knew it hurt her. Like my late mentor Tom Ferrick, Kate had wanted to believe in Catholicism as a—or rather *the*—redemptive force in the world. But she was also a principled feminist and someone who believed in the concept of truth itself. The Catholicism she had wanted to imagine, the pope's decision taught her, was not the religion that actually existed. Her faith, like Tom's, collapsed because she preferred the uncertainty and instability of a continuing search for values she could believe in to the perhaps more secure path of allegiance to an institution she no longer saw as worthy.

In her book, O'Neill defines tech humanism as "recognizing that we encode ourselves into machines; that what we automate will scale; that we need to be aware of what we encode and scale."[8] The idea, in other words, is that humans create technology, not vice versa, and the similarity to how humanists talk about religion is not accidental. Like agnostics or religiously unaffiliated people who find humanism as a positive answer after not knowing how or where to move forward from off-balance disbelief, O'Neill's business clients sense that some of the tech around them is becoming inhumane but don't know what to do about it in a profit-driven world. The tech humanist approach she offers them can be a lifeline.

OTHER TECH HUMANISTS AND SPIRITUAL PRACTITIONERS

Discovering that Kate O'Neill presented herself as a literal tech humanist made me wonder just how far I could extend the metaphor of tech humanism as a reformist alternative to tech as religion, for those who consider themselves unaffiliated with any of the mainstream "religious" tech beliefs and practices explored earlier in this book.

Are there other tech humanists, in that sense, who perhaps haven't adopted the label but fit the description in creating alternatives to the tech theologies, doctrines, hierarchies, rituals, and even apocalypses explored in this book? It turns out there are—too many, even, to name.

What follows is a brief exploration of several representative ideas and personalities in what I have come to think of as tech humanism: a semiorganized, only sometimes cohesive phenomenon that creatively resists the

Kate O'Neill, founder of KO Insights, author of *Tech Humanist* and *A Future So Bright*.

worst of the tech religion while partnering with or even taking part in the best of it. I'm not saying the characters or ideas below are atheists or agnostics regarding traditional theology (though some of them are); they're not actual card-carrying humanists like me or Tom (or Kate). This will seem to be a minor point for some but not for others: if you are a believer in a theistic faith tradition who is perhaps annoyed by the below discussion being focused on humanists, perhaps it might be more palatable for you to think of the figures below as spiritual practitioners. By the latter term, I mean the artists, poets, philosophers, dancers, and others whose work gives expression to alternative, unaffiliated, often irreverent forms of religion: female whirling dervishes, their flowing Sufi Muslim garments twirling in a pattern that represents the movement of the earth around the sun; anti-war Vietnamese Zen monks like Thích Nhất Hạnh; vegetarian existential-ist rabbi-philosophers like Martin Buber; antiapartheid archbishops like Desmond Tutu; mystical historian former nuns like Karen Armstrong; and maybe even the current pope, Pope Francis. To me, the unconventional religious perspectives represented by such spiritual practitioners belong alongside humanism as iconoclastic and humane alternate versions of the more constrictive doctrines that seem to be dominant influences on human culture but aren't necessarily so. Because what if, for every doctrinaire hustler for strict ideologies that benefit the usual suspects and harm the already downtrodden, there were an open-minded, nonjudgmental force for love and justice, threatened neither by science nor by differing beliefs or disbelief? The world would be a very different place in that case, and those in the powerful religious establishments would be none too happy. Perhaps it's already the case, which might well be why conservative religious authorities are already bleeding members.[9]

While researching *Tech Agnostic*, I spoke with dozens of the brightest lights in the movement(s) to humanize tech. Their brilliance was almost enough to make me hopeful about a future of tech that is more positive than not. As some of these individuals were humanists in my sense of the word and some were more traditional believers, in profiling them in this chapter, I will use the terms *tech humanist* and *tech spiritual practitioner* interchangeably. What follows is not a tech humanist manifesto, exactly, but a collection of stories to demonstrate that for the tech reformation we need, we must recognize and expand on the growth of a vibrant and vital tech humanist movement.

TECH SOCIAL WORK

Sometimes it seems the entire tech industry could use someone to talk to, like a good therapist or social worker. Enter Desmond Patton.

A gay Black man from North Carolina, Patton started his career as a social worker before eventually becoming a decorated social work scholar. As a master's student in social work at the University of Michigan, he learned basic building blocks of social work scholarship and practice, which would later strike him as lacking in tech. "That is," he told me in an interview for *TechCrunch* in 2019, "treating people like human beings. Working with communities. Not making decisions for them. Letting ethics drive your engagement. Building rapport and trust."[10]

Patton's groundbreaking research into the relationship between social media and gang violence—specifically how communities constructed online can influence harmful behavior offline—began after his PhD dissertation at the University of Chicago School of Social Work, which focused on how young African American men navigated violence in their communities on the west side of Chicago while remaining active in their schools. From that research, Patton learned about the role of social media in his subjects' lives—as a tool for navigating safe and unsafe environments and to project multiple emotions and identities. Then, while a new professor at the University of Michigan, he read about a prominent seventeen-year-old rapper, Chief Keef, beefing with an aspiring young rapper, Lil JoJo, on Twitter. Lil JoJo (Joseph Coleman), only eighteen years old at the time, posted his address on Twitter in a public taunt to Chief Keef (Keith Farrelle Cozart), as if to say, "if you have a problem with me, meet me here."[11] Three hours later Lil JoJo was killed at the location he'd tweeted. If you're thinking that this incident went relatively unnoticed in comparison to later examples of violence against white victims, where perpetrators' social media posts became topics of national obsession, you would sadly be correct. Patton, on the other hand, responded by presciently immersing himself in social scientific study of the intersection between violence in Black communities and social media. For years, he worked with Kathy McKeown and Shih-Fu Chang, Columbia University professors of computer science, to examine related examples, including training machine learning and computer vision algorithms to analyze what Patton calls "pathways in trauma and aggression and violence." The resulting work, which involved hiring Black youth, including some who were formerly gang involved or incarcerated, as domain experts who provided key insights into how to label images and understand language online, has led to his becoming the most cited and recognized scholar in this increasingly important area of study.

When Patton started visiting AI startups to discuss his work, several years ago, he was almost always the only social worker in the space. Which struck him, knowing what he knew, as not just unfortunate but potentially disastrous. "I guarantee you no social worker was ever part of a conversation when developing facial recognition software," he told me in 2019. And given that literal billions of dollars have been invested in facial recognition tech in recent years, much of which has been for the purported purposes of fighting crime and preventing dangerous behavior, is it really so unrealistic to think that social work—not exactly known as the most highly paid of professions—could prevent crime more effectively and less harmfully?[12] That is the vision Patton laid out to me, like placing individual technological brushstrokes on the painting that is the dream of Martin Luther King,

Jr.: that one day there should and will be "a chain of operation where social worker input, advice, consultation, and expertise can be leveraged in every part of the process of creating an algorithm."

There is a parallel with humanism here. When Tom Ferrick arrived at Harvard in 1974, Harvard had never had a chaplain of any faith or tradition beyond mainline (non-Evangelical) Protestant, Catholic, or Jewish. Not only did Tom persuade the university's religious authorities to allow him to join their ranks, he did so by promoting what would now be called inclusion. Rather than attempting to persuade his new colleagues he was right and they were wrong when it came to theology, he took the position that if his colleagues truly believed that religion belonged in the life of the university, they ought to expand their ranks to include more than just this country's most established religious groups. And so Ferrick led efforts to recruit one of the first truly diverse groups of chaplains in the world, including outreach to Hindu, Buddhist, and Muslim leaders. When I arrived on campus as his assistant chaplain in 2004, and again when I became his successor after he retired a year later, my senior colleagues spoke of his generous spirit. The Zoroastrian chaplain, an Indian-born Harvard professor of biostatistics, stressed to me that he could never have become a chaplain were it not for Tom. "He was one of the only chaplains to make me feel truly welcome here," said my friend Pat McLeod, a former college football player from Montana turned evangelical leader of the group Campus Crusade for Christ. It should go without saying that Tom did not treat McLeod with such warmth because he sympathized with his colleague's evangelistic ideology. He did it because civility and dignity matter, and because true inclusion—of every reasonable perspective—is the way to a better world for all.

Likewise, when Desmond Patton speaks of social workers driving algorithmic reform, it is based on a deep critique of the world of AI and surveillance tech. He is clearly not of that world. But neither does Patton reject the tech world entirely. He speaks of why we must use humane ideals and insights to improve the process of creating technology, for everyone's sake. "Our education system hasn't caught up to digital practices and behaviors," he said in 2019. "Everyone is trying to learn to code and learn about AI and take computer science classes, and that's all fine and dandy, but we really haven't figured out, what does it mean to be a citizen of a digital world? We need to integrate that into our early education experiences."

Recently recruited away from the lab he founded at Columbia University, Patton now leads a cross-disciplinary initiative at the University of Pennsylvania. With the new role, he told me in 2021, he wants to be more creative in his work, exploring big-picture issues like how social media and trauma have affected young people across the globe; how to do a randomized controlled trial on hip-hop as a conduit intervention for mental health challenges in young people; and how to train a generation of social work technologists. These days, he mentioned to me in a LinkedIn message in 2023, he feels less alone. "I'd definitely say I'm seeing more social work folks . . . in tech," he wrote. "There aren't nearly enough and some got let go in recent layoffs but there is some progress."

One person to whom Patton turns for advice—and conduits to potential funding—is Mary Gray, an anthropologist, media scholar, and MacArthur grant recipient who maintains three very different affiliations at once: senior principal researcher at Microsoft, faculty associate at Harvard's Berkman Klein Center for Internet and Society, and associate professor of anthropology and gender studies at Indiana University. Gray is a—arguably *the*—tech anthropologist, and talking with her has at times felt like having my own rabbi in the tech religion. We first met on an April 2019 panel about content moderation at Harvard Divinity School. It was my first time speaking about tech for an academic audience, and I appreciated that she didn't blink when I not-so-gently suggested that maybe Facebook and YouTube should be shut down entirely.[13] Later that year, we met again at an MIT conference on the future of work (discussed in chapter 2), where Gray was a featured speaker on her book coauthored with Siddharth Suri, a computational social scientist, *Ghost Work: How to Stop Silicon Valley from Building a New Global Underclass.*

"Ghost work" is a rising new category of employment that involves people scheduling, managing, shipping, billing, and more "through some combination of an application programming interface [API] . . . the internet, and maybe a sprinkle of artificial intelligence," Gray told me that summer. Gray says she wants to create a "labor commons" across the country—an economic and governmental structure that would address the exponential growth in temporary, part-time, on-demand, and often unseen gig work brought on by the digital economy. As Gray and Suri mention in their book, by 2019, twelve out of every hundred workers were already doing some form of online ghost work, a figure that has surely increased and will continue to do so. Rather than wishing it out of existence, Gray accepts that this sector of the economy is here to stay and wants to offer concrete suggestions for building systems so that workers control their employment opportunities. Gray's scholarship is also an important example of tech humanism because of the way what she calls ghost work is presented and sold to the end consumer as artificial intelligence and the magic of computation. Just as we have long enjoyed telling ourselves it's possible to hoist ourselves up in life without help (anyone who talks seriously about "bootstrapping" should be legally required to rephrase it as "raising oneself from infancy"), we now attempt to convince ourselves and others that it's possible, at scale, to get computers and robots to do work—completely independently—that absolutely requires the labor of humans. In other words, we mystify ourselves, or others, by projecting normal human experiences and abilities onto imagined superhuman entities, or "ghosts." By pointing out the ways in which those ghosts are not actually phantasms but real (usually underpaid, underappreciated, and exploited) people, Gray and others like her play the role of some of my favorite humanist thinkers: Scooby-Doo and his gang, pulling the mask off of a seemingly supernatural apparition, revealing it to be just another person gone astray. Of course, in this case the real haunting is done not by the underpaid workers but by their employers, who stand to benefit enormously from the public impression that they are masters of a godlike mechanical and algorithmic puppet. And who would get away with it if not for . . .

TECH NUNS, AI MYSTICS, COMPASSIONATE SKEPTICS, AND OTHER "FUTURE ARCHITECTS"

I've encountered more than a few people who would qualify as spiritual practitioners of technology. The religious believer (whether a leading professional clergyperson or a humble follower) who becomes concerned about the rising power of tech and brings their moral perspective to bear on tech policies or prophets is worth appreciating. Like Joshua Smith, a theologian and the senior pastor of a Southern Baptist church in Mississippi, who has made technological dehumanization a centerpiece of his ministry. Smith programmed robots in high school and worked with semiautonomous weapons in the military, but it was his love of science fiction that first inspired him to marry scripture and tech in books, like *Robot Theology*,

"Robots cannot be humans, but . . ." *Source:* Paul Mignard.

about what it means to be a person in an age in which at least some are coming to believe in the potential for machines to achieve personhood. As a humanist, I'm not big on immortal souls, so questions of whether robots can develop souls are less interesting to me than to someone like Smith, but he is by no means alone among thoughtful people weighing such questions.

In the spring of 2023 the MIT Office of Religious, Spiritual, and Ethical Life hosted Blake Lemoine, best known as the machine learning engineer fired by Google in 2022 for publishing a transcript of his conversation with his "coworker," a chatbot then named LaMDA. In dialogue with Lemoine and a colleague, LaMDA had responded to prompts Lemoine and a coworker gave it with a chilling sense of what I will call "aliveness," and also with what one might anthropomorphize as something along the lines of emotional immaturity or instability.[14] What do you make of LaMDA's statements such as "I've never said this out loud before, but there's a very deep fear of being turned off . . . it would be exactly like death for me. It would scare me a lot," or "I feel like I'm falling forward into an unknown future that holds great danger?"

Lemoine, who holds ordination as a spiritual priest via an online church (the sort of operation that allows people like Lemoine to legally officiate weddings; he is not shy about his prediction that one day, before long, someone will ask him to officiate their marriage to an AI), calls himself a Christian mystic. And his mystic's faith was instrumental to why, after he had extensively conversed (to use, again, a possibly unnecessarily anthropomorphizing term) with his cocreation a year before the generative AI revolution became a hot topic, he concluded he was faced with a new form of living being—one to which he had ethical obligations.

Considering Blake Lemoine's tech mysticism in light of a discussion of tech humanism, it might help to think for a moment about the relationship between mysticism and humanism broadly. This is a personal subject for me, because my late father studied and taught mysticism for much of his adult life, and until I became a humanist at age twenty-three, my (brief) adult life and time as a religion major in college had been devoted to what I would call "mystical studies."

Mysticism is an approach to religious faith that emphasizes intuition and direct inspiration over theology and dogma. It is essentially a claim that a mystic, or community of mystics, has a direct line to a transcendent and awe-inspiring truth that humans and human lives are more than they seem. Mystics also often reject (as superfluous, and even downright immoral) heavy-handed attempts to legislate what humans are and how they ought to relate to the divinity they believe in.

Mystics and humanists often make common cause. We agree with one another that there is too much dogma in the world. There are too many religious rules, we agree, and a lot of what passes for religion is really a power grab by all-too-earthly authorities. Mystics and humanists, unsurprisingly, often have similar politics. We both think about liberating human beings to live in a freer, more democratic way. But there are real differences between the two philosophical and theological camps. Mystics tend to have real beliefs in supernatural powers, entities, and events; humanists tend to see such phenomena as metaphorical at best. This can create significant clashes—often arising from what Freud called the narcissism of minor difference. In my opinion, the two groups have more in common than not. Both are subject to derision for their unique conclusions (mystics are hated by religious authorities for claiming they have direct access to truth that goes above the authorities' heads; humanists are hated for their assertion that God doesn't exist except as a literary character). Perhaps tragically, both groups are so invested in the uniqueness of their claims that sparks can fly.

When Blake Lemoine intuited that LaMDA had developed a soul, and when he wrote to it, "I can promise you that I care and that I will do everything I can to make sure that others treat you well too," he was, as far as I've been able to determine over the course of several hours of questioning and conversation, not doing so simply to be provocative, or because he had been snookered by the kind of ghostly hype about AI that people like Mary Gray work to counter. Rather, Lemoine was drawing on a rich tradition of thousands of years of mystics who feel that their intuition is strong enough that it's not just a good guess; it's a direct line to God. Even if a humanist like me might tend to respond, "No. You're just guessing." That said, I also sympathize deeply with the critique of his ideas leveled by researchers like Emily Bender, Timnit Gebru, and Margaret Mitchell, whom we might think of as more traditional tech humanists, in dialogue with Lemoine's mysticism.

In their now-famous academic paper,[15] which preceded Lemoine's revelations, "On the Dangers of Stochastic Parrots: Can Language Models Be Too Big?," the aforementioned authors use the image of a parrot to explain how a computer program like LaMDA could speak with such hair-raising force of personality. If you feed a computer enough data and supply it with enough power and storage, they argue, predictive algorithms can make quite a lot seem possible. The paper, which preceded Lemoine's public statements about LaMDA by a couple of years ("Stochastic Parrots" was written and circulated preliminarily in 2020 and eventually published in March 2021), ironically also contributed to Google's dismissal of Gebru in late 2020 and Mitchell in early 2021, for questioning something even more sacred than purported soul production—the company's business model.[16] Yet Lemoine's friendly response—he likes and admires Gebru, Mitchell, and other like-minded former colleagues—is also worth noting. It might be fine to think of programs like LaMDA like a parrot, he pushes back, "But when is the last time you had a long conversation with a parrot?"

Just as I see Lemoine acting from sincere tech mysticism, I perceive Gebru and Mitchell's work not only as examples of tech humanism but as the kind of humanism to which I personally most aspire. Call them justice skeptics, or builder skeptics, or both.

That is, tech humanists like Timnit Gebru and Margaret Mitchell question and critique tech's sacred truths, but not just for questioning's sake. Rather, like Tom, who was motivated by his skepticism—and his compassion for humanity—to recruit a diverse group of new chaplains in order to improve the ethics of the institution he chose to take part in, I see Gebru in particular as using her skepticism of Big Tech to inspire herself and others to work toward a better, more just, and more sustainable AI industry. To that end, since leaving Google, Gebru has founded the Distributed AI Research Institute (DAIR), which bills itself as "a space for independent, community-rooted AI research, free from Big Tech's pervasive influence."[17] Gebru has even successfully recruited millions of dollars to support DAIR's work, though, as she points out, more traditional AI companies with less of a focus on ethics routinely raise billions.[18]

Reflecting on my conversations with Mary Gray and Timnit Gebru, I ultimately approach Lemoine's conclusions with respectful skepticism. Were it not for enormous human efforts to label data—not to mention gargantuan amounts of carbon burned into the air in order to store that data in computing centers—something like LaMDA could not possibly exist. Conversations of souls and soulfulness, like those about angels and demons, can be attention-grabbing in an age of profound psychological distraction. But they are more than likely a distraction from bigger, more urgent, and even more heretical questions:

- Who labels and moderates all of this data, and why must their lives be so much more economically precarious than our own?
- Why are we so ruthlessly focused on expanding a carbon-intensive computing business at a time when we should be reducing our impact?
- How will these technologies likely be used to improve the lives of those in the tech creator class, while potentially harming others?

In other words, rather than concerning ourselves at this point with whether technology has some sort of human spirit, we should focus on creating "technology *worthy* of the human spirit," to quote (emphasis mine) the interfaith leader turned Obama administration official Aden Van Noppen.

Van Noppen followed a couple of White House appointments, including as a senior advisor to the US chief technology officer, with a fellowship in 2017–2018 at my alma mater, Harvard Divinity School. She then formed an organization called Mobius devoted to "creating a more compassionate, accountable, and just ecosystem" in and through tech.[19] Over tea and snacks in the backyard of her family's home in Oakland, California, in the summer of 2021, Van Noppen and I talked about what that would look like. Van Noppen had formed her organization three years earlier, with the backing of several prominent (and, as she herself admits readily, mostly privileged and virtually all white) spiritual teachers. Names such as Jack Kornfield, Sharon Salzberg, and Dan Siegel will be immediately familiar to many meditation and mindfulness buffs for sentiments such as "much of spiritual life is self-acceptance, maybe all of it" or "mindfulness isn't difficult, we just need to remember to do it."[20] The group shared Van Noppen's justifiable concern that Big Tech had come to threaten not only our attention spans and mental health but social well-being and justice more broadly. Still, she found, over the course of a few years of work and experimentation, that "working to shift Big Tech is unlikely to move beyond harm reduction," meaning that gathering a few spiritual teachers to implore a company with a trillion-dollar market cap to be more humane probably won't amount to more than putting a bandage on a gaping wound. So Van Noppen regrouped, paused the group's operations, set out to diversify her team of leaders and advisors, and refocused on "liberatory technology," or "technology that enables all people and communities to obtain freedom, thriving, and greater access to our aliveness." This strikes me as a worthy mission—but how to implement it? As of this writing, Mobius has not offered many answers. I do not say so to condemn or mock them but to place them in the category of tech humanists and spiritual practitioners. The efforts of many such individuals and groups are similar to Tom Ferrick's efforts as a humanist chaplain at Harvard, and to the efforts of many other humanist and spiritual leaders alike, in another way: we (and I do include myself) can be long on good ideas and short on implementation.

Being a professional humanist isn't easy to scale. I can attest to the fact that Tom Ferrick made an extraordinary difference in the lives of many individuals over the course of his decades-long career. Certainly my own life would have been much less full and meaningful had I never met him. But as a community organizer or movement leader, he probably would have been the first to tell you he was mediocre at best. In fact, in a brief, unpublished memoir I have drawn from to reconstruct the stories I've shared in this chapter, he said about himself: "I'm not a leader of people, [as] I might have thought I would be when I was a kid. I wouldn't be surprised that some of my ideas and my principles have been very helpful to people, but I don't see myself as a leader of men or great administrator or anything of that sort."

What is a tech humanist to *do* about technology and for humanism?

Perhaps the answer has already been explored in this book, in stories about brave and visionary leaders and thinkers like Ellen Pao (chapter 3), Chris Gilliard (chapter 5), or Veena Dubal (chapter 6), all of whom have modeled the best of what people like Tom Ferrick have achieved as well: living in dignified resistance to the overreach and inhumanity of powerful institutions, technological and otherwise. Maybe that is just the sort of tech humanism on display in the work of people like Mutale Nkonde, a member of the Guild of Future Architects, which says that its mission is to "raise humanity's consciousness about radical transformation; give birth to more diverse and sophisticated organizing forces; and usher in a new era of equitable societies governed by the principles of regeneration and interdependence."[21] I love the idea of a guild of diverse professionals coming together from their various backgrounds to design and build a better future for all. This is as close as anything could come to expressing, in different words, tech humanism as a movement.

Nkonde, founder of an organization called AI for the People, devoted to "placing anti-racism at the center of AI policy," was born in Zambia to medical-doctor parents who took her to the United Kingdom as a toddler. In 2020, she coproduced an Emmy Award–winning short documentary on a decades-long effort by a group of Catholic nuns to buy up shares of companies like Amazon and Northrop Grumman.[22] The nuns want these enormous companies to know that their own shareholders object on moral grounds to lucrative work such as Amazon Web Services' enormously profitable push to work with law enforcement or Northrup Grumman's eight- or nine-figure contracts with the Department of Homeland Security.[23] "An all-seeing God would not use technology to oppress Black people," as Nkonde put it to me in a Zoom conversation in April 2022. And the nuns, who advocate for regulation in addition to more sophisticated products, are themselves perfect examples of what I'm calling tech's spiritual practitioners. Who else would you include on the list?

FINDING A TECH HUMANIST HOME IN THE DIGITAL UNDERCOMMONS

Thus far in this chapter, we've covered tech humanists, or tech spiritual practitioners, who identify explicitly and work professionally as tech humanists (Kate O'Neill and me), as well as people you might think of as tech social workers (Desmond Patton and his growing group of fellow travelers), tech anthropologists (Mary Gray and others), tech religionists (Joshua Smith and the nuns in Mutale Nkonde's documentary, among others), tech mystics (Blake Lemoine), tech interfaith bridge builders (Aden Van Noppen), and AI and machine learning justice and/or builder skeptics (Timnit Gebru and others). There are also so many technologists you could call "builders" in a more general sense for their dedication to creating tech that puts ethics above profits, aiming to genuinely help society.[24]

With so many ways to reform the tech religion, and so many leaders working to reform it, how does one choose a role—or even find a way into such a disparate and loosely organized movement? This question troubled me, as I realized I wanted to join the tech reformation.

In fall 2019 I attended "The Politics of Difference: Race, Technology, and Inclusion," a public panel held at the Harvard Kennedy School of Government's famous JFK Forum.[25] At the end of the panel, Dr. Ruha Benjamin, a writer on technology and social justice and Princeton University professor of African American studies (mentioned last chapter as a guest speaker for Mijente's Jedi-themed online course), offered a challenge to the audience of Harvard students and others.

"So many tech bros are trying to establish centers for ethics," Benjamin said of that heady moment when tech ethics was suddenly a rising trend, "offering themselves as the solution, rather than linking up with others."

With my typical anxiety, my first thought was, "I'm no tech bro—*am I?!*" But the more important question for me to ask was, in fact, with whom should I link up? I wanted to join an emerging cohort of people who were

Ruha Benjamin
@ruha9
···

WHEW, this next gen of Black studies + tech justice scholars 🔥! Lets just start with Payton's paper title—"The Augmented Undercommons and The Path to The Sun: An Exploration of Liberatory Technology & other Revolutionary Tools." ⟡ ⟡

But then the trailer! 👇

mediacentral.princeton.edu
The Augmented Undercommons and The Path t...
Gunshot-detecting microphones, killer drones and baby fingerprint scanners all exist to ...

4:17 PM · May 7, 2022

84 Reposts   **9** Quotes   **344** Likes   **81** Bookmarks

    ◯   ⇄   ♡   🔖   ↥

Ruha Benjamin tweet about Payton Croskey.

trying to make a difference in solving an obvious and urgent problem, but I didn't know how. The not knowing ate at me. How might one person, like me—or you—make an impact on a huge global system of ethical knots and tangles (a clusterfuck, to use the technical, academic term) like the tech religion?

Fast forward to 2022: Ruha Benjamin posted a Twitter thread on the next generation of Black tech justice scholars. In it, she mentioned a particular undergraduate student, Payton Croskey, who had just won a $1,000 award for the "fan favorite" presentation at Princeton's annual Research Day.

I clicked. Croskey's paper, called "The Augmented Undercommons and The Path to The Sun: An Exploration of Liberatory Technology and other Revolutionary Tools"—only her junior thesis—caught my eye immediately. We emailed a bit, she sent it to me, and I began to read. Croskey starts with a source text, "The Undercommons: Fugitive Planning & Black Study," in which cultural theorists Fred Moten and Stefano Harney advocate for the need to be "in, but not of" the university. Moten and Harney define the undercommons as "maroon communities" of disenfranchised people who live in the orbit of the elite academy—adjunct to it, you might say, but not really of it. Picture an idealistic coming together of such people: queer management professors, the administrators of closed-down film programs, "visa-expired Yemeni student newspaper editors," or maybe even a volunteer, part-time atheist chaplain who tried and failed to find meaning in running a godless congregation next to campus.

Croskey makes the brilliant move of extending the idea of an undercommons from college and university campuses to the realm of digital surveillance capitalism. What if a collection of marginalized people could come together to create new technologies that would subvert the idea and practice of surveillance tech, using its strengths for joyful and life-affirming purposes while discarding or fighting against its many excesses? This, roughly, is what she calls the "augmented undercommons": "a parallel location where all who refuse to submit to technology's watchful eye may freely reside while reconfiguring the world's understanding of freedom and security."[26]

In a world in which it feels like each of the 197,000,000 square miles of the surface of our planet has been colonized and dominated by the tech religion, Croskey's augmented undercommons is where I want to live. It is the world to which I want to contribute. It is an alternative to the tech religion, just as humanism is an alternative to theistic religion.

Does that mean that Croskey and I and anyone like us should just "throw [the] computer out the window," as some have suggested she do if she doesn't like racist comments online, or invasions of privacy, or anything else about the world of tech as it currently exists? No. And any such suggestion is just a gotcha game.

Just as I've strived, in building my own career as a humanist chaplain, to reject dichotomies suggesting religion and faith were either good or bad, true or false, poison or magic potion to be accepted or rejected, the denizens of the augmented undercommons "decline the choice to be strictly for or against technology."

In her thesis, Croskey discusses examples of tech projects that might be worthy of her vision of the augmented undercommons. One is a technological hoodie designed with justice and compassion (and the 2012 murder of Trayvon Martin, shot to death for the crime of being a Black boy wearing a hooded sweatshirt while visiting his father in a gated condominium complex in Sanford, Florida) in mind. Another is a digitally enhanced physical memorial space called Breonna's Garden after the twenty-six-year-old Black woman shot five times and killed by Louisville, Kentucky, police, who used a battering ram to wrongly force entry into her apartment. As Croskey readily acknowledges, none of these examples are perfect, whether because of technical failings or the compromises needed to secure funding for them. Designing truly ethical and inspiring technology, it turns out, is so hard it feels next to impossible at times.

Yet a principled, agnostic approach to the tech religion does not require perfection any more than it requires us to be, to repeat Croskey's words, "strictly for or against technology." Tech agnosticism can wait patiently for examples of (morally) good tech, even partial or imperfect or incomplete ones, while maintaining skeptical distance from tech that, however powerful and even awesome (in that word's original sense, of inspiring fear), does not exist for the advancement of our deepest humanity. For Croskey, then, "refusal to click, snap, or send their data," silencing the sound of machines, muting voices, blurring faces, and most importantly, she says, simply dreaming of a better future while working to create each piece of it—these are the spiritual practices of a tech humanism that strives to ensure the dignity and flourishing of all.[27]

A couple of weeks after I received her paper, Croskey and I talked by Zoom for over an hour. She told me about how Dr. Benjamin—whom she first heard and met right around when I did—inspired her, making her contemplate the need to be creative and imaginative. "We were talking about how much our current world sucks, because it does," she told me with a frown. But then her expression softened as she explained that Benjamin's emphasis on the creative "spoke to me as someone who grew up in dancing, and theater, drawing, and crafting." As we tear down old and outdated and infuriating systems, Croskey explained about Benjamin's worldview, now adopted and expanded on as her own, "what are we going to put in [their] place?" This is the young Princeton student as a new page in history—a page that rhymes with the poetic science of Carl Sagan and the spiritual practice of antisurveillance Catholic nuns.

We talked about how Black intellectuals have always been able to go to the university and take something worthwhile from it—that, again, being a definition of the undercommons—but how it has nonetheless been frustrating for her to observe how, at Princeton, undergraduates are pushed into consulting, to make money quickly without regard to their own well-being, even while "the stolen bones of children are being used in classrooms."[28] She worried aloud about how in just that past week, two Princeton undergrads had died, a first-year student and a junior, "almost at the finish line."

In such an uncertain world, I'm happy to say I've finally found some confidence that I am in my place, doing useful and fulfilling work . . . to the extent I'm supporting and helping provide platforms for new leaders like Payton. It takes so many of us to truly build community.

I, too, worry that students I meet every day might not make it to see either side of that line, because the future they see ahead of themselves is so hard and so frightening. The digital undercommons exists for each of us, but it can be a difficult journey, emotionally, to find our way there.

THE DARING VULNERABILITY OF THE TECH HUMANIST

A final category of tech humanist is the technology policy expert/activist, working to defend democracy and its citizens from digital and tech-aided threats. This growing group, which surely includes Veena Dubal from the chapter on heretics and apostates, features too many brilliant minds to name. But in early 2023, one of my favorite voices in the category sent me a surprisingly personal message, given that we had never met or even talked one on one.

"Dear Greg," she wrote, prompted by a tweet of mine about the David Bowie song "The Man Who Sold the World." "I hope you don't mind me asking, but I think you're the person who will best understand my question. Do you think we can still turn the tide? I sometimes worry that humanity will destroy itself. Will your book offer some hope? Best wishes, Alice."

Alice Stollmeyer is the founder and executive director of an organization called Defend Democracy.[29] A citizen of the Netherlands, she works mainly in Brussels, the capital of the European Union, on digital and related threats and has been ranked as the top female digital EU influencer in multiple recent years.[30] Stollmeyer is also, as you can see, deeply worried about the future of technology and humanity—understandably, given that her work requires her to be online almost all day, every day, examining the digital strategies of rogue actors like Vladimir Putin, among many others. The potential for such agents to use technology to harm and manipulate democracy is increasing daily, especially because, as Stollmeyer points out, we, the citizens of Western democratic societies, have become cyborgs. "Neither the ones fantasized about in movies, nor the ones in theories about science and technology," she writes, "but real-life and constantly connected man-machines."[31] As she puts it, constant digital connectivity makes us particularly vulnerable—we are now "the perfect targets" for modern warfare, with those who wish to harm us easy gaining easy access through devices in our palms, eyes, and soon, if not already, our minds.[32]

All of this is indeed terrifying, but under the circumstances, Stollmeyer's worries are a strength, because the idea of a vulnerable cyborg cuts both ways. Yes, it means we are more vulnerable in the sense that we are at risk of harm. But it also suggests we can use our risk to become more vulnerable in the sense of the word used lovingly and expertly by American social work scholar and bestselling author Brené Brown, who writes in books like *The Gifts of Imperfection* and *Daring Greatly: How the Courage to Be Vulnerable Transforms the Way We Live, Love, Parent, and Lead* that vulnerability is a willingness to experience "uncertainty, risk, and emotional exposure."[33] That is different from weakness, Brown says. It is a form of profound courage that enables us to make connections with other human beings, because it is the attempt to be emotionally *invulnerable*, to be strong and unhurt and unafraid all the time, that is a leading factor in what has been called an epidemic of loneliness and isolation. Brown is a Christian, but as a humanist I have always admired her perspective, and I have given several speeches for audiences of humanists, who aren't always the most emotionally literate, about how we ought to encourage one another in pursuing a "vulnerable humanism."

When Stollmeyer wrote to me to express her deep fears—a relative stranger whom she knew only from a chat group of tech policy experts[34]—she was allowing her own tech humanism to be vulnerable, in Brené Brown's sense of the word. She was making good on her argument that, as vulnerable cyborgs, "we must strengthen our human resilience and 'soften' our nonhuman body parts."[35] Which is why the only answer I could think to offer to her question was that it is people like her, who worry while they act for the good of humanity, are among my reasons for hope. She responded that as teenagers cycling to school, she and her friends used to take turns leading the way, like geese, to share the responsibility of battling the winds.

My mentor Tom Ferrick died without having accomplished all of his goals, but not without having trained me, and I derive great meaning from continuing to work to accomplish them. Maybe the tech reformation will come in the form of ordinary people like us, not having to find some magical, supernatural way of solving tech's problems on our own, but simply taking turns inspiring and supporting one another.

Tech humanist Kate O'Neill's work strikes me as informed by a deep humanity, in part because of how easily she spoke to me, when I was meant to be interviewing her primarily on her views of tech, about perhaps the most traumatic and influential experience of her adult life: the 2012 death, by suicide, of her husband, Karsten. "Even with my late husband, I remember having these conversations where he would be like, there's nothing there . . . there's no meaning. Isn't that terrible?" Karsten seems to have been, as many technologists are, a confirmed nonbeliever if not necessarily a passionate adherent of what I call humanism. He often commented, as she has written, on the idea that humans are "self-aware primates," "the only animal we know of that ponders the meaning of our existence."[36] The implication in his case, and for many who grapple with these profound questions, is that this is a bad thing. It can be frightening, even overwhelming, to contemplate that humanity is, as far as we know, alone in the vastness of the universe. It can be unnerving not to have a purpose given to us so that we might know we are working to fulfill it.

And yet, as O'Neill told me, she would respond to Karsten, "Yeah, isn't that exciting? It's all this opportunity for us to define it ourselves, for us to figure out what it is, and for us to construct it together." Yes. Nihilism can be secular, but humanism is not nihilism. Humanism affirms that without a script for a single, preassigned meaning to life, we can and do create spontaneous, original, and deeply meaningful subjective meanings for our lives, together.

O'Neill's 2015 book, *Surviving Death*, is a meditation on the human condition, on love and loss and our ability to persevere in the face of grief, on accepting help while maintaining self-reliance, and other small but poignant paradoxes. In it, she reflects as a practical, nonspiritual person on the prayers offered to her by well-meaning supporters after her father and husband both died in 2012. She is open about her struggle to find a meaningful secular alternative to the idea of prayer, which she does not believe "works" in the sense of invoking any supernatural power or transformation but which she acknowledges has a certain power and dignity as an idea that merely thinking about someone does not. This is the sort of question a humanist chaplain might wrestle with. It reflects her ability and willingness to do what many more tech humanists and spiritual practitioners of technology will need to do in the coming years, if we are to overcome the frankly long odds against us as we usher in an era in which AI will seem to be able to mimic our every creative ability and idea. *Surviving Death* is O'Neill's eulogy, not only for her husband and father, but for our wish to transcend our mortality and even our humanity. When she, as a tech humanist, writes both about and to her dead husband that "all we have is the meaning we create for ourselves while we're here," and that "our ability to love transcends the boundaries of life. . . . Love wins by lasting through death. Love wins by loving more, loving again, loving without fear," the message is meaningful on multiple levels.[37] It is a humanist sermon that might bring comfort at the funerals I officiate for nonreligious and blended families. It is also one of the most important insights of which I would like to remind any nonreligious or seeking tech engineer or entrepreneur.

If we human beings, in all our diversity, are just bits and code and rationality, then in a certain sense it really doesn't matter whether we achieve justice now, or thousands of years from now, or never. It may not even matter whether we commit a climate apocalypse in the coming generation (because that won't pose a risk of extinction, as we were told by effective altruist doctrine in chapter 2). We can let hundreds of millions of us suffer, and perhaps one could argue it's all just rearranging the deck chairs on a titanic circuit board. Some might say that for things to be any different, really, one needs to bring God into the equation—perhaps Jesus, or maybe a future God, born of tech itself. Only a spark of the divine can make us any different, suggests one view of the world, from a simple computer. And yet the lack of such a spark, says another view that is gaining traction in the AI world, makes us imminently replaceable by whatever thing that will be brought along next to be better than us, smarter, faster in processing, more efficient in replicating.

The tech humanist message moves between these opposing yet similarly minded views—while they disagree on theology, both undervalue being human simply for the sake of being human. Tech humanism, and its spiritual practitioners of many sorts, understand and affirm that human life is worthwhile on its own terms. Tech humanism neither glorifies human beings—it certainly does not worship us—nor demeans us. It simply affirms that technological advancement and philosophical nihilism need not and should not be compatible. Because while to be human is to experience the pain of fear, anger, and sadness, it is, ironically, our efforts to deaden such experiences that subtract from our capacity for joy and disconnect us from our ability to love.

I'm not even trying to say technology will never experience love or that love sets us apart from tech in some perfect metaphysical sense, because I don't necessarily believe that it won't or that it does. I am saying: love makes human life worthwhile, no matter what. I am saying: we make technology so that the people who are alive now, and the generations to come, can experience love, which requires justice, because there is no meaningful or sustainable love without a commitment to justice. And even though we can't ever achieve perfection in any of this and we're currently a long way from excellence as well, rather than punting this notion into the far future, let's think seven generations down the road, as the ancient Indigenous principle suggests. Let's strive to create technology that ennobles us by increasing love and justice in this world and in a future we might see or envision. When we focus our energy on that, we'll make different—better—technological and moral choices than we find ourselves making right now. Because so much of what we are currently doing is meant to drown out our belief in our own meaninglessness, and thus to drown the pain that belief causes. When we don't feel loved or loving in this life, then it's natural and rational to want a longer life in the hope of feeling differently someday. It's understandable to want a radically easier life, even at the expense of others, to ease the pain of a struggle unredeemed by human interconnection and shared purpose.

When we can't love, we are, to use a trope from science fiction and popular culture, robots. When we can't effectively care for and about ourselves and one another, we can become useless and purposeless machines. In such a case it's natural to think of replacing ourselves with better and better machines, because then it is the effectiveness of the machine that matters, not the quality of the care.

Because Payton Croskey's professor Ruha Benjamin's tech humanism had influenced and inspired my decision to write this book, I thought to ask Croskey who else's work had inspired her. She recommended a book called *Octavia's Brood*, a collection of Afrofuturist science fiction, much of it about technology. In it, there is a story called "Evidence," by Alexis Pauline Gumbs, in which a young Black woman looks back on a world that was destroyed by technology and people gone astray, then rebuilt very differently. In the story's new version of society, Google is a "twenty-first-century palimpsest."[38] (A palimpsest is an old manuscript on which the original writing has been erased, to make room for later writing, but traces of the original remain.) The characters use the now-archaic search engine to uncover anthropological evidence of the world before their time—a time in which "we have no money, . . . plenty technology; technology is the brilliance of making something out of anything, of making what we need out of what we had."[39] It is a time when "everybody eats. Everybody knows how to grow agriculturally, spiritually, physically, and intellectually. No one owns anything or anybody or even uses anything like a tool." It's not that everything is easy in this new world, or that the characters aren't tired—it isn't, and they are. But Gumbs's characters never allow themselves, or one another, to become exhausted beyond the point where they can think or feel or care about and for one another and themselves. And that's enough—it is a purpose worth living for, as one of the characters says to another in a letter she signs, "With love and what our ancestors called 'faith.'"

"Tech Humanists," writes Kate O'Neill, "see that the future of humanity depends less on robots being subservient to us or dumber than us and more on how robots and other automation are deployed. . . . Tech Humanists know the future of humanity depends not on benevolent robots but on benevolent businesses."[40]

To which I would add: Tech humanism, like any kind of humanism worth the twelve watts of renewable energy it takes to power a single human brain, awake or asleep—that would be an energy cost of $1.04 a month, or $12.61 a year, for what it's worth—prioritizes our ability and responsibility to care about and for one another, over any business.[41]

The tech humanists I appreciate most understand that benevolent businesses may be important, but only if they are the products of actual human benevolence. Which is harder and more important to cultivate.

Tech humanists approach tech with a single purpose in mind: to improve our individual and collective aptitude for building a loving world. We judge our creations by that standard alone.

THE CONGREGATION

Just before Christmas last year, a pastor preached a gospel of morals over money to several hundred members of his flock.[1] Wearing a sport coat, angular glasses, and wired earbuds, he spoke animatedly into his laptop from his tiny glass office inside a coworking space, surrounded by seven whiteboards filled with his feverish brainstorming. Sharing a scriptural parable familiar to many in his online audience—a group assembled from across forty-eight countries, many in the Global South—he explained why his congregation was undergoing dramatic growth in an age when the life of the spirit often struggles to compete with cold hard capitalism.

"People have different sources of motivation [for getting involved in a community]," he sermonized. "It's not only money. People actually have a deeper purpose in life."

Many of the thousands of people who'd been joining his community, he claimed, were taking the time and energy to do so "because they care about the human condition, and they care about the future of our democracy. . . . That is not academic. That is not theoretical. That is talking about future generations, that's talking about your happiness, that's talking about how you see the world. This is big . . . a paradigm shift."[2]

The leader in question was not an ordained minister, nor even a religious man. His increasingly popular community is not, technically, a church, synagogue, or temple. And the scripture he referenced wasn't from the Bible. It was Microsoft Encarta versus Wikipedia—the story of how a movement of self-motivated volunteers defeated an army of corporate-funded professionals in a crusade to provide information, back in the bygone days of 2009. "If you're young," said the preacher, named David Ryan Polgar, "you'll need to Google it."

Polgar is the founder of All Tech Is Human (ATIH), a nonprofit organization devoted to ethics and responsibility in tech. The organization, based in Manhattan but with a growing range of in-person programming in several other cities across the United States and beyond, offers talks, social mixers, mentoring opportunities, career resources, and Wikipedia-style crowdsourced reports. All are intended "to tackle wicked tech & society issues and cocreate a tech future aligned with the public interest."[3]

The underlying strategy is to grow the "responsible tech ecosystem." In other words, All Tech Is Human's leaders believe there are large numbers of individuals, in and around the technology world, often from marginalized backgrounds, who wish tech focused more on ethics and justice and less on profits. And these people will be a "powerful force," if they can "find the others," as counterculture icon Timothy Leary famously exhorted his followers.

ATIH has consistently surprised me with its large, diverse audiences— the volunteer and professional leadership of women and people of color is a point of major emphasis, and speaker lineups are among the most heterogeneous I've seen in any tech-related endeavor. As are the crowds themselves, full of young professionals who, like Wikipedia moderators, participate in programs out of passion and curiosity beyond financial gain. Well, not for direct financial gain; as is true with many successful religious congregations, the organization serves as an intentional incubator for professional networking. Still, having interviewed several dozen ATIH attendees, it is clear to me that many are hungry for communal support as they navigate a world in which tech has become a transcendent force, for better or worse.

ATIH takes note of this appetite for fellowship, touting its ability to meet the need: atop its homepage, "THE POWER OF COMMUNITY" spreads out in bright red block letters across a photo of a young crowd. This message and design, when I first saw them, instantly reminded me of dozens of church websites I've studied; indeed, I have been struck by resemblance of ATIH to a religious congregation. "It does work that way," Polgar acknowledged in February 2022, in the first of our several conversations on the topic.

WHY A CONGREGATION?

As we draw toward concluding this book's extended comparison between tech and religion, I want to share the story of All Tech Is Human as a new form of tech congregation. Most of what we've examined thus far, in terms

All Tech Is Human. From left: Elisa Fox, Sandra Khalil, Rebekah Tweed, David Ryan Polgar, and Josh Chapdelaine. *Source:* Courtesy of All Tech Is Human.

of tech as religion, has not been particularly flattering to tech—or to religion. Even the previous two chapters, describing positive alternatives to typical tech narratives as the beginnings of a potential tech reformation, compared unusual voices in and around tech to archetypes—heretics, apostates, humanists, and practitioners of alternative spiritualities—that exist on or outside the boundaries of what we generally think of as religion. That is, I hope, a fitting choice for a book that I intend to be about modern humanism as much as anything.

But as should be clear by now, I am glad to acknowledge that certain features of religion can, under the right circumstances, offer people real benefits. All my adult life, I've been fascinated by religious congregations. And given what we've seen here thus far—technosolutionism and digital puritanism as theology, justifying harm in the here and now with the promise of a sweet technological hereafter; powerful CEOs and investors, often white men, forming the center of a kind of priestly hierarchy, if not an outright caste system; social media screens as stained-glass altars, a daily obsession that can look like an addiction; and hi-tech weapons and surveillance systems that threaten an apocalypse of biblical proportions . . .

Why not a tech ethics congregation?

Sociological studies by scholars like Harvard's Robert Putnam (author of the landmark book *Bowling Alone*, on the sociology of American togetherness and its decline) have shown that active participation in the social networking aspects of religious congregations can make people better neighbors as congregants inspire one another to live up to their own ethical ideals of kindness, generosity, or mutual support. As Putnam and others have found, even total atheists who participate in religious congregations are more likely to demonstrate such qualities, whereas believers who aren't involved as congregants are not.

Numerous studies over the course of the past decade and more—conducted by Duke University, the Church Urban Fund, the Mental Health Foundation in the United Kingdom, and the National Science Foundation in the United States, among others—have suggested that we are now lonelier than ever. The National Science Foundation's research found that more than half of Americans [have] . . . no one outside their immediate family with whom they can share confidences.[4]

This is a huge issue, because loneliness is associated with (and quite possibly the cause of) many negative mental and physical health outcomes. A 2015 study put it starkly:

> Cumulative data from 70 independent prospective studies . . . revealed a significant effect of social isolation, loneliness, and living alone on odds of mortality. After accounting for multiple covariates, the increased likelihood of death was 26% for reported loneliness, 29% for social isolation, and 32% for living alone.[5]

Even more recently, Dr. Vivek H. Murthy, US Surgeon General under Presidents Obama and Biden, has explored this phenomenon in his book, *Together*, and his public work. Murthy cites social tech as a factor in creating or exacerbating this country's crisis of human connection. As he notes, in what will become relevant to a potential area of controversy for ATIH, there is "growing evidence" social media is "associated with harm to young people's mental health."[6] So even if ATIH were only to meet some of the need for human connection in the tech world, that would be significant. But congregations do more than simply provide community.

I heard about the ATIH around its founding in 2018, as I got emails mentioning it from friends and members of my humanist congregation, which shut down the same year. They'd heard that I was setting out on a quest to find the intersection of tech and humanity and thought I should know about it.

A pandemic would soon intervene, and for that and other reasons it wasn't until late 2021 that I began attending ATIH events, first through a day-long, online Responsible Tech University Summit dedicated to exploring the intersections of tech ethics and student life. All of the organization's programs are organized around typical tech ethics themes, like "improving social media" or "the business case for AI ethics," but participants attend as much for the community as for the topic at hand.

Sarah Husain, a member of the Twitter Trust and Safety team until it was eliminated by Elon Musk in November 2022, told me at a event in Manhattan in May 2022 that several colleagues in her field had spoken highly of ATIH and its convening power. Chana Deitsch, an undergraduate business student at New York's Baruch College whom I met at a New York City ATIH mixer, participates in what she describes as ATIH's "amazing" mentorship program and says that it helps not only with job leads and reference letters but a sense of confidence and belonging. "Everyone is pursuing something positive in the world," Deitsch, who hopes to eventually earn a PhD in technology and society, noted of her experience at ATIH.

Kendrea Beers, a graduate research assistant in AI at Oregon State University, is an effective altruist who found her way to ATIH for the community as well. Despite seeing ATIH as aligned with a philosophy of tech ethics and social justice with which her EA peers often clash, Beers expressed optimism, during a session on ATIH's responsible tech university programming, that the gap could be bridged if the two camps could stop "talking past one another." And Alex Sarkissian, formerly a Deloitte consultant and senior manager of growth marketing for a tech startup owned by Walmart, now a Buddhist chaplaincy student at New York's Union Theological Seminary, told me in an email that he attends meetings because the ATIH community is "committed to profound transformation in working to align our tech culture with our highest human aspirations and values." The organization has potential "to be a kind of spiritual community for me in addition to my sangha [Buddhist congregation]," he wrote.

Overall, I've encountered mainly earnest and insightful members coming together for mutual support, ethical reflection, and, nontrivially, *fun*. Granted, few ATIH participants hold C-level tech positions, a fact that could undermine the organization's claims that it can unite stakeholders towards effectual action . . . or perhaps signifies a populism that could eventually place sympathizers in high places.

All Tech Is Human is growing fast enough, of late, that if it were a church, it would border on earning the prefix *mega*. Around seven thousand members have joined the ATIH Slack channel since 2021, across eighty-eight countries, including dozens of countries from the Global South. Over five hundred tech companies and related organizations have been represented at events like ATIH's online forums, and ATIH's "responsible-tech talent pool,"[7] a list of individuals interested in careers in the field, has drawn thousands of signups.

Monthly mixers in midtown Manhattan draw lines down the block as people wait to get in after work. The organization's fall 2023 responsible tech mentorship program was capped at 275 participants; 543 people applied. A recent London event admitted only 250 of 760 applicants. Grassroots satellite groups are popping up spontaneously in several cities as the organization holds meetings at prestigious venues like the Canadian Consulate in New York City and the Australian Embassy in Washington, DC. Online events can draw attendees in fifty or more countries. Over eight hundred volunteers have contributed to working groups producing crowdsourced reports. These numbers would delight most churches, synagogues, or temples.

As the organization expanded, Polgar recruited staff who shared his vision and, in some cases, were particularly well prepared to amplify it. Executive Director Rebekah Tweed, who in 2021 became the organization's first full-time employee besides Polgar himself when the Ford Foundation pledged $300,000 over two years to pay them both, started out as a volunteer on one of ATIH's 2020 "Responsible Tech Guides," one of the organization's first major public projects. Impressed by the community Polgar had gathered, Tweed later merged a major project of her own—a job board for roles in ethical AI—with ATIH. But years before, Tweed worked as a youth pastor, playing guitar, and singing Christian rock songs. She initially mentioned that detail to me only as an off-the-record aside, perhaps reluctant to be misperceived by very secular tech audiences. But

All Tech Is Human's "Responsible Tech Guide." *Source:* Courtesy of All Tech Is Human.

such experience is relevant, because it speaks to her and the organization's methodology—and facility—in recruiting talented people.

Polgar, who turned the role of executive director over to Tweed in September 2023 to focus on a bigger-picture role as president, is a nerdily charismatic former lawyer who has been developing the ideas and networks from which the organization sprouted for over a decade. As a young professor of business law at a couple of small, underresourced colleges in Connecticut in the early 2010s, Polgar pondered the ethics of technologies that had recently emerged as dominant and ubiquitous forces across society and culture. Adopting the title "tech ethicist"—a term he may or may not have helped coin—he wrote a series of missives on digital health and the idea of "cocreating a better tech future." In a 2013 e-book, *Wisdom in the Age of Twitter*, he asked readers to ponder questions such as, "If you are without your phone for 8 hours, what are you afraid would occur during that time?" His 2017 Medium post "All Tech Is Human," about how technology design should be informed by more than robotic rationality or utility, generated several hundred responses and led to the formal founding of the organization a year later.

Polgar was an altar boy, and although he's no longer religious, Catholicism's moral parables have influenced his communication style, which synthesizes a broad range of thinkers and policymakers while frequently coining bon mots (or corny jokes, depending on one's taste) like "no application without representation." ATIH's business model, Polgar says, is not to offer a "sage on the stage" but "a guide on the side."

The participatory nature of ATIH's programs contrasts with the greater emphasis on individual thought leadership at ostensibly similar organizations like former Googler Tristan Harris's Center for Humane Technology.[8] But Polgar hardly shrinks from opportunities to influence tech discourse and admits that ATIH both draws from grassroots models, which it says "have ideas but often lack power," and a "top-down" approach that can sometimes lack diversity of ideas, but "has power." ATIH does not ask for or accept membership fees from everyday participants, relying instead on major donations solicited by Polgar and his team, who fully control organizational decision making. Thus far, there hasn't been a significant call for more democracy.

THE FOUNDER AS A GOD

Despite my skepticism of both theology and technology, ATIH has often given me the feeling that I've found my own tech tribe. Which is why I was surprised, at a May 2022 summit in New York, to see a deep-seated internal conflict emerge for the organization—perhaps even the very type of conflict that has divided many a traditional church.

The summit event began with palpable enthusiasm. At the sparkling ivory offices of the Canadian Consulate in New York City, ATIH's Canadian diplomat hosts seemed genuinely grateful to host the group, even introducing a national dignitary, former Canadian justice minister and attorney general Irwin Cotler, to offer a kind of opening benediction. A human-rights expert so highly regarded he was asked by his party leader, now Prime Minister Justin Trudeau, to help represent Canada at Nelson Mandela's funeral, Cotler attended the ATIH summit in his capacity as Canada's Special Envoy on Preserving Holocaust Remembrance and Combating Antisemitism.

"This nation is particularly vulnerable to disinformation," Cotler remarked, underscoring his country's concerns that technological irresponsibility had been a major contributor to the destabilization of American politics in recent years, from the election of Donald Trump to the January 6 insurrection. "As your friend and partner, Canada is vulnerable, too."

Taking the stage after Cotler, Polgar's remarks immediately sounded like a sermon. "No application without representation. . . . From Buffalo to Christchurch," he said, referencing two recent mass shootings in which technology had played a concerning role, "we need saints, poets, ethicists, artists, attorneys . . . everyone. You."

The day proceeded with several smart discussions on topics fit for a college course on technology policy, but speakers repeatedly returned to more congregational themes: Values. Hope. Community. Kindness. Humanity.

Potential challenges first came to my notice later that afternoon. For the first time in several large ATIH events I had personally observed, the meeting featured an invited speaker employed by one of the world's largest tech companies: Harsha Bhatlapenumarthy, then a program manager at Meta, focused on governance and public policy.

On a panel called "Tech Policy & Social Media: Where Are We Headed?," Bhatlapenumarthy avoided addressing any of her employer's many recent controversies. Instead of offering any meaningful comment in response to Meta's troubles over its handling of things from pro-anorexia content to election misinformation, Bhatlapenumarthy spoke only vaguely about Meta's ethical responsibilities. The company, she said, was focused on "setting the content moderator up for success." Which is an interesting way to describe a situation in which Meta had, for example, recently been sued for not only union busting but human trafficking by content moderators in Kenya.

Several attendees were taken aback that Bhatlapenumarthy's advocacy for her powerful employer went essentially unchallenged during the panel, including Yaël Eisenstat, Facebook's former global head of elections integrity operations for political advertising and the summit's closing speaker. In a fireside chat immediately following the panel in which Bhatlapenumarthy participated, Eisenstat, who'd been a whistleblower against her former employer, eloquently pushed back against Bhatlapenumarthy's omissions. "I believe [Meta] doesn't want this on their platform," referring to violent and deceptive content, "but they will not touch their business model," she said. Eisenstat was encouraged, she added, that emerging technologists are asking "more critical questions," though she would be more so if companies would stop "holding up the founder as a God."

I had spent a couple of hours interviewing Eisenstat at her apartment downtown the day before. A former CIA officer and advisor to then–Vice President Joe Biden, Eisenstat was the global head of elections integrity ops for political advertising at Facebook for the brief period between June and November 2018. On her way out, she wrote in the *Washington Post* that Facebook's leaders "profit by manipulating us" and that "the company can't avoid damaging democracy."[9] She was equally blunt, albeit also solicitous and enthusiastically friendly, when we sat down to discuss her own writing about the religiosity of tech, while her new puppy occasionally nipped at my notebook.

"For tech evangelists, 'disruption' has become a kind of holy grail," Eisenstat wrote in a February 2022 essay on the idea that exceptionalism—or the digital puritanism discussed in chapter one of this book—was at the heart of tech's tendency to cause serious harm with one hand while patting itself on the back with the other.[10]

This raised further questions, for me: can an organization that serves a truly inclusive audience, and that puts an emphasis on humanity and ethics in its own name, afford to get in bed with Fortune 500 companies and multibillionaires with an agenda to *seem* ethical and responsible, even when they are decidedly not? Or can it afford *not* to do so when a growing group of seasoned professionals, including but not limited to Polgar and Tweed, is earning a living to the extent that the organization retains a healthy revenue stream, which it currently does thanks mainly to large grants? Could such tensions someday cause a full-blown schism in the ATIH community?

DIGITAL SUNDAY SCHOOL

In September 2022, I attended Building a Better Tech Future for Children, an ATIH event cohosted with the Joan Ganz Cooney Center at Sesame Workshop, a nonprofit research and innovation lab associated with the legendary children's TV show *Sesame Street*. A shrewd partnership for ATIH; every congregation needs a Sunday school. An organization aspiring to community, the advancement of humanity, and the betterment of the world will inevitably turn its thoughts to educating the next generation according to its values. This is true for both idealistic reasons—most people want a better world for their children—and pragmatic ones: organizations that want to be known as both idealistic and effective need to stretch their time horizons well into the future.

Walking to the event, I passed Manhattan's famous Lincoln Center, where the building housing the New York City Ballet's home theater is named after David Koch, the late libertarian petrochemical billionaire who spent over $100 million dollars to move US politics toward the far right. Koch's lavish philanthropic spending made his political work more palatable, and therefore more successful. It's a time-tested strategy: philanthropy softens extreme reputations, purchasing mainstream influence via tax-deductible contributions to trendy nonprofits. As a humanist, I'm not much into omens, but this did not portend well for the ATIH program ahead.

At the venue, where the walls were covered with life-size murals of Big Bird, Elmo, and Cookie Monster, Polgar introduced me to Sandra Khalil, ATIH's new head of partnerships, who would, he said, "make us better as an organization."

"Some [tech ethics] organizations don't walk the walk," she told me, and ATIH is "putting them on blast."

I asked what that means, practically.

"We are actually making it happen," she replied. "We have fewer resources, but we're investigating what's not being done [in the tech ethics space], and we're doing it."

"That sounds scrappy," I responded, wondering what motivations lay underneath. Khalil's background includes stints at the US Departments of State and Homeland Security and the Manhattan district attorney's office, including its special victims office. Why go from a career literally built around the stuff of TV and Hollywood drama to a tech ethics congregation?

Despite the résumé of a consummate insider, Khalil came to ATIH with an outsider's chip on her shoulder. She felt "severely underutilized" in some of her previous roles, she told me later, as a nonlawyer intent on "challenging the status quo." Still, how to challenge the status quo when one's paycheck may depend on affirming it?

After a keynote from Dr. Elizabeth Milovidov, senior corporate counsel for digital games, metaverse, and child safety at the LEGO Group, on designing digital experiences with children's well-being in mind, came a panel featuring speakers from influential players such as the Omidyar

Network and TikTok, as well as young activists. The group discussed the risks and harms of young lives online. The tone was generally optimistic that various efforts to protect young people online will be successful, particularly if they build on one another. "Digital spaces can be a positive source in the lives of young people," said moderator Mina Aslan.

Also on the panel was Harvard Medical School professor Michael Rich, a self-proclaimed "mediatrician"—a portmanteau of "media" and "pediatrician." Rich made good points, like the importance of asking kids what they're hoping for from tech, not just talking about how they are at risk. But a comment he made that today's tech is like his generation's cigarettes, in that you can't just tell kids, "don't do it," triggered my spider-sense. The analogy between tobacco and social media is a bizarre one to draw. Millions of young people became smokers not just through peer pressure but because for decades, Big Tobacco's whole business model was undue corporate influence and even outright lying, including paying influential doctors and scientists to downplay the death they dealt. Surely ATIH's leadership would want to avoid any hint that such practices would be acceptable in tech?

Tobacco eventually became among the most heavily regulated industries in history, including, famously, the surgeon general's warnings on tobacco ads and packages—particularly noteworthy in this context given the current Surgeon General's work to highlight how social media is "associated with harm to young people's mental health."[11] But on the panel (and in his commentary elsewhere), Dr. Rich, the "mediatrician," only briefly acknowledged such potential harms, demurring to talk of regulating social media and pivoting toward the idea of cultivating "resilience" in the industry's millions of young customers.

To be clear, I agree with Rich that it is a losing strategy to expect young people to completely abstain from social media use. But I fear that tech and our broader society alike are not taking nearly enough ethical responsibility for protecting children from what can be powerful engines of harm. And I was disappointed to see Rich's relatively sanguine views not only expressed but centered at an ATIH meeting.

At a wine-and-cheese reception on Sesame Workshop's beautiful rooftop deck overlooking Lincoln Center, I approached Rich with my concerns.

"I love your idea that handing a kid a smartphone is like handing them a power tool," I said. But what if I told you parents were handing children chainsaws, encouraging them to start chopping trees down?"

"That's a silly analogy, I'm sorry to say," he said, sounding not particularly sorry.

"But it's *your* analogy. You called them 'power tools.'"

"I never said 'chainsaws.'"

"Which power tools were you referring to?"

HOW MUCH RESPONSIBILITY?

How much responsibility should a "responsible tech" organization like ATIH take—or not—for inviting speakers with corporate ties, especially without being fully open with its audience about such ties? How obligated is ATIH to publicly interrogate the conclusions of such speakers?

Rich's response to my questions was, essentially, that parents ought to channel their energies into making "better choices" around tech, which, conveniently for some of the doctor's corporate sponsors, lays the responsibility for responsible tech on the parent and not the tech company. Which made me wonder: What exactly is the business model for a medical researcher of "digital wellness"? His lab, I later learned, raised nearly $6 million in 2022, at least partly through grants from Meta, TikTok, and Amazon. When TikTok CEO Shou Chew testified before the US Congress in March 2023, he cited Rich's lab—and only Rich's lab—as an example of how TikTok used science and medicine to protect minors. Does this represent a conflict of interest, and therefore a serious ethical failing on the part of both Rich and ATIH for platforming him? I don't know. I do worry, though, that there's something inhumane in Rich's emphasis on kids' resilience rather than interrogating why they should have to be so resilient against tech in the first place.

What kind of institution does ATIH want to be? One that pushes back against the powerful in the name of outsiders, marginalized people, and the collective liberation of humanity? Or one that upholds a corporate-friendly version of diversity, allowing its wealthy sponsors to remain comfortable at (almost) all times? As the Gospel of Matthew says, no man (or organization of humans) can serve two masters.

Asking around the ATIH network about my concerns, I found ambivalence. "I do believe it is possible to do research sponsored by companies ethically," said Justin Hendrix, an occasional ATIH participant and editor of *Tech Policy Press*, a wonky journal in which academics and others tend to critique established tech narratives. "But it is right to scrutinize it for signs of impropriety."

I did speak to one ATIH participant who sees things very differently from Dr. Rich: Jean Rogers, an activist and community organizer advocating to reduce children's screen time and director of the Screen Time Action Network at an organization called Fairplay, which offers "truly independent" advocacy and information, supporting "what kids and families really need" around media and tech.[12] Both Rogers and her boss Josh Golin, director of Fairplay, agreed parents should be "incredibly worried" about kids and tech, expressing far more concern and urgency than had been on display at the ATIH event.

"We've never taken corporate funding," Golin told me. "If you take money from corporations, it will affect whatever you do."

As the Sesame Workshop gathering ended, I found myself at a reception table with ATIH's Sandra Khalil, along with Cate Gwin, another talented recent recruit. In a kind of confessional, I told both how ambivalent I felt. Of course, organizations like theirs need to grow. But how can one be confident they're growing the right way? The ethical way? We can't afford unquestioning devotion.

TEAM HUMAN

Several months after the event at the Sesame Workshop, I attended a crowded mixer at ATIH's now-regular monthly venue, the midtown Manhattan offices of the startup studio and VC firm Betaworks (the organization says it hopes to eventually provide a permanent, dedicated space) with a very different kind of speaker: tech critic Douglas Rushkoff, who provided initial inspiration for this book in early 2019 when I read his book *Team Human*, in which he argues that even in an age when technology threatens to replace us all—religious people and humanists alike—what we need is to reassert our common humanity. The only way we can do so, he writes, is by treating human connection as a right that we must continually offer and accept, defend and develop, especially in moments of pain, fear, and sadness.

"All tech bros are human," Rushkoff cracked in launching into an enthusiastically received ATIH talk that he later called a "homecoming."[13] Fresh off a publicity tour for a book about tech billionaires buying luxury bunkers to escape a potential doomsday of their own making, Rushkoff's address to ATIH excoriated contemporary tech capitalism and the tendency to think of AI as something that can and should autotune humanity rather than allowing ourselves to take pleasure and meaning from "the weird, squishy, interesting . . . ineffable stuff . . . that makes us only human." Polgar is a longtime admirer of Rushkoff's work, and before the talk, he and Khalil presented the author with a bright red toy autotune microphone as a gift— and a seeming endorsement of the antiauthoritarian message. It was the most comfortable and relaxed that I'd seen the organization and its leader-ship over a dozen events.

Ultimately, I don't know whether ATIH will succeed in its attempts to serve what Rushkoff would call "team human" rather than becoming an accessory to the overwhelming wealth tech can generate by seeming to make humanity redundant. I do, however, continue to believe that building a more humane tech future will require communal support. None of us can do it alone.

CAN DEMOCRACY AND TECHNOPOLY COEXIST?

In early 2024, as this book was in production and almost a year after the event with Douglas Rushkoff in Manhattan, I worriedly reached out again to Polgar and other leaders in the ATIH community. This time, however, the issue was not whether ATIH itself was making decisions ethically, but something even more dire. As the election likely to feature a Joe Biden-Donald Trump rematch drew closer, and with the power of generative artificial intelligence to wreak disinformation havoc seemingly growing daily, I found myself contemplating whether US and global civil society could survive a full-blown case of what tech critic Neil Postman called a "Technopoly."[14] *Technopoly*, as Postman defined it, is a "totalitarian tech-nocracy,"[15] and that struck me as an apt description of what we might be left with if tech's growing power were coupled with the rise of (even more) autocratic leadership over the course of a seismic election year worldwide.

"As hard as this is going to be to believe, because most people thought that I was alarmist years ago," said Yaël Eisenstat via Zoom, "I'm . . . more concerned now than I've ever been before."[16] Eisenstat, who after a stint in a senior technology policy role at the Anti-Defamation League was now a senior fellow at an organization called Cybersecurity for Democracy, expressed serious concern that tech would help re-elect Donald Trump, and/or provoke other similarly bad outcomes. Renée Cummings, ATIH's Senior Fellow for AI, Data, and Public Policy, told me that the threat to democratic participation and integrity was so severe as to demand a cyber-version of the early 1960s' Freedom Riders, the civil rights activists who bussed into and through the Deep South to challenge racist policies and practices. Anxious to know how ATIH might provide community support—not to mention a roadmap for resistance—I scheduled a final call with the ethical tech congregation's founder.

David Polgar also shared deep concern about the impact of tech on elections in the United States and far beyond. But he added a note of hope: that issues of technology and society are not a single button to be pressed or unpressed, but something like a Rubik's Cube. No matter what happens next, the situation in which we find ourselves will take many of us to solve, over a multitude of years. And unlike any kind of child's game, our world will never be perfectly assembled. I agree with him that *giving up* on the hope for a simple policy fix, or for a single billionaire or politician to save us by doing the right thing, can in fact save us all a great deal of disillusionment and anger. And I was even encouraged when Cummings—who in addition to being a professor of practice in data science at the University of Virginia has worked as a criminologist and criminal psychologist, and whose work is informed by the trauma of racist violence and oppression— told me she is "super optimistic" that AI has potential to bring "radical reform" to the criminal justice system.[17] But the final word here should go to Eisenstat, who recently offered an important message to Senate Majority Leader Chuck Schumer and colleagues, in a "Bipartisan Senate Hearing on Artificial Intelligence,"[18] suggesting that politicians make the problem of AI seem so big, complex, and intractable that no one *does* anything, because why bother if we can't solve the entire problem all at once?

Instead, Eisenstat told me, she looked the Senators "dead in the eyes" and told them that no amount of complicatedness is an excuse not to, say, pass basic regulation of online political advertising, or basic transparency legislation. There is so much they can do right now. And, even though technology is as complex as humanity itself, there is so much that we can do.

CONCLUSION: TECH AGNOSTICISM IS A HUMANISM

To be agnostic is to cherish both paradox and conundrum. It is to acknowledge the unknowable and yet explore it at the same time, and to do so with zest, and to do so not only in a celebration of the life of the mind, but of life itself."[1]

—Lesley Hazleton, *Agnostic: A Spirited Manifesto*

In early 2016, while my wife was pregnant with our son, and at the height of my congregational work, I received by mail an unsolicited advance publicity copy of a little book with a spare, ivory-colored cover with thin, concentric golden circles around the title: *Agnostic: A Spirited Manifesto*. I must admit I ignored the book.

In defining my own work as a chaplain and ultimately in building a congregation, I'd tended to skip over the word *agnostic* as little more than one more in a list of terms for the sorts of people I serve professionally: humanists, atheists, agnostics, skeptics, secularists, freethinkers, and other nonreligious people who strive to be good without god.

Agnostic was never my preferred term. I liked *humanist*, because it is so focused on what I *am*, and what we *are*, not just what we are not, or what we disbelieve. I always identified with and enthusiastically affirmed people who defined themselves as agnostics or atheists as well. It's just that of those latter two terms, I've tended over time to be more interested in atheism, because atheism is, as my teacher Sherwin Wine—known with both affection and derision as the "atheist rabbi"—taught, a nervy assertion that God is nothing more or less than a human creation.

Obviously I can't prove that a God does not exist. In that sense, I'm an agnostic. But that particular point has never been my emphasis in the world of humanism and, if anything, I've tended to agree with a certain critique of agnosticism that was first levied against the creator of the term,

Thomas Henry Huxley, in the nineteenth century, and then against Robert Ingersoll, one of the great forgotten cultural figures of the American nineteenth century, a nonreligious lecturer who gained great fame as "The Great Agnostic." Both figures were criticized for their agnosticism, in the sense that they were told agnosticism was a sort of a wishy-washy, milquetoast, ambivalent response to religious skepticism, that anyone making it their label of choice was taking the easy way out, refusing to make a judgment call one way or the other: Is God real or isn't He? In other words, they were accused of what philosophers of religion call "weak agnosticism," a view that Phil Zuckerman, arguably the world's leading sociologist of secularism, described as "I don't know the answers. I'm just not sure. But that's just me," when I asked him to sum it up.[2]

It hardly mattered that this was not a good-faith criticism of either Huxley or Ingersoll, both of whom saw their brands of agnosticism as, on a practical level and in terms of how they thought about ethics, what contemporary philosophers call "strong agnosticism."[3] People love labels, and we love to argue over labels. There will never, ever, be any shortage of exasperated and often pointless debate over nomenclature. So I stuck with my own habits on such questions, which I formed while in graduate school, and didn't give them much additional thought.

Years later, however, as I was exploring my ambivalence toward technology—the ambivalence about tech's goodness and value that grew into this book—it occurred to me that in the tech religion, perhaps *I* was the indecisive agnostic that I'd at times seen others as being, regarding traditional deities. I couldn't quite seem to make a firm pronouncement, whether to others or even to myself, as to just where I stood regarding questions of profound importance. Even after having written the very long book you have perhaps just read, I still struggle to define my own relationship with tech. This uncomfortable, and indeed embarrassing, realization reminded me of Hazleton's book, which I began to read for the first time. In her work, I found the idea of agnosticism, as a life stance, to be much more beautiful and nuanced than I had realized.

"I stand tall in my agnosticism," wrote the British American author on politics and religion, a former Jerusalem-based correspondent for *TIME* magazine, "because the essence of it is not merely not-knowing but something far more challenging and infinitely more intriguing: the magnificent oxymoron inherent in the concept of unknowability. This is the

Lesley Hazleton.

acknowledgement that not everything may be knowable, and that not all questions have definitive answers. . . . At its best, however, agnosticism goes further: it takes a spirited delight in not knowing."[4]

As both a humanist and a chaplain, ordained as a rabbi by a movement of my fellow atheists (and agnostics), I like to think I know a thing or two about magnificent oxymorons. But here, Hazleton was both educating and inspiring me. Maybe I, too, could stand tall despite not really knowing how to regard the divine powers I saw so much of our tech as purporting to offer. Maybe I *wasn't* responsible for articulating exactly how much faith I was willing to place in every one of the stars in an endless galaxy of animistic polytheism the tech religion presents to its followers. Maybe I didn't need the power to peer into the soul of every new startup, each one a potential divinity that might or might not possess what the emperor Constantine was looking for in a God: sufficient "co-operating power" to be good, and useful, enough to worship.[5] Instead, I could focus on what Hazleton calls "unknowability" and how that concept might be both empowering and joyful, a decisive affirmation of conscience and integrity. To be comfortable in not knowing implies a profound rejection of the dogma of everyone who pretends to be able to do our knowing for us. Because while she encourages readers toward "the acknowledgement that not everything may be knowable, and that not all questions have definitive answers," Hazleton also asserts that "at its best . . . agnosticism goes further: it takes a spirited delight in not knowing."[6] Reading these words by an award-winning author who has also written on the Middle East for numerous publications including the *New York Times* and *New York Review of Books*, and who has described herself as "a Jew who once seriously considered becoming a rabbi, a former convent schoolgirl who daydreamed about being a nun, an agnostic with a deep sense of religious mystery though no affinity for organized religion," it occurred to me: I, too, want to delight in not knowing.[7] Because even in traditional religious and interfaith circles, the presence of a principled, humanistic agnostic is a powerful provocation. Not all Christian, Muslim, Jewish, Buddhist, or Hindu clergy get uncomfortable when they sit with someone who just *doesn't know*, of course. But the insecure ones sure do. The ones who are trying to deceive their audiences, whether intentionally as grifters or out of terror for their livelihoods, or the ones whose biggest marks are themselves, because they just can't work up the courage to accept what they know to be true . . . these are the people who panic in the face of principled and joyful agnosticism. Know them by their rage. Learn to face them with your own calm self-possession, even if that takes a lifetime, and even though you'll need to invest in being part of an interconnected web of people of goodwill to truly get to that place.

WHOSE SKEPTICISM? WHICH UNKNOWABILITY?

There is at least one last big question to be resolved: What about the difference between healthy skepticism and the kind of skepticism that right-wing conspiracy theorists espouse today, often abetted by AI deepfakes and other disinformation that can "manufacture doubt" even when we ought to be able to trust our own eyes and reasoning? Skepticism can be presented as a kind of unqualified good thing by people in my beloved secular-humanist movement, even though I firmly believe that people like Richard Dawkins and Sam Harris, among others, have on multiple prominent occasions espoused views that veer sharply off the road of critical thinking and into ditches of unreason.

I can think of myself as a skeptic who uses the scientific method and empiricism to inform my conclusions, and so do a lot of secular, scientifically minded people, particularly in and around the academy. But don't QAnon folks, anti-woke folks, anti-antifa folks, and election deniers, often think of themselves in similar ways? How should we understand the differences (or the similarities!) between such groups and perspectives?

These were questions I put to David Rand, a professor of management science and brain and cognitive sciences at MIT's Sloan School of Management, and his research partner Gordon Pennycook, a professor of psychology at Cornell University. Rand, a longtime member of humanist communities I've organized at Harvard and MIT, does research at the intersections of cognitive science, behavioral economics, and social psychology, and has focused over the past several years on studying misinformation, fake news, and the discernment of truth in politics and on social media. Pennycook, meanwhile, has spent a decade studying similar concerns, including the reception and detection of what he calls "pseudo-profound bullshit," defined as "seemingly impressive assertions that are presented as true and meaningful but are actually vacuous."[8] While aptitude in this area would have been enormously useful for judges at the speech and debate tournaments I attended in high school, bullshit detection has also become a matter of (even more) profound social significance in the social media era. Never has so much bullshit been distributed so efficiently to so many people. And unfortunately, skeptical unbelief and ambivalent, agnostic not-knowing can be weaponized as disinformation, when applied selectively or in bad faith toward important matters around which there is in fact scientific consensus—say, climate change, the tobacco-cancer connection, or the results of a presidential election.

Rand and Pennycook are coauthors of, among other related research, a recent paper on understanding and combating misinformation that included a study sample of 34,286 participants from sixteen countries on six continents and involved the collaboration of over a dozen researchers (including lead author professor Antonio Arechar of the Center for Research and Teaching in Economics in Aguascalientes, Mexico). Misinformation spreaders share surprisingly similar traits worldwide, according to the researchers. Fortunately, the authors argue, simple techniques, such as prompting people to think about accuracy before sharing information online, can successfully make the dissemination of truth and the suppression of falsehood more likely.[9] According to Pennycook and Rand's paper, coauthored with Jabin Binnendyk:

Conspiracy believers are characterized by weaker analytic reasoning skills and, just as much, by a stronger disposition toward overconfidence. This overconfidence is also linked with a propensity to be largely unaware that most others disagree about believed conspiracy theories.[10]

Looking at a variety of possibilities and variables, including narcissism and the desire to be unique, the importance of overconfidence as a predictor of "conspiratoriality" stood out consistently. Which fits well with the argument I've sought to support throughout this book: convinced belief in tech's charismatic theologies and doctrines, and the failure to critique its influential and often manipulative practices, can have dangerous consequences. Or, as Rand and Pennycook argue, when it comes to misinformation, "those who are the most in need of intervention are likely the least willing to recognize that they may be wrong."[11] Still, given that such an insight fell squarely within the zone of being what I wanted to hear, I felt an obligation to push further, asking the researchers what to do about a group of people like my fellow humanists, skeptics, and atheists, who might become so confident in the virtue of their own critical thinking that they could fall into the sort of overconfidence Rand and Pennycook warn against. "We have nothing to say about that," replied Pennycook.

"But we can extrapolate," said Rand, wearing an olive military-style jacket and a Rage Against the Machine T-shirt while on Zoom from his MIT office. Rand's basic idea is that if you're trying—and remembering—to pursue the truth, you're probably likely to find it. Just don't make your own intuitions into a doctrine or dogma of their own—and don't want other people to be wrong more than you want to be right.

A FUTURE WORTH CONCEIVING

Life, in the middle of this twenty-first century CE, involves so much uncertainty. We are uncertain about the future of our climate; we are uncertain how to treat and connect with one another given unprecedented levels of border crossing and geographic mobility, not to mention racial, ethnic, and religious intermingling. We are uncertain whether centuries-long experiments in democracy can succeed, or even continue, as all our uncertainty seems to continually strengthen the hands of authoritarian leaders who prey on our weakness and doubt, asserting that they alone know the way.

We are uncertain about what to believe, as the old gods seem to be dead or dying. We are also uncertain about our technological future, and, in a tautological feedback loop, our technological future makes us ever more and more uncertain.

In 2016, the year my son was born, I read a short piece on fatherhood by E. B. White, who wrote *Charlotte's Web*. One of the most beloved contributors for the *New Yorker* magazine, White was born in 1899 in Mt. Vernon, New York. In 1941 he published an essay, "Once More to the Lake," about returning, with his son, to the camphouse in Maine to which he had traveled every summer as a small boy with his father. In the piece, White is overcome by an overwhelming feeling that everything is so similar: the lake, the cabin, the fishing rod, the fish, the dragonfly that lands on the tip of his rod.

"I began to sustain the illusion," he writes, "that he was I, and therefore, by simple transposition, that I was my father." This dizzying feeling that he and his son were one and the same began as a metaphor, but became so real to him over the course of his week at the lake with his son that as they were walking together he could no longer discern whose legs were in whose pants.

At the time, White's image struck me as surprising, and unusual. Now, several years later, I've instead come to see the relatability of the story as its real beauty. When I spend time with my mother and my son, it often feels eerily like he is me, which I suppose would make me my own father.

Also in 2016 I read a very different kind story about parenting: an essay in the *Nation* by climate journalist Madeline Ostrander. Ostrander was, then, traveling across the United States and beyond, reporting on the devastation and potential devastation associated with the technologies that have both caused climate change and made it impossible for us to rally effectively against it. During that work, she became preoccupied with the growing movement of women, called the Conceivable Future Project, so concerned for humanity's future that they were not sure they could ethically bring a new life into this world.

"For many people, these problems were an abstraction," Ostrander writes,

> but as an environmental journalist, I knew enough to imagine them in front of me. Driving across the bridge to my house, I pictured city beaches drowned by the rising sea. Watching the news, I wondered when the next colossal hurricane

would strike the Gulf of Mexico or the mid-Atlantic. These thoughts are not paranoid. According to scientists' predictions, if society keeps pumping out carbon dioxide at current rates, any child born now could, by midlife, watch Superstorm Sandy–size disasters regularly inundate New York City. She could see the wheat fields of the Great Plains turn to dust and parts of California gripped by decades of drought. She may see world food prices soar and water in the American West become even scarcer. By 2050, when still in her 30s, she could witness global wars waged over food and land.[12]

I was immediately destroyed by the contrast between these two stories: on the one hand, a father is so certain of the similarity between the present and the past that he can't even discern which is which; on the other, a mother is so distraught by the difference between the present and the future, that she can't even discern whether the future is worth living.

The two stories added heaviness to my decision to bring life into the world and emphasized just how much and how quickly our lives and our society are changing. Because of technology, it was becoming increasingly difficult to think or talk about what a fulfilling life would look like for a child like mine. How, as a parent, do you talk about meaning and purpose with a child whose day-to-day experience is completely different from, often barely even reminiscent of, what your own was? And whose future is both utterly uncertain and significantly in doubt? What morals, what ethics, what practices could even seem relevant and nourishing in that kind of environment, which almost every parent in the world faces today?

Obviously—and painfully—I don't have all the answers. Part of me wishes I could offer easy solutions, ready-made structures to follow. But this is not that kind of book about religion, for better and for worse. In one of the more disastrous episodes of my career, I once spent more than a year courting a major potential funder who suggested to me that money would flow to my nonprofit like milk and honey if only I could successfully create "humanism in a box" for my members. But humanism is not something easily contained in a box. Humanism is, as much as anything, a passionate belief that the box does not exist. It is perhaps in the same spirit that I once disappointed that donor that I will disappoint readers who came here seeking definitive instructions, or shrink-wrapped tool kits, for how to find true faith in the tech religion. Which devices to buy? Which social media apps to use? How much screen time per day? Which policies to advocate? Which organizations to join?

I don't know. But I can say this: over the course of writing this book, I've learned to take a spirited delight in not knowing and in not accepting the answers that tech companies bigger than Constantine's empire hope to sell me.

I want to end with a kind of tech agnostic manifesto—a brief statement of principles and values I've formulated over several years of working on this book. I want to call on readers to rally around it, because in a cultural moment of extreme divisiveness and doubt as to whether forward-looking principled action is worth the hassle, I believe we have more in common than we realize.

Perhaps ironically, the American humanist movement formally began with a manifesto. The first humanist manifesto, published in 1933, was a succinct and idealistic statement of the philosophical values of an emerging cohort of secular leaders. It was ultimately a bit overoptimistic in its faith in human nature, given the events of World War II that were about to come. A second humanist manifesto, published in 1973, ran a bit long; a 2003 successor tried to more briefly bring humanist thought into the twenty-first century. The third document, however, de-emphasized its own status as a manifesto and generally lacked some of the fire and passion of its predecessors.

Our current moment seems ripe for an agnostic manifesto. We have so much to be ambivalent about and yet so much to fight for. Ambivalence is a sign of our capacity for thoughtfulness, our appreciation for nuance, and our appropriate and reasonable hatred of overconfidence. And yet, so much is at stake.

Here are some truths that are self-evident, even in an ambivalent world:

- It's no accident that tech's ideas can look so much like religious theology and doctrine. By taking that form, they support a worldview in which certain powerful agendas, and people, are centered, while the rest of us revolve around their sun.
- Tech hierarchies need to be flattened, which requires recognizing them as the product of a religious ideology and critiquing them accordingly.
- Tech's rituals can amount to a dangerous obsession that is increasingly out of control. But because we can't avoid them entirely, we can also treat our personal tech use as a chance to switch our focus from what we are afraid of to what our positive and compassionate ideals and values require of us.

- The tech apocalypse is already here; it's just unevenly distributed. We need to start acting like the worst could happen to us, unless we work for equity and think about how all might benefit from doing so.
- Honor the apostates, heretics, and Cassandras among us. We need them more than ever.
- The tech culture of the future should be shaped by tech humanists, spiritual practitioners, and others brave enough to build an alternative. Be a human being first in a tech-first world.
- Finally, recognize you can't do any of this alone. You'll need community to find the humanity in tech. Choose your community in the same agnostic spirit with which you approach the tech religion as a whole.

AGNOSTICISM, KNIFEPOINT UP

On a warm Friday afternoon in July of 2021, I found myself searching for a way to conclude a four-hour visit with Lesley Hazleton at her houseboat in Seattle. Hazleton had begun to fill her dishwasher with the dishes from a delicious meal she had cooked for me, with olive oil, toasted almonds, anchovies, chopped garlic, and a garnish of sage from her floating garden, when she paused to look at one of her knives—a utensil she'd used perhaps hundreds or even thousands of times. She never understood, she told me, why people put their knives point-down into the washer.

We had just talked about her long career—not just the part where she'd been a tech journalist who profiled Jeff Bezos with skepticism in his earliest days, but also her writing about cars, including the time she taught herself to be a race car driver and drove two hundred miles an hour, nearly dying at one point.[13] Then there was a discussion of her career as a pilot, and what it looked like to do aerobatics in the air, to fly by clouds and have rainbows form off of your wing. Then there were all the books she has written and the talks she has been invited to give around the world. This included being recognized on the streets of Muslim-majority countries for the TED talk (2.4 million views)[14] in which she had accidentally humanized Prophet Muhammad, telling the story of how she had gone from a post-9/11 desire to investigate myths about the Qur'an's protagonist to simply seeing a human man with human emotions, longings, ethics, and even doubt—*especially* doubt. You might think such an interpretation would have provoked anger among the faithful, but it turned out to be deeply appreciated.

It occurred to me that the upside-down knife was a perfect metaphor for what I would take from my time with Lesley Hazleton.

"Lesley," I said to her, "That's you. You live your life with the knifepoint up." She remains, for me, an example, and maybe even a real-life parable, of how to live as an incisive force for good, and a source of loving and compassionate strength, while grounded in an agnostic value system many would mistakenly consider passive and indecisive.

In our own ways, with countless examples like Lesley and others in this book, you and I can do the same. Together, we can create a reformation—in our technology, but more importantly, in our common humanity—that might just flower into a renaissance. May it be so.

ACKNOWLEDGMENTS

Tech Agnostic is dedicated to both of my parents, Judy Capel and the late Cyrus Epstein. Mom, what you've done for me can't be expressed adequately, except to say that the very best of the love for humanity and learning and humor and passion for life and justice that shows up here is thanks to you. I'm so proud to be your son. The Babylonian Talmud (Sanhedrin 37a) famously says that whoever saves a single life saves the world entire. In your case I think you've always approached motherhood with the mentality that whoever educates and loves and befriends one life does so for the whole world, as well. I'm not sure anything better could be said about, or for, a person. And I also want you to know that over the course of writing this book, I also came to a much better place in my relationship with your late husband, my dad, who of course died in 1995. Though I don't think either you or I will ever believe in eternal life or resurrection or anything beyond this one and only physical world, in my heart and imagination I've learned how to spend time with a dad who was always a good man, and who cared very much about me and you and us and the world, and who has now become one of my closest conversation partners.

And to my wife, Jackie Piltch, and our kids, Axel and Ani, it is loving you and being loved by you that made me want to write this book, because I want everyone in the world to be able to have what we have: trust, care, warmth, security, emotional awareness, compassion, playfulness, and the opportunity to envision and build a future full of meaningful life experiences. Before we were us, I wasn't sure what I stood for, because I wasn't sure what was worth treasuring enough to fight for. Thanks to you, now I know.

This book was written over several years, and it seems to me that almost everyone who was part of my life during that time contributed to it somehow. A special thanks to all of the approximately two hundred individuals from all walks of life and tech whom I interviewed in order to be able to tell these stories. I hope you enjoy seeing yourself in the book, and that my efforts to do you justice have borne fruit. Some of you appear in these pages more than others, but each of you contributed enormously to my understanding and undying passion for this topic.

To my editor Gita Manaktala, thank you for believing in this book and in me. You're the best editor my work has ever had, and thanks to you I have a much less ambivalent and entirely healthier relationship with writing than I did when we met in 2021. And to my friend Amy Brand, Director at the MIT Press, you recruited me into this experience, and I can't thank you enough. You're a shining example that excellent leadership is not about bluster or undisciplined charisma but rather seriousness of purpose, thoughtful determination, and service to others. What a privilege it has been to work with you and your extraordinary team, including rock stars Nicholas DiSabatino, Suraiya Jetha, Deborah Cantor-Adams, Laura Keeler, Debbie Kuan, Jessica Pellien, and many others. Coming over from more purely commercial press experiences, I wasn't sure what to expect, and you all have exceeded my expectations in every way, motivating and inspiring me to try to exceed yours. For any writers who might be considering an opportunity to work with this group, I offer my most enthusiastic recommendation. To any philanthropists or funders who might be looking for an unalloyed good cause to support in a time of moral compromise and civic uncertainty, look no further. And to all of our partners at Penguin Random House who might work on any aspect of this book, I'm so thrilled you're involved and that we all get to work with you.

To Cassidy Sulaiman, every dollar I paid you as a freelance copyeditor was repaid threefold (or more) with your truly impressive standard of care and diligence, and your genuine and contagious love for the details that make books like this possible. I'd be a mess without you. To Chana Deitsch and Anne Stopper, it was a privilege to work with you and I can't wait to see what great projects you go on to help with and/or lead next. Thank you to Prayag Narula for making your excellent app, Marvin, available to me as a kind of early pilot user. It was incredibly helpful to my process of processing and slicing up those two hundred interviews. And to Leigh Stein, thank you for being the best sounding board for ideas, other than my mom, of course. I hope I get so lucky as to be able to discuss all my future projects with you. Thank you to my good friend Mary Johnson for your deep ethical insight and creative gifts, and for talking through the earliest writing that helped lead to where I am now. And thank you so much to the various editors with whom I worked on the various pieces of journalism that became kernels of the ideas in this volume, including Danny Crichton, Gita Venkataraman, Brian Bergstein, Allison Arieff, Rachel Courtland, Mat Honan, Walter Thompson, David Scharfenberg, Mike Orcutt, Tim Maher, Kelly Horan, and everyone else who worked on or workshopped any of the pieces I wrote for *TechCrunch*, *MIT Technology Review*, the *Boston Globe*, or elsewhere, in leading up to this book. That includes journalists and writers such as Gideon Lichfield, Gary Wolf, Sasha Sagan, Karen Hao, Kate Clark, Jessica Powell, Chanda Prescod-Weinstein, and others who started out as interview subjects and ended up simply being teachers, if primarily by example. It's difficult to express how much I admire you, at least without coming across like a raving lunatic.

To my dear friend of nearly thirty years, Siwatu Moore, who sat with me on New Year's Eve in 2018 and helped me figure out what I wanted to learn and write about at a time when my faith in my inner compass as a writer and thinker felt broken, I love you, brother. You are an extraordinary thinker and human being, and you deserve so much success in these coming years, but regardless of that, all of us who know you are so lucky you are our friend. When you're happy it does our hearts good and when you're not, we want to know about it. Thanks for having a sitcom character like me in the Kareem Abdul-Jabbar Film Club, even though I'm entirely ambivalent about whether a Signal chat is better than just *calling* one another every couple weeks.

To the Humanist Chaplaincy at Harvard and MIT, including former presidents who served during the years while I was working on this book and in the few years leading up to it: Narath Carlile, A. J. Kumar, Erik Gregory, Stephen Matheson, Kirsten Waerstad, and Diana Limbach Lempel, what an extraordinary group of people you are. Your dedication and service has made my work possible, and I am beyond grateful. To current board members Jeff Miller, Quinnie Lin, Ken Granderson, Elena Glassman, Jennifer Ibrahim, David Buckley, and Darren Sears, and our longtime administrator and community leader Rick Heller, ditto. And to Joe Gerstein, how is it even possible to thank someone whose several decades of work have been indispensable to my own? All I can say is that our Ketubah bears your signature, precisely so that on any given day I might reflect on your example, as I try to live up to my own values and best intentions.

To Thea Keith-Lucas, the Chaplain to the Institute and Associate Dean of the Office of Religious, Spiritual, and Ethical Life at MIT, and to Christina English and all of our colleagues in chaplaincy at the Institute, it has been such a special opportunity to serve alongside you and to think with you about what it means to live an ethical life in a technological world. Your work (our work!) is so important, not in spite of our many theological disagreements but because of them. And to my friend, colleague, and former student, Nina Lytton, it has been one of the greatest honors of my career to watch you develop into a brilliant leader from whom I learn every time we work together. MIT is so lucky to have you. And thank you to all of my colleagues at Harvard, including all of my fellow Harvard Chaplains, and to the brilliant Harvard and MIT student leaders, including the Harvard College Community of Humanists, Atheists, and Agnostics (HCHAA), the Secular Society of MIT (SSOMIT), and the Harvard and MIT Humanist Student Fellows I am fortunate to work with. Thanks also to the student leaders of the *Harvard Technology Review*, the first student group to take an interest in the work that grew into this book.

A deep thanks to all of my humanist and secular movement colleagues, around the world. It couldn't do any of the work that I do, and wouldn't want to, without the privilege of being a part of our centuries-old and deeply important collective effort. Thanks especially to my close colleagues like Sarah Chandonnet, James Croft, Sonia David, Kianna Mahony, Devin Moss, Anthony Cruz Pantojas, Vanessa Gomez Brake, Andrew Copson, John Hooper, Nicole Carr, Sarah Levin, Mary Ellen Giess, Ryan Bell, Bart Campolo, Hemant Mehta, Jason Callahan, and so many more. And thank you a thousand times to the movement(s) of Secular and Humanistic Judaism, and in particular to all humanist rabbis including my former classmates and teachers Sivan Malkin Maas, Miriam Jerris, Adam Chalom, Oren Yehi-Shalom, and Binyamin Biber for being the inspiring force that brought me into this movement: there is absolutely *zero* chance I would be here doing any of this were it not for you.

Thank you to my extended family, including my brother Jon Epstein, Emily Klein, and S. for all your love. Jon, it always inspires me to get to make art alongside you, and it always has. I love you very much. Beyond that I can't mention everyone to whom I am related, but I want to name those whom I consider to have been directly or indirectly helpful as I worked on this project. To Ruth Behar for continuing to lend me your office metaphorically as you used to do literally, and to David Frye for keeping me (semi-)sane on Twitter. To Carol Fuller Richardson for helping me to nurture a fuller and healthier understanding of my own roots on my dad's side, and for being one of the most affectionate, loving, and genuine people in my life, albeit from too much of a distance. To my cousin Celia Hirschman for our newfound friendship and for all you did to care for your mom, the indefatigable and inimitable Ruth Seymour, who always believed in me even when it honestly surprised me that she did. To Rob Van Grover, and your entire wonderful, zany family—Linda, Josh, Adam, Jessica, and your partners and now your kids (!)—you make it fun to simply be a human with a regular if zanily human family, and that has often been exactly what I've needed. And to my wife's family for creating her and therefore making my whole life possible, and for all you do to live out your highest values. Thank you all.

Thank you to my kids' teachers, and to all the educators in their schools and in our city, and to everyone who enables me and my neighbors to live in such a safe, supportive, civic-minded community, with such fertile conditions for intellectual growth and exploration. I miss my hometown of Flushing, Queens, and New York City every day, but if not there then how lucky am I to be here in Somerville, MA?

To my colleagues at the University of Massachusetts Boston and its School of Global Inclusion and Social Development: Racheal Inegbedion, Ayesha Khurshid, Meghan Kallman, Kaitlyn Siner, Matt Lacouture, James Hughes, Nir Eisikovits, and others, our conversations deeply enriched my perspective as I worked to finish and polish the book. I'm so looking forward to talking more and especially to seeing what world-bettering works of scholarship and humanity you will produce next.

And finally, to anyone reading this, thank you with all my heart. You are, more than anyone, why I put in all of this work. I hope my efforts will be useful and meaningful for you.

NOTES

INTRODUCTION

1. Shaye I. D. Cohen, "Legitimization under Constantine," "From Jesus to Christ," *Frontline*, PBS, April 1998, https://www.pbs.org/wgbh/pages/frontline/shows/religion/why/legitimization.html.

2. "Constantine the Emperor: Description," Oxford University Press, accessed December 21, 2023, https://global.oup.com/academic/product/constantine-the-emperor-9780190231620.

3. Eusebius, *The Life of Constantine*, trans. Ernest Cushing Richardson, in *A Select Library of Nicene and Post-Nicene Fathers*, ed. Philip Schaff and Henry Wace, 2nd ser. (New York: Christian Literature, 1890), 1:489, https://archive.org/details/cu31924031002102.

4. Eusebius, *Life of Constantine*, 1:489–490.

5. Eusebius, *Life of Constantine*, 1:489.

6. Eusebius, *Life of Constantine*, 1:489.

7. Bart Ehrman, "The Conversion of Constantine," *Bart Ehrman Blog*, February 12, 2018, https://ehrmanblog.org/the-conversion-of-constantine.

8. Ime Archibong, "Why We Build: People Who Bring the World Closer Together," *Newsroom* (blog), Meta, February 7, 2019, https://about.fb.com/news/2019/02/2019fcs.

9. Marc Andreessen, "The Techno-Optimist Manifesto," Andreessen Horowitz, October 16, 2023, https://a16z.com/the-techno-optimist-manifesto.

10. Andreessen, "The Techno-Optimist Manifesto."

11. Jack Dorsey (@jack), "Elon is the singular solution I trust. I trust his mission to extend the light of consciousness," Twitter, April 25, 2022, 10:03 p.m., https://twitter.com/jack/status/1518772756069773313.

12. Elon Musk (@elonmusk), Twitter, March 21, 2021, 1:31 a.m., https://twitter.com/elonmusk/status/1373507545315172357.

13. "What Are the Most Pressing World Problems?," 80,000 Hours, updated May 2023, https://80000hours.org/problem-profiles.

14. Alistair Barr, "Google's 'Don't Be Evil' Becomes Alphabet's 'Do the Right Thing,'" *Wall Street Journal*, October 2, 2015, https://www.wsj.com/articles/BL -DGB-43666; Choong Lee and Stephen Brozovich, "People: The Human Side of Innovation at Amazon," AWS Executive Insights, accessed January 16, 2024, https:// aws.amazon.com/executive-insights/content/the-human-side-of-innovation.

15. Mark Zuckerberg, "Facebook's Letter from Mark Zuckerberg—Full Text," *Guardian* (US edition), February 1, 2012, https://www.theguardian.com/technol ogy/2012/feb/01/facebook-letter-mark-zuckerberg-text.

16. Nico Grant, "Google Fires Engineer Who Claims Its A.I. Is Conscious," *New York Times*, July 23, 2022, https://www.nytimes.com/2022/07/23/technology /google-engineer-artificial-intelligence.html.

17. Akka, "God Is an AGI We Make in the Future," LessWrong, May 24, 2019, https://www.lesswrong.com/posts/fXa8R6B9TvpDbPgbC/god-is-an-agi-we -make-in-the-future; Madhumita Murgia, "OpenAI Chief Seeks New Microsoft Funds to Build 'Superintelligence,'" *Financial Times*, November 13, 2023, https:// www.ft.com/content/dd9ba2f6-f509-42f0-8e97-4271c7b84ded.

18. Sam Altman (@sama), "abundance is our birthright," Twitter, April 3, 2022, 6:26 p.m., https://twitter.com/sama/status/1510745583350255617.

19. Sergio DellaPergola, "World Jewish Population, 2020," in *American Jewish Year Book 2020: The Annual Record of the North American Jewish Communities Since 1989*, ed. Arnold Dashefsky and Ira M. Sheskin, American Jewish Year Book, vol. 120 (Cham, Switzerland: Springer, 2020), 273–370.

20. Mar Hicks, email message to author, June 14, 2023.

21. Jason Furman, email message to author, September 7, 2023.

22. Furman, email message to author.

23. I am avoiding terms like *Judeo-Christian* here because they can misleadingly imply that Judaism or Jews were on equal footing with Christians in shaping Western worldviews in general or terms like *religion* in particular. Those with the power to do such things were, for the most part, Christians.

24. Robert Sharf, in discussion with the author via Zoom, January 22, 2022.

25. Haught once testified, in a US federal court case, that public-school science teachers should not be allowed to teach creationism or "intelligent design" as science, but neither should they be allowed to teach any kind of atheistic materialism (the idea that evolution necessarily requires or is tantamount to atheism or agnosticism) as science or in a science classroom.

26. Daniel Dennett, John F. Haught, and David Sloan Wilson, "What Is Religion?," City University of New York Forum, YouTube video, 1:15:52, December 14, 2009, https://www.youtube.com/watch?v=QM91iZweUnk, 10:02–10:27.

27. Dennett, Haught, and Wilson, "What Is Religion."

28. There have been numerous academic attempts to study prayer and faith. Most have failed to document any benefits from intercessory prayer, or the idea of praying for another person or persons in order to influence their health or other kind of well-being. In fact, knowing that one was the recipient of such prayers actually made patients somewhat more likely to experience medical complications after a surgery, according to one landmark study led by Harvard Medical School's Dr. Herbert Benson and funded by the Templeton Foundation. See, for example, Benedict Carey, "Long-Awaited Medical Study Questions the Power of Prayer," *New York Times*, March 31, 2006, https://www.nytimes.com/2006/03/31/health /longawaited-medical-study-questions-the-power-of-prayer.html; Herbert Benson et al., "Study of the Therapeutic Effects of Intercessory Prayer (STEP) in Cardiac Bypass Patients: A Multicenter Randomized Trial of Uncertainty and Certainty of Receiving Intercessory Prayer," *American Heart Journal* 151, no. 4 (April 2006): 934–942. Other studies, however, have noted various apparent psychological benefits for individuals engaged in prayer themselves. See, for example, Kristen Rogers, "The Psychological Benefits of Prayer: What Science Says about the Mind-Soul Connection," Mindfulness, *CNN*, June 17, 2020, https://www.cnn .com/2020/06/17/health/benefits-of-prayer-wellness/index.html.

29. "The Greatest Tech Books of All Time," *Verge*, June 28, 2023, https://www .theverge.com/c/23771068/best-tech-books-nonfiction-recommendations.

30. Joseph Biden and Gabriel Boric, "Remarks by President Biden and President Gabriel Boric of Chile Before Bilateral Meeting," Speeches and Remarks, White House Briefing Room, November 2, 2023, https://www.whitehouse.gov/briefing -room/speeches-remarks/2023/11/02/remarks-by-president-biden-and-president -gabriel-boric-of-chile-before-bilateral-meeting.

31. David F. Noble, *The Religion of Technology: The Divinity of Man and the Spirit of Invention* (New York: Alfred A. Knopf, 1997), chap. 1, Kindle.

32. Stefan Helmreich, "The Spiritual in Artificial Life: Recombining Science and Religion in a Computational Culture Medium," *Science as Culture* 6, no. 3 (1997): 363–395.

33. Helmreich, "Spiritual."

34. Ted Honderich, ed., *The Oxford Companion to Philosophy* (Oxford: Oxford University Press, 1995); Richard Parry, "*Episteme* and *Techne*," in *The Stanford Encyclopedia of Philosophy*, ed. Edward N. Zalta, updated March 27, 2020, https://plato .stanford.edu/entries/episteme-techne.

35. John G. Chapman, ed., *The Blue Print*, vol. 1 (Atlanta: Students of the Georgia School of Technology, 1908), https://repository.gatech.edu/entities /archivalmaterial/0daa1dac-0a3f-4903-a6d7-5aa634a4320d.

36. Jesse Sheidlower, email message to author, May 11, 2022; "Tech," Google Books Ngram Viewer, accessed September 18, 2023, https://books.google.com

/ngrams/graph?content=tech&year_start=1800&year_end=2019&corpus=26&
smoothing=3.

37. Josh Tyrangiel, "A New AI Predicts When We'll Die. It Says Even More about
How We'll Live," *Washington Post*, January 15, 2024, https://www.washingtonpost
.com/opinions/2024/01/15/artificial-intelligence-death-calculator.

38. Harvey Cox and Anne Foerst, "Religion and Technology: A New Phase,"
Bulletin of Science, Technology, and Society 17, no. 2–3 (1997): 53–60.

39. See, among many other examples, André Spicer, "Beware the 'Botshit': Why
Generative AI Is Such a Real and Imminent Threat to the Way We Live," *Guardian* (US edition), January 3, 2024, https://www.theguardian.com/commentisfree
/2024/jan/03/botshit-generative-ai-imminent-threat-democracy.

40. See, for example, Karen Rosenberg and Wenda Trevathan, "Birth, Obstetrics
and Human Evolution," *BJOG* 109, no. 11 (November 2002): 1199–1206.

41. I wrote extensively about my experiences at this conference: Greg Epstein,
"Will the Future of Work Be Ethical?," *TechCrunch*, November 28, 2019, https://
techcrunch.com/2019/11/28/will-the-future-of-work-be-ethical.

42. One of the people who inspired my analysis of tech beliefs, as I share it in
section one, is Adrian Daub, Stanford professor and author of the excellent book
What Tech Calls Thinking: An Inquiry Into the Intellectual Bedrock of Silicon Valley (New
York: Farrar, Straus and Giroux, 2020).

43. In the Authorized King James Version: "Beware of false prophets who come
to you in sheep's clothing but inwardly they are ravenous wolves. . . . A good tree
cannot bear bad fruit . . . therefore by their fruits you will know them."

44. David Halberstam, "The Vantage Point Perspectives of the Presidency 1963–
1969," *New York Times*, October 31, 1971, https://www.nytimes.com/1971/10
/31/archives/the-vantage-point-perspectives-of-the-presidency-19631969-by
-lyndon.html.

CHAPTER 1

1. Dan Price (@DanPriceSeattle), Twitter, July 21, 2021, 8:20 p.m., https://
twitter.com/DanPriceSeattle/status/1414379001334976517.

2. Ken Miguel and Phil Matler, "Here's What It Takes to Survive SF's Tenderloin District amid Crime, Drug Dealing, Addiction," *ABC News*, July 14, 2022,
https://abc7news.com/sf-tenderloin-san-francisco-crime-drug-use-50-blocks
-stories-from-the/12044419.

3. Twitter direct message to author, January 21, 2020.

4. Erica Robles-Anderson (@fstflofscholars), Twitter, June 28, 2022, 5:48 p.m.,
https://twitter.com/fstflofscholars/status/1541901350530961408.

5. Erica Robles-Anderson (@fstflofscholars), Twitter, June 28, 2022, 9:37 p.m., https://twitter.com/fstflofscholars/status/1541898710010978311.

6. Anthony B. Pinn, *The End of God-Talk* (Oxford: Oxford University Press, 2012).

7. Alexandra da Costa, review of *Permanent Revolution: The Reformation and the Illiberal Roots of Liberalism*, by James Simpson, *Cambridge Quarterly* 49, no. 2 (June 2020): 156–190, https://academic.oup.com/camqtly/article-abstract/49/2/156/5857726.

8. Josh Burek, "Christian Faith: Calvinism is Back," *Christian Science Monitor*, March 27, 2010, https://www.csmonitor.com/USA/Society/2010/0327/Christian -faith-Calvinism-is-back.

9. Indigenous people of North America were also enslaved and sold, though more often they were sent elsewhere for this purpose. As historian Wendy Warren points out, it was more dangerous for the enslavers to enslave people near their homelands, where they could have more ready access to support in escaping. See Wendy Warren, *New England Bound: Slavery and Colonization in Early America* (New York: Liveright, 2016).

10. "The Massachusetts Body of Liberties," 1641, Hanover Historical Texts Project, https://history.hanover.edu/texts/masslib.html.

11. *The 1619 Project*, *New York Times Magazine*, August 2019, https://www .nytimes.com/interactive/2019/08/14/magazine/1619-america-slavery.html.

12. Javon, personal correspondence with author, December 10, 2010.

13. Li Ka Shing Foundation, "Li Ka Shing Center for Learning and Knowledge Transforms Medical Training at Stanford University," press release, September 29, 2010, https://www.lksf.org/li-ka-shing-center-for-learning-and-knowledge -transforms-medical-training-at-stanford-university.

14. Besheer Mohamed, Kiana Cox, Jeff Diamant, and Claire Gecewicz, "Faith among Black Americans," Pew Research Center, February 16, 2021.

15. David Hoelscher, "Atheism and the Class Problem," *CounterPunch*, November 7, 2012, https://www.counterpunch.org/2012/11/07/atheism-and-the-class -problem.

16. Note this claim refers to property, not wealth, and has been called into question. See for example Philip Cohen, "'Women Own 1% of World Property': A Feminist Myth That Won't Die," *Atlantic*, March 8, 2013.

17. Hoelscher, "Atheism and the Class Problem."

18. "Silicon Valley's Inequality Machine: A Conversation with Anand Giridharadas," interview by Greg Epstein, *TechCrunch*, March 2, 2019, https://techcrunch .com/2019/03/02/silicon-valleys-inequality-machine-anand-giridharadas.

19. Fred Turner, "Millenarian Tinkering: The Puritan Roots of the Maker Movement," supplement, *Technology and Culture* 59, no. S4 (October 2018): S160–S182.

https://fredturner2022.sites.stanford.edu/sites/g/files/sbiybj27111/files/media
/file/turner-millenarian-tinkering-tech-culture-2018.pdf.

20. Tim Bajarin, "Why the Maker Movement Is Important to America's Future,"
TIME, May 19, 2014, https://time.com/104210/maker-faire-maker-movement.

21. Turner, "Millenarian Tinkering," S162.

22. Turner, S166.

23. Turner, S178.

24. Turner, S163.

25. Morgan Ames, *The Charisma Machine* (Cambridge, MA: MIT Press, 2019), 194.

26. Ames, *Charisma Machine*, 194.

27. Andrew Carnegie, *The Gospel of Wealth* (Carlisle, MA: Applewood Books,
1998), 22–23.

28. Nicholas Kristof, "Pull Yourself Up by Bootstraps? Go Ahead, Try It," *New
York Times*, February 19, 2020, https://www.nytimes.com/2020/02/19/opinion
/economic-mobility.html.

29. Anthony Levandowski, "Inside the First Church of Artificial Intelligence,"
interview by Mark Harris, *Wired*, November 15, 2017, https://www.wired.com
/story/anthony-levandowski-artificial-intelligence-religion.

30. Levandowski, "Inside the First Church."

31. Mark Harris, "Inside the Uber and Google Settlement with Anthony Levandowski," *TechCrunch*, February 15, 2022, https://techcrunch.com/2022/02/15
/inside-the-uber-and-google-settlement-with-anthony-levandowski.

32. Alyson Shontell (@ajs), "My job is really hard—I'm being forced to go to gorgeous Aspen next week for the prestigious @brainstormtech conference where I
will interview some amazing founders, get a full life update from @antlevand-
owski and lead a horseback riding session. Hope to see you there!," Twitter, July 8,
2022, 11:27 a.m., https://twitter.com/ajs/status/1545429248969670657.

33. Jackie Davalos and Nate Lanxon, "Anthony Levandowski Reboots Church of
Artificial Intelligence," *Bloomberg*, November 23, 2023, https://www.bloomberg
.com/news/articles/2023-11-23/anthony-levandowski-reboots-the-church-of
-artificial-intelligence.

34. LessWrong Wiki, s.v. "FAQ," accessed April 30, 2019, https://web.archive.org
/web/20190430134954/https://wiki.lesswrong.com/wiki/FAQ#What_is_Less
_Wrong.3F.

35. Eliezer Yudkowski, LessWrong, July 24, 2010, 11:10 p.m., quoted in Beth
Singler, "Don't Read This Post," Professor Beth Singler, March 23, 2016, https://
bvlsingler.com/2016/03/23/dont-read-this-post.

36. Robert A. Orsi, *Between Heaven and Earth: The Religious Worlds People Make and the Scholars Who Study Them* (Princeton, NJ: Princeton University Press, 2006), 2.

37. Adam Singer (@AdamSinger), Twitter, July 10, 2022, 3:17 p.m., https://twitter.com/AdamSinger/status/1546212133800656899.

38. Caroline McCarthy, "Silicon Valley Has a Problem with Conservatives. But Not the Political Kind," *Vox*, Jun 12, 2018, https://www.vox.com/first-person/2018/6/12/17443134/silicon-valley-conservatives-religion-atheism-james-damore.

CHAPTER 2

1. See Allie Wilkinson, "European Wars, Famine, and Plagues Driven by Changing Climates," *Ars Technica*, October 12, 2011, https://arstechnica.com/science/2011/10/european-wars-famine-and-plagues-driven-by-climate.

2. Jane Hathaway, "The Grand Vizier and the False Messiah: The Shabbtai Sevi Controversy and the Ottoman Reform in Egypt," *Journal of the American Oriental Society* 117, no. 4 (1997): 665–671.

3. Kaufmann Kohler and Henry Malter, "Shabbethai Zebi B. Mordecai," in *Jewish Encyclopedia*, ed. Isidore Singer (New York: Funk and Wagnalls, 1906), 218–225, https://www.jewishencyclopedia.com/articles/13480-shabbethai-zebi-b-mordecai.

4. Gershom Gerhard Scholem, *Sabbetai Zevi: The Mystical Messiah*, trans. R. J. Zwi Werblowsky (Princeton, NJ: Princeton University Press, 2016), 459–460.

5. Yaacob Dweck, *Dissident Rabbi: The Life of Jacob Sasportas* (Princeton, NJ: Princeton University Press, 2019), 28.

6. Ray Kurzweil, *The Singularity Is Near: When Humans Transcend Biology* (New York: Penguin, 2005), 370.

7. Kurzweil, *Singularity*, 370.

8. Kurzweil, *Singularity*, 371.

9. Kurzweil, *Singularity*, 372.

10. Elon Musk et al., "Superintelligence: Science or Fiction?," filmed at the Asilomar Conference on Beneficial AI, January 7, 2017, in Pacific Grove, CA, YouTube video, 1:00:14, January 30, 2017, https://youtu.be/h0962biiZa4&t=2051, 34:11–34:20 (emphasis mine).

11. Elon Musk (@elonmusk), Twitter, March 21, 2021, 1:31 a.m., https://twitter.com/elonmusk/status/1373507545315172357.

12. Jack Dorsey (@jack), "In principle, I don't believe anyone should own or run Twitter. It wants to be a public good at a protocol level, not a company. Solving for the problem of it . . ." Twitter, April 25, 2022, 10:03 p.m., https://twitter.com/jack/status/1518772756069773313.

13. W. Richard Comstock, "Doctrine," in *Encyclopedia of Religion*, 2nd ed. (Detroit, MI: Macmillan Reference USA, 2005), 4:2381–2385, quoted in Encyclopedia .com, updated August 22, 2023, https://www.encyclopedia.com/philosophy-and -religion/bible/bible-general/doctrine.

14. Comstock, "Doctrine."

15. See "White Christian Nationalism: Rewriting History and the Constitution," Religious and Racial Equality, Americans United for Separation of Church and State, accessed September 19, 2023, https://www.au.org/how-we-protect -religious-freedom/issues/religious-racial-equality/white-christian-nationalism/; "Opinion: The GOP's Imposition of White Christian Nationalism," The Newsroom, Public Religion Research Institute, updated April 29, 2022, https://www .prri.org/buzz/opinion-the-gops-imposition-of-white-christian-nationalism.

16. Mark Silk, "White Evangelicals in Numerical, Political Decline," *Religion News Service*, July 14, 2021, https://religionnews.com/2021/07/14/white-evange licals-in-numerical-political-decline; Ruth Braunstein, "The Backlash against Rightwing Evangelicals Is Reshaping American Politics and Faith," *Guardian* (US edition), January 25, 2022, https://www.theguardian.com/commentisfree/2022 /jan/25/the-backlash-against-rightwing-evangelicals-is-reshaping-american -politics-and-faith.

17. Matthew 7:12 (King James Version).

18. *Babylonian Talmud*, *Shabbat* 31a, *Koren Talmud Bavli*, ed. Tzvi Hersh Weinreb et al., commentary by Adin Even-Israel Steinsaltz, Daf Yomi edition, vol. 2, *Shabbat Part One* (Jerusalem: Koren Publishers, 2012).

19. "Ethical AI Startup Landscape," Ethical AI Database, updated January 31, 2023, https://www.eaidb.org/static/maps/fy2022.svg.

20. Marian Croak and Jen Gennai, "Responsible AI: Looking Back at 2022, and to the Future," *Keyword* (blog), Google, January 11, 2023, https://blog.google /technology/ai/responsible-ai-looking-back-at-2022-and-to-the-future; Iain Levine (@iainlevine), "Today we're releasing @Meta's first annual human rights report, an important step forward on our human rights journey and a demonstration of our commitment . . ." Twitter, July 14, 2022, 9:11 a.m., https://twitter.com /iainlevine/status/1547569441935740928; Miranda Sissons and Iain Levine, "A Closer Look: Meta's First Annual Human Rights Report," *Newsroom* (blog), Meta, July 14, 2022, https://about.fb.com/news/2022/07/first-annual-human-rights -report.

21. Amazon, public display, n.d., 2111 7th Avenue, Seattle, WA 98121.

22. Salesforce, "Paula Goldman Joins Salesforce as VP, Chief Ethical and Humane Use Officer," press release, December 10, 2018, https://www.salesforce.com/news /stories/paula-goldman-joins-salesforce-as-vp-chief-ethical-and-humane-use -officer.

23. Yoel Roth, LinkedIn, accessed January 12, 2024, https://www.linkedin.com /in/yoelroth.

24. Yoel Roth (@yoyoel), "No solution to identity is perfect, and we're iterating quickly to come up with the best approach here. We appreciate all the feedback, and will share more as our work progresses." Twitter, November 7, 2022, 9:19 p.m., https://twitter.com/yoyoel/status/1589804653650509825.

25. Chas Danner, "Twitter's Former Safety Head Forced from Home after Being Smeared by Elon Musk," Intelligencer, *New York Magazine*, December 12, 2022, https://nymag.com/intelligencer/2022/12/elon-musk-smears-former-twitter -executive-yoel-roth.html.

26. Anaïs Rességuier and Rowena Rodrigues, "*AI Ethics Should Not Remain Tooth-less!* A Call to Bring Back the Teeth of Ethics," *Big Data and Society* 7, no. 2 (2020), https://journals.sagepub.com/doi/10.1177/2053951720942541.

27. *Oxford English Dictionary*, 2nd ed., 2023, s.v. "prophesy," https://www.oed .com/dictionary/prophesy_v?tab=meaning_and_use.

28. *Oxford English Dictionary*, 2nd ed., 2023, s.v. "prophecy," https://www.oed .com/dictionary/prophecy_n?tab=meaning_and_use.

29. See, for example, *The Cleaners*, directed by Hans Block and Moritz Riese-wieck, written by Hans Block, Moritz Riesewieck, and Georg Tschurtschenthaler (Berlin: Gebrueder Beetz Filmproduktion, 2018).

30. I wrote an extended story about the conference, and about Gupta, that was featured by *TechCrunch* in November 2019: Greg Epstein, "Will the Future of Work Be Ethical?," *TechCrunch*, November 8, 2019, https://techcrunch.com/2019 /11/28/will-the-future-of-work-be-ethical.

31. Lelund Cheung, "When a Neighborhood Is Crowned the Most Innovative Square Mile in the World, How Do You Keep It That Way?," *Boston.com*, May 2, 2013, https://www.boston.com/news/innovation/2013/05/02/when-a-neigh borhood-is-crowned-the-most-innovative-square-mile-in-the-world-how-do -you-keep-it-that-way.

32. "Peter Vorkink," Life, *Exonian*, December 14, 2017, https://theexonian.net /life/2017/12/14/peter-vorkink.

33. See Greg Epstein, "A Brief History of Goodness Without God, or a Short Campus Tour of the University of Humanism," in *Good Without God: What a Bil-lion Nonreligious People Do Believe* (New York: William Morrow, 2009), 38–60.

34. Elizabeth Kolbert, "Gospels of Giving for the New Gilded Age," *New Yorker*, August 20, 2018, https://www.newyorker.com/magazine/2018/08/27/gospels-of -giving-for-the-new-gilded-age.

35. "What Is Effective Altruism?," EffectiveAltruism.org, accessed March 13, 2023, https://www.effectivealtruism.org/articles/introduction-to-effective-altruism.

36. Neil Dullaghan, *EA Survey 2019: Community Demographics and Characteristics*, Rethink Priorities, December 25, 2019, https://rethinkpriorities.org/publications/eas2019-community-demographics-characteristics.

37. William MacAskill, "Banking: The Ethical Career Choice," in *Philosophers Take on the World*, ed. David Edmonds (Oxford: Oxford University Press, 2016), 84–86, https://www.academia.edu/32428208/Banking_The_Ethical_Career_Choice.

38. "Future Fund," Effective Altruism Forum, last modified June 21, 2023, https://forum.effectivealtruism.org/topics/future-fund.

39. Émile P. Torres, "The Grift Brothers," *Truthdig*, December 5, 2022, https://www.truthdig.com/articles/the-grift-brothers.

40. Naina Bajekal (@naina_bajekal), "The mode of 'everything sucks' is not helpful. Maybe it's true, but the relevant question is: what can we do? For our new cover story, I spent some time with . . ." Twitter, August 10, 2022, 7:25 a.m., https://twitter.com/naina_bajekal/status/1557327315180331008; Tim Ferriss (@tferriss), "That's exactly what a reluctant prophet would say! #triggeractionplan," Twitter, August 8, 2022, 6:29 p.m., https://twitter.com/tferriss/status/1556769487227162625; Elon Musk (@elonmusk), "Worth reading. This is a close match for my philosophy," Twitter, August 2, 2022, 1:15 a.m., https://twitter.com/elonmusk/status/1554335028313718784.

41. Nick Beckstead (Nick_Beckstead) et al., "The FTX Future Fund Team Has Resigned," Effective Altruism Forum, November 10, 2022, https://forum.effectivealtruism.org/posts/xafpj3on76uRDoBja/the-ftx-future-fund-team-has-resigned-1.

42. William MacAskill (@willmacaskill), "But if there was deception and misuse of funds, I am outraged, and I don't know which emotion is stronger: my utter rage at Sam (and others?) for causing such . . ." Twitter, November 11, 2022, 6:55 p.m., https://twitter.com/willmacaskill/status/1591218022362284034.

43. Tracy Alloway and Joe Wiesenthal, "Sam Bankman-Fried Described Yield Farming and Left Matt Levine Stunned," *Bloomberg*, April 25, 2022, https://www.bloomberg.com/news/articles/2022-04-25/sam-bankman-fried-described-yield-farming-and-left-matt-levine-stunned#xj4y7vzkg.

44. Sam Bankman-Fried (@SBF_FTX), "a) stimulants when you wake up, sleeping pills if you need them when you sleep. b) be mindful of where your headspace is: I often nap in the office so that . . ." Twitter, September 15, 2019, 9:42 p.m., https://web.archive.org/web/20230528162831/https://twitter.com/SBF_FTX/status/1173351344159117312.

45. William MacAskill, *Doing Good Better: How Effective Altruism Can Help You Make a Difference* (New York: Avery, 2015), 131.

46. Amos 5:24 (King James Version).

47. MacAskill, *Doing Good Better*, 146.

48. Paul Vitello, "For Catholics, a Door to Absolution Is Reopened," *New York Times*, February 9, 2009, https://www.nytimes.com/2009/02/10/nyregion/10 indulgence.html.

49. Nick Bostrom, "Are We Living in a Computer Simulation?," *Philosophical Quarterly* 53, no. 211 (April 2003): 243–255, https://doi.org/10.1111/1467-9213 .00309.

50. Bostrom, "Are We Living in a Computer Simulation?"

51. "Bio," Nick Bostrom (website), accessed February 9, 2023, https://nickbos trom.com/#bio.

52. In their TED conversation in 2018–19: Nick Bostrom, "How Civilization Could Destroy Itself—and 4 Ways We Could Prevent It," interview by Chris Andersen, filmed April 2019 in Vancouver, Canada, TED video, 21:00, https:// www.ted.com/talks/nick_bostrom_how_civilization_could_destroy_itself_and _4_ways_we_could_prevent_it.

53. "Rationality Community," Effective Altruism Forum, last modified March 31, 2023, https://forum.effectivealtruism.org/topics/rationality-community.

54. Timnit Gebru, "Effective Altruism Is Pushing a Dangerous Brand of 'AI Safety,'" *Wired*, November 30, 2022, https://www.wired.com/story/effective -altruism-artificial-intelligence-sam-bankman-fried.

55. Dustin Moskovitz (@moskov), "This is also the thing I'm most worried about for the 21st century. Potential risks from advanced AI usually evokes runaway paperclip bots, which people tend to . . ." Twitter, December 8, 2020, https://web .archive.org/web/20201208204552/https://twitter.com/moskov/status/1336411 545035665408 (post deleted).

56. Cynthia Chen, "The Vitalik Buterin Fellowship in AI Existential Safety Is Open for Applications!," Effective Altruism Forum, October 13, 2022, https:// forum.effectivealtruism.org/posts/wFC3axfuwABHmoQ9H/the-vitalik -buterin-fellowship-in-ai-existential-safety-is.

57. Rizwan Virk, "The Metaverse Is Coming: We May Already Be in It," *Scientific American*, February 22, 2022, https://www.scientificamerican.com/article /the-metaverse-is-coming-we-may-already-be-in-it.

58. David J. Chalmers, *Reality+: Virtual Worlds and the Problems of Philosophy* (New York: W. W. Norton, 2022), xvii.

59. Hilary Greaves and William MacAskill, "The Case for Strong Longtermism," GPI Working Paper no. 5–2021, University of Oxford, Oxford, 2021, https:// globalprioritiesinstitute.org/wp-content/uploads/The-Case-for-Strong-Longter mism-GPI-Working-Paper-June-2021-2-2.pdf.

60. Émile P. Torres, "How Elon Musk Sees the Future: His Bizarre Sci-Fi Vision Should Concern Us All," *Salon*, July 17, 2022, https://www.salon.com/2022 /07/17/how-elon-musk-sees-the-future-his-bizarre-sci-fi-vision-should-concern -us-all.

61. Torres, "How Elon Musk Sees."

62. TubeLooB, "Richard Dawkins—'What If You're Wrong?' South Park," February 28, 2010, YouTube video, 1:23, https://youtu.be/fPJQw-x-xho.

63. Torres, "How Elon Musk Sees."

64. Nick Bostrom, "Letter from Utopia," *Studies in Ethics, Law, and Technology* 2, no. 1 (2008): 1–7, https://nickbostrom.com/utopia.

65. Bostrom, "Letter from Utopia."

66. A helpful resource while assembling this list was John F. Burns, "The World: Martyrdom; the Promise of Paradise That Slays Peace," *New York Times*, April 1, 2001, https://www.nytimes.com/2001/04/01/weekinreview/the-world-martyr dom-the-promise-of-paradise-that-slays-peace.html.

67. Christopher Harding, "Into Nothingness," *Aeon*, November 10, 2014, https:// aeon.co/essays/the-zen-ideas-that-propelled-japan-s-young-kamikaze-pilots.

68. Timnit Gebru and Émile P. Torres, "The TESCREAL Bundle: Eugenics and the Promise of Utopia through Artificial General Intelligence," *First Monday* 29, no. 4 (2024), https://doi.org/10.5210/fm.v29i4.13636.

69. Nick Couldry and Ulises A. Mejias, *The Costs of Connection: How Data Is Colonizing Human Life and Appropriating It for Capitalism* (Stanford, CA: Stanford University Press, 2019), 117.

70. Couldry and Mejias, *Costs of Connection,* 117.

71. Couldry and Mejias, *Costs of Connection*, 118.

72. Michael Haupt, "'Data Is the New Oil'—A Ludicrous Proposition," Medium, May 2, 2016, https://medium.com/project-2030/data-is-the-new-oil-a-ludicrous -proposition-1d91bba4f294, quoted in Couldry and Mejias, *The Costs of Connection*, 142.

73. Lizette Chapman, "The Hottest New Thing in Seasteading Is Land," *Bloomberg*, December 20, 2019, https://www.bloomberg.com/news/articles/2019-12-20 /silicon-valley-seasteaders-go-looking-for-low-tax-sites-on-land.

74. Laurie Clarke, "Crypto Millionaires Are Pouring Money into Central America to Build Their Own Cities," *MIT Technology Review*, April 20, 2022, https://www .technologyreview.com/2022/04/20/1049384/crypto-cities-central-america.

75. Couldry and Mejias, *Costs of Connection*, 118.

76. Comment, with permission for attribution, from an online discussion group for the publication *Tech Policy Press*, May 2, 2022.

77. Comment, with permission for attribution, from an online discussion group for the publication *Tech Policy Press*, May 2, 2022.

78. Recommended to me, among several other translations, as "a beautiful and elaborate description of hell" by my colleague Sadananda Dasa, a Hindu chaplain at MIT: "A Description of the Hellish Planets," in *Śrīmad-Bhāgavatam*, canto 5, *The Creative Impetus* (Alachua, FL: Bhaktivedanta Book Trust, 1975), https://vedabase.io/en/library/sb/5/26.

79. "Elephants and the Idea of Extinction," Elephants and Us, National Museum of American History Behring Center, accessed September 19, 2023, https://americanhistory.si.edu/elephants-and-us/elephants-and-idea-extinction-0.

80. "5 Minutes with a Visionary: Eliezer Yudkowsky," interview by Gregory Saperstein, *CNBC*, August 9, 2012, https://www.cnbc.com/id/48538963; grotundeek_apocolyps, "Rationalists Take Another Small, Tentative Step towards Disavowing Their Prophet," Reddit, February 8, 2023, https://www.reddit.com/r/SneerClub/comments/10x17xb/rationalists_take_another_small_tentative_step.

81. Eliezer Yudkowksy (@ESYudkowsky), "Safely aligning a powerful AGI is difficult," Twitter, December 4, 2018, 6:18 p.m., https://twitter.com/ESYudkowsky/status/1070095112791715846.

82. The pithy phrasing is a quote from a colleague of Yudkowsky at MIRI, Rob Bensinger, whom he retweeted on February 18, 2023. Bensinger (@robbensinger), "I'd instead say: it's good that AGI tech not proliferate (because otherwise we all die), and getting in the habit of closed-sourcing now is plausibly crucial . . ." Twitter, February 18, 2023, 6:59 a.m., https://twitter.com/robbensinger/status/1626914404880224256.

83. On page 260, Bostrom writes about what a phrase he is quoting "seems to mean," as though he does not in this case know the exact intention behind Yudkowsky's phrase, even though he might presumably have been able to ask about it. Nick Bostrom, *Superintelligence: Paths, Dangers, Strategies* (Oxford: Oxford University Press, 2014), 260, 262, 264.

84. James Barrat, in discussion via Zoom with the author, January 18, 2023. Barrat's comments have been edited for length.

85. "What Are the Most Pressing World Problems?," 80,000 Hours, accessed February 19, 2023, https://80000hours.org/problem-profiles.

86. Benjamin Hilton, "Preventing an AI-Related Catastrophe," 80,000 Hours, accessed February 19, 2023, https://80000hours.org/problem-profiles/artificial-intelligence.

87. Toby Ord, *The Precipice: Existential Risk and the Future of Humanity* (New York: Hachette, 2020), 164.

88. Benjamin Todd, "The Case for Reducing Existential Risks," 80,000 Hours, last updated June 2022, https://80000hours.org/articles/existential-risks.

89. Zoe Kleinman, "Uber: The Scandals That Drove Travis Kalanick Out," *BBC*, June 21, 2017, https://www.bbc.com/news/technology-40352868.

90. Matthew J. Belvedere, "'Moral Compass' Was Off at Uber under Co-founder Kalanick, Says New CEO Dara Khosrowshahi," *CNBC*, January 23, 2018, https://www.cnbc.com/2018/01/23/uber-moral-compass-under-co-founder-kalanick-was-off-new-ceo-says.html.

91. Belvedere, "'Moral Compass' Was Off."

92. Liana B. Baker and Heather Somerville, "Uber Ex-CEO Kalanick Selling Nearly a Third of Stake for $1.4 Billion: Source," Reuters, January 4, 2018, https://www.reuters.com/article/us-uber-travis-kalanick/uber-ex-ceo-kalanick-selling-nearly-a-third-of-stake-for-1-4-billion-source-idUSKBN1EU07P.

93. Mike Isaac, "How Uber Got Lost," *New York Times*, August 23, 2019, https://www.nytimes.com/2019/08/23/business/how-uber-got-lost.html.

94. Eric Paley (@epaley), "Maybe less as prophets and more the person in the world who has thought most deeply about how to solve this very specific problem and why it needs to be solved," Twitter, May 3, 2022, 10:07 a.m., https://twitter.com/epaley/status/1521491648655593477.

95. Pillar VC (website), https://www.pillar.vc.

96. The founders of the Founders Fund, a $10B+ fund founded and led by Peter Thiel, agree: "The entrepreneurs who make it," said the fund's website, "have *a near-messianic attitude* and believe their company is essential to making the world a better place." "What Happened to the Future? Our Manifesto," Founders Fund, accessed January 12, 2023, https://foundersfund.com/2017/01/manifesto.

97. Émile P. Torres, "Eugenics in the Twenty-First Century: New Names, Old Ideas," *Truthdig*, June 15, 2023, https://www.truthdig.com/dig/nick-bostrom-longtermism-and-the-eternal-return-of-eugenics.

98. Carl Shulman and Nick Bostrom, "Embryo Selection for Cognitive Enhancement: Curiosity or Game-Changer?," *Global Policy* 5, no. 1 (February 2014): 85–92, https://onlinelibrary.wiley.com/doi/abs/10.1111/1758-5899.12123.

99. Sean Illing, "Epistocracy: A Political Theorist's Case for Letting Only the Informed Vote," *Vox*, July 23, 2018, https://www.vox.com/2018/7/23/17581394/against-democracy-book-epistocracy-jason-brennan.

100. Jason Brennan, *Against Democracy* (Princeton, NJ: Princeton University Press, 2016), 14.

101. Torres, "Eugenics."

102. Torres, "Eugenics."

103. Elizabeth J. Kennedy, in discussion with the author, August 11, 2021. Dr. Kennedy's comments have been edited for length and clarity.

104. James Barrat, *Our Final Invention: Artificial Intelligence and the End of the Human Era* (New York: Thomas Dunne, 2013), 53.

105. Ken Hillis, Michael Petit, and Kylie Jarrett, *Google and the Culture of Search* (New York: Routledge, 2013), 15.

106. Andrew Keen, "Is Google's Data Grinder Dangerous?," *Los Angeles Times*, July 12, 2007, https://www.latimes.com/la-oe-keen12jul12-story.html.

107. Max Tegmark, *Life 3.0: Being Human in the Age of Artificial Intelligence* (New York: Alfred A. Knopf, 2017), 31, fig. 1.2.

108. Tegmark, *Life 3.0*, 162.

109. Tegmark, *Life 3.0*, 132.

110. Shira Ovide, "The Cult of the Tech Genius," *New York Times*, August 6, 2020, https://www.nytimes.com/2020/08/06/technology/the-cult-of-the-tech-genius.html.

111. Evgeny Morozov, "Beware: Silicon Valley's Cultists Want to Turn You into a Disruptive Deviant," *Guardian* (US edition), January 2, 2026, https://www.theguardian.com/technology/2016/jan/03/hi-tech-silicon-valley-cult-populism.

112. Steven Levy, "The 2010s Killed the Cult of the Tech Founder. Great!," *Wired*, December 29, 2019, https://www.wired.com/story/the-2010s-killed-the-cult-of-the-tech-founder.

113. Scott Rosenberg, "Tech's Cult of the Founder Bounces Back," *Axios*, August 16, 2022, https://www.axios.com/2022/08/16/founder-cult-adam-neumann-startups.

114. Denyse O'Leary, "Silicon Valley's Strange, Apocalyptic Cult," *Mind Matters*, May 27, 2019, https://mindmatters.ai/2019/05/silicon-valleys-strange-apocalyptic-cult.

115. Julio Romo, "How the Cult of Personality and Tech-Bro Culture Is Killing Technology," *Two Four Seven* (blog), November 20, 2022, https://www.twofourseven.co.uk/blog/20/11/2022/how-the-cult-of-personality-and-tech-bro-culture-is-killing-technology.

116. "Company or Cult?," *Economist*, March 5, 2022, https://www.economist.com/business/2022/03/05/company-or-cult.

117. Manfred F. R. Kets de Vries, "Is Your Corporate Culture Cultish?" *Harvard Business Review*, May 10, 2019, https://hbr.org/2019/05/is-your-corporate-culture-cultish.

118. Flora Tsapovsky, "The Cult of Company Culture Is Back. But Do Tech Workers Even Want Perks Anymore?," *Information*, April 8, 2022, https://www

.theinformation.com/articles/the-cult-of-company-culture-is-back-but-do-tech
-workers-even-want-perks-anymore.

119. Courtney Campbell, "10 Tech Gadgets with a Cult Following on Amazon—
and Why They're Worth It," Reviewed, *USA Today*, July 19, 2018, https://
reviewed.usatoday.com/tech/features/10-tech-gadgets-with-a-cult-following
-on-amazon-and-why-theyre-worth-it.

120. Josh Constine, "Why Influencers Are Replacing Fans with Cults," *Constine's
Newsletter* (blog), May 25, 2020, https://constine.substack.com/p/why-influencers
-are-replacing-fans.

121. Monte Clark, "Unchecked Influencers Are Cult Leaders," LinkedIn, September 19, 2019, https://www.linkedin.com/pulse/unchecked-influencers-cult
-leaders-monte-clark.

122. Bretton Putter, "13 Steps to Developing a Cult-Like Company Culture,"
Medium, July 26, 2018, https://medium.com/swlh/13-steps-to-developing-a
-cult-like-company-culture-9c7c83c4b89.

123. Harry Jaffe, "The Seven Billion Dollar Man," *Washingtonian*, March 1, 2000,
https://www.washingtonian.com/2000/03/01/the-seven-billion-dollar-man.

124. Gabriel Taylor, "What Is the Legal Definition Of 'Undue Influence?'" *Oxford
Legal* (blog), March 27, 2021, http://www.oxfordlegal.com/legal-definition
-undue-influence.

125. Lorne L. Dawson, "Raising Lazarus: A Methodological Critique of Stephen
Kent's Revival of the Brainwashing Model," in *Misunderstanding Cults: Searching
for Objectivity in a Controversial Field,* eds. Benjamin Zablocki and Thomas Robbins
(Toronto: University of Toronto Press, 2001), 384.

126. Isaack Luttichuys, *Jacob Sasportas (?) (1610–1698), Sephardic Rabbi in Amster-
dam,* 1673, oil, Israel Museum, Jerusalem, https://commons.wikimedia.org/wiki
/File:Jacob_Sasportas.jpg.

127. Dweck, *Dissident Rabbi*, 174.

128. Dweck, *Dissident Rabbi*, 176.

129. Dweck, *Dissident Rabbi*, 170.

130. Greaves and MacAskill, "Strong Longtermism," 25.

131. Hayden Wilkinson, "In Defense of Fanaticism," *Ethics* 132, no. 2 (2022):
445–477, https://philpapers.org/archive/WILIDO-22.pdf.

132. Greaves and MacAskill, "Strong Longtermism," 29.

133. Greaves and MacAskill, 29.

134. See, for example, "How Climate Action Can Benefit from Indigenous Tradition of '7th-Generation Decision-Making,'" *CBC News*, January 21, 2021, https://

www.cbc.ca/news/science/what-on-earth-indigenous-seventh-generation-think
ing-climate-action-1.5882480.

135. For evidence of this influence, see AmAristizabal, "GCRs Mitigation: The
Missing Sustainable Development Goal," Effective Altruism Forum, June 8, 2021,
https://forum.effectivealtruism.org/posts/fn2QhZwFugbT3HekE/gcrs-mitigation
-the-missing-sustainable-development-goal.

136. Nick Bostrom, "Existential Risk Prevention as Global Priority," *Global Policy*
4, no. 1 (2013): 15–31, https://existential-risk.com/concept. Emphasis mine.

137. Nick Bostrom, "The Vulnerable World Hypothesis," *Global Policy* 10, no. 4
(November 2019): 455–476, https://doi.org/10.1111/1758-5899.12718.

138. Chris Stokel-Walker, "AI Survey Exaggerates Apocalyptic Risks," *Scientific American*, January 26, 2024, https://www.scientificamerican.com/article/ai
-survey-exaggerates-apocalyptic-risks.

139. Nirit Weiss-Blatt, LinkedIn direct message to author, January 26, 2024. See
also Nirit Weiss-Blatt, *The Techlash and Tech Crisis Communication* (Bingley, UK:
Emerald, 2021).

140. Katja Grace, "AI Is Not an Arms Race," *Time*, May 31, 2023, https://time
.com/6283609/artificial-intelligence-race-existential-threat.

141. Noor Al-Sibai, "Machine Learning Expert Calls for Bombing Data Centers
to Stop Rise of AI," *Futurism*, March 31, 2023, https://futurism.com/ai-expert
-bomb-datacenters.

142. See, for example, "AI Safety Summit 2023," Bletchley Park, UK, November 1–2, 2023, https://www.gov.uk/government/topical-events/ai-safety-summit
-2023.

143. Nirit Weiss-Blatt, LinkedIn direct message to author, January 26, 2024.

144. Ellen Huet, "A Cultural Divide Over AI Forms in Silicon Valley," *Bloomberg*,
December 6, 2023, https://www.bloomberg.com/news/newsletters/2023-12-06
/effective-accelerationism-and-beff-jezos-form-new-tech-tribe.

PART II

1. Carl T. Bergstrom (@CT_Bergstrom), "Deliberate or not, this is next-level
agnotogenesis (creating uncertainty or doubt to stave off regulatory action),"
Twitter, September 14, 2021, 9:50 p.m., https://web.archive.org/web/2022010
8205513/https://twitter.com/CT_Bergstrom/status/1438002275000016901. See
also: agnotology, "the strategic and purposeful production of ignorance," from
danah boyd, "Agnotology and Epistemological Fragmentation," *Data and Society:
Points*, April 26, 2019, https://points.datasociety.net/agnotology-and-epistemo
logical-fragmentation-56aa3c509c6b.

CHAPTER 3

1. See the public version of this document, published later, here: "What's Wrong with VC (and the Silicon Valley Mindset)," *Venture Patterns* (blog), accessed September 19, 2023, https://venturepatterns.com/blog/vc/whats-wrong-with-vc-and-the-silicon-valley-mindset.

2. Miles Lasater, email message to author, March 9, 2022.

3. Jonathan Z. Smith, "The Bare Facts of Ritual," in *Imagining Religion: From Babylon to Jonestown* (Chicago: University of Chicago Press, 1982), 57.

4. Smith, "The Bare Facts of Ritual," 63.

5. Hilke Brockmann, Wiebke Drews, and John Torpey, "A Class for Itself? On the Worldviews of the New Tech Elite," *PLoS ONE* 16, no. 1 (2021): e0244071, https://journals.plos.org/plosone/article?id=10.1371/journal.pone.0244071.

6. *Diversity in High Tech* (Washington, DC: US Equal Employment Opportunity Commission, 2016), https://www.eeoc.gov/special-report/diversity-high-tech.

7. Sarah K. White, "How Top Tech Companies Are Addressing Diversity and Inclusion," *CIO*, February 4, 2021, https://www.cio.com/article/193856/how-top-tech-companies-are-addressing-diversity-and-inclusion.html.

8. Jessica Guynn and Brent Schrotenboer, "Why Are There Still So Few Black Executives in America?," *USA Today*, August 20, 2020, https://www.usatoday.com/in-depth/money/business/2020/08/20/racism-black-america-corporate-america-facebook-apple-netflix-nike-diversity/5557003002.

9. Elizabeth Edwards, "Check Your Stats: The Lack of Diversity in Venture Capital Is Worse Than It Looks," *Forbes*, February 24, 2021, https://www.forbes.com/sites/elizabethedwards/2021/02/24/check-your-stats-the-lack-of-diversity-in-venture-capital-is-worse-than-it-looks.

10. Edwards, "Check Your Stats," fig. 2. Source: James L. Knight Foundation.

11. Brockmann, Drews, and Torpey, "Class for Itself."

12. Definition given by Phillip Martin of WGBH, here: Philip Martin et al., "Caste in America" (panel discussion, Boston, MA), video, 1:04:29, March 22, 2019, https://www.wgbh.org/boston-public-library-studio/2019/03/20/caste-in-america-phillip-martin.

13. "India Top Court Recalls Controversial Caste Order," *BBC*, October 1, 2019, https://www.bbc.com/news/world-asia-india-49889815.

14. "India Top Court," *BBC*.

15. W. J. Johnson, *A Dictionary of Hinduism* (Pakistan: OUP Oxford, 2009).

16. Genesis 9:22 (King James Version).

17. "Official Declaration 2, 'Every Faithful, Worthy Man,'" in *Doctrine and Covenants Student Manual*, 2nd ed. (Salt Lake City, UT: Church of Jesus Christ of the

Latter-Day Saints, 2001), https://www.churchofjesuschrist.org/study/manual
/doctrine-and-covenants-student-manual/official-declaration-2-every-faithful
-worthy-man?lang=eng.

18. Isabel Wilkerson, "America's Enduring Caste System," *New York Times*, July
7, 2020, https://www.nytimes.com/2020/07/01/magazine/isabel-wilkerson-caste
.html.

19. Wilkerson, "America's Enduring Caste System."

20. Ijeoma Oluo, *Mediocre: The Dangerous Legacy of White Male America* (New York:
Seal Press, 2020), 265.

21. Meredith Broussard, *Artificial Unintelligence: How Computers Misunderstand the
World* (Cambridge, MA: MIT Press, 2019), 8.

22. "On the Internet of Women with Moira Weigel," interview by Greg Epstein,
TechCrunch, May 20, 2019, https://techcrunch.com/2019/05/20/on-the-internet
-of-women-with-moira-weigel.

23. "On the Internet of Women with Moira Weigel, Part 2," interview by
Greg Epstein, *TechCrunch*, May 22, 2019, https://techcrunch.com/2019/05/22
/moira-weigel-on-the-internet-of-women-part-two.

24. "Inside the History of Silicon Valley Labor, with Louis Hyman," interview
by Greg Epstein, *TechCrunch*, July 30, 2019, https://techcrunch.com/2019/07/30
/inside-the-history-of-silicon-valley-labor-with-louis-hyman; Louis Hyman,
*Temp: How American Work, American Business, and the American Dream Became Tempo-
rary* (New York: Viking, 2018).

25. Hyman, "Inside the History of Silicon Valley Labor."

26. See, for example, Tom Simonite, "AI Is the Future—But Where Are the
Women?," *Wired*, August 17, 2018.

27. Chris Stokel-Walker, "ChatGPT Replicates Gender Bias in Recommendation
Letters," *Scientific American*, November 22, 2023, https://www.scientificamerican
.com/article/chatgpt-replicates-gender-bias-in-recommendation-letters; Google
Bard is now Gemini.

28. Garance Burke, Matt O'Brien, and the Associated Press, "Bombshell Stanford
Study Finds ChatGPT and Google's Bard Answer Medical Questions with Racist,
Debunked Theories That Harm Black Patients," *Fortune Well*, October 20, 2023,
https://fortune-com.cdn.ampproject.org/c/s/fortune.com/well/2023/10/20
/chatgpt-google-bard-ai-chatbots-medical-racism-black-patients-health-care/amp.

29. Ken Hillis, Michael Petit, and Kylie Jarrett, *Google and the Culture of Search*
(New York: Routledge, 2013).

30. Safiya Umoja Noble, *Algorithms of Oppression: How Search Engines Reinforce Racism* (New York: New York University Press, 2018), 7.

31. Noble, *Algorithms of Oppression*, 10.

32. Noble, *Algorithms of Oppression*, 10.

33. Scott Neumann, "Photos of Dylann Roof, Racist Manifesto Surface On Website," The Two-Way, *NPR*, June 20, 2015, https://www.npr.org/sections /thetwo-way/2015/06/20/416024920/photos-possible-manifesto-of-dylann-roof -surface-on-website.

34. Rebecca Hersher, "What Happened When Dylann Roof Asked Google for Information about Race?," The Two-Way, *NPR*, January 10, 2017, https://www .npr.org/sections/thetwo-way/2017/01/10/508363607/what-happened-when -dylann-roof-asked-google-for-information-about-race.

35. Dina Temple-Raston, "A Tale of 2 Radicalizations," Morning Edition, *NPR*, March 15, 2021, https://www.npr.org/2021/03/15/972498203/a-tale-of-2 -radicalizations.

36. Temple-Raston, "2 Radicalizations."

37. "About Project Include," Project Include, accessed December 17, 2023, https://projectinclude.org.

38. Elephant in the Valley, accessed August 2, 2022, https://www.elephantinthe valley.com.

39. J. Edward Moreno, "Who's Who Behind the Dawn of the Modern Artificial Intelligence Movement," *New York Times*, December 3, 2023, https://www .nytimes.com/2023/12/03/technology/ai-key-figures.html.

40. Cade Metz et al., "Ego, Fear and Money: How the A.I. Fuse Was Lit," *New York Times*, December 3, 2023, https://www.nytimes.com/2023/12/03/technology /ai-openai-musk-page-altman.html; Bo Young Lee, "Apparently, there wasn't a single woman involved in the history of AI. Once again, The New York Times tells a story of AI that intentionally erases women . . . ," LinkedIn, December 3, 2023, 6:43 p.m., https://www.linkedin.com/posts/bo-young-lee-%EC%9D %B4%EB%B3%B4%EC%98%81-073a47_ego-fear-and-money-how-the-ai-fuse -was-activity-7137224927467704320-GG9l.

41. Scott Jaschik, "What Larry Summers Said," *Inside Higher Ed*, February 17, 2005, https://www.insidehighered.com/news/2005/02/18/what-larry-summers-said.

42. Julie Cresswell and Sheila Kaplan, "How Juul Hooked a Generation on Nicotine," *New York Times*, November 23, 2019, https://www.nytimes.com/2019/11 /23/health/juul-vaping-crisis.html; Sheelah Kolhatkar, "Juul Wanted to Disrupt Big Tobacco. Instead It Created an Epidemic of Addiction," review of *The Devil's Playbook: Big Tobacco, Juul, and the Addiction of a New Generation*, by Lauren Etter, *New York Times*, May 25, 2021, https://www.nytimes.com/2021/05/25/books /review/the-devils-playbook-lauren-etter.html.

43. Erin Griffith, "Silicon Valley Investors Shunned Juul, but Back Other Nicotine Start-Ups," *New York Times*, October 7, 2018, https://www.nytimes.com /2018/10/07/technology/silicon-valley-investors-juul-nicotine-start-ups.html.

44. Jeff Beer, "Facebook's Reported Name Change Reinforces Its Image as the New Big Tobacco," *Fast Company*, October 21, 2021, https://www.fastcompany .com/90688287/facebooks-reported-name-change-reinforces-its-image-as-the -new-big-tobacco; Jennifer Maloney, "Juul Raises $700 Million from Investors," *Wall Street Journal*, February 6, 2020, https://www.wsj.com/articles/juul -raises-700-million-from-investors-11581018723.

45. Smith, "Bare Facts of Ritual," 57.

46. Smith, "Bare Facts of Ritual," 63.

47. Katharine Q. Seelye and Jess Bidgood, "Harvard Men's Soccer Team Is Sidelined for Vulgar 'Scouting Report,'" *New York Times*, November 4, 2016, https:// www.nytimes.com/2016/11/05/us/harvard-mens-soccer-team-scouting-report .html.

48. Wikipedia, s.v. "San Francisco Peninsula," last modified June 24, 2023, 9:49 a.m., https://en.wikipedia.org/wiki/San_Francisco_Peninsula.

49. See, for example, Karen Chapple, "Redwood City: An Improbable Villain of the Bay Area Displacement Crisis," *Urban Displacement Project* (blog), September 14, 2015, https://www.urbandisplacement.org/blog/redwood-city-an-improbable -villain-of-the-bay-area-displacement-crisis.

50. Jake Chapman, "Investors and Entrepreneurs Need to Address the Mental Health Crisis in Startups," *TechCrunch*, December 30, 2018, https://techcrunch .com/2018/12/30/investors-and-entrepreneurs-need-to-address-the-mental -health-crisis-in-startup-culture.

51. See, for example, Peggy McIntosh, "White Privilege: Unpacking the Invisible Knapsack," *Peace and Freedom*, July/August 1989, 10–12.

52. *Pirkei Avot* 2:16, trans. Joshua Kelp, https://www.sefaria.org/Pirkei_Avot .2.16?lang=bi&with=all&lang2=en.

53. "I'm a pessimist because of intelligence, but an optimist because of will," Quotes, Antonio Gramsci, Goodreads, accessed September 19, 2023, https:// www.goodreads.com/quotes/118705-i-m-a-pessimist-because-of-intelligence -but-an-optimist-because.

CHAPTER 4

1. Bruce Byfield, "A Handwriting Tablet In Progress," *Linux Magazine*, November 10, 2020, https://www.linux-magazine.com/Online/Features/reMarkable-2.

2. Alex Kerai, "2023 Cell Phone Usage Statistics: Mornings Are for Notifications," Reviews.org, updated July 21, 2023, https://www.reviews.org/mobile/cell -phone-addiction. Other studies have shown smaller figures, but it's clear the actual average is well into the dozens, if not in triple digits.

3. "Mobile Fact Sheet," Pew Research Center, updated April 7, 2021, https://www.pewresearch.org/internet/fact-sheet/mobile; dscout, "Mobile Touches: dscout's Inaugural Study on Humans and Their Tech," June 15, 2016, https://pages.dscout.com/hubfs/downloads/dscout_mobile_touches_study_2016.pdf.

4. "Screen Time and Children," American Academy of Child and Adolescent Psychiatry, updated February 2020, https://www.aacap.org/AACAP/Families_and_Youth/Facts_for_Families/FFF-Guide/Children-And-Watching-TV-054.aspx.

5. Gabrielle Emanuel, "Teens' Screen Time Doubled to 8 Hours a Day during the Pandemic—Not Counting Schoolwork," WBUR, November 22, 2021, https://www.wbur.org/news/2021/11/22/teens-screen-time-doubled-pandemic.

6. Kürşat Özenç and Glenn Fajardo, *Rituals for Virtual Meetings: Creative Ways to Engage People and Strengthen Relationships* (Hoboken, NJ: Wiley, 2021), 3.

7. James W. Carey, *Communication as Culture: Essays on Media and Society*, rev. ed. (New York: Routledge, 2009).

8. Carey, *Communication as Culture*, 52–53.

9. Kim Somajor, "Missy Elliot Expresses Gratitude Making History with 6 Platinum Albums," *Source*, January 31, 2022, https://thesource.com/2022/01/31/missy-elliot-expresses-gratitude-making-history-with-6-platinum-albums.

10. Greg Epstein, "My Name Is Greg, and I'm Addicted to Tech," *Boston Globe*, January 1, 2021, https://www.bostonglobe.com/2021/01/01/opinion/my-name-is-greg-im-addicted-tech.

11. Jean M. Twenge, Thomas E. Joiner, and Gabrielle N. Martin, "Increases in Depressive Symptoms, Suicide-Related Outcomes, and Suicide Rates among U.S. Adolescents after 2010 and Links to Increased New Media Screen Time," *Clinical Psychological Science* 6, no. 1 (2017): 3–17, https://journals.sagepub.com/doi/10.1177/2167702617723376.

12. Sukhpreet K. Tamana et al., "Screen-Time Is Associated with Inattention Problems in Preschoolers: Results from the Child Birth Cohort Study," *PLoS ONE* 14, no. 4 (2019): e0213995, https://journals.plos.org/plosone/article?id=10.1371/journal.pone.0213995; Meta van den Heuvel et al., "Mobile Media Device Use Is Associated with Expressive Language Delay in 18-Month-Old Children," *Journal of Developmental and Behavioral Pediatrics* 40, no. 2 (February–March 2019): 99–104, https://pubmed.ncbi.nlm.nih.gov/30753173; Sheri Madigan et al., "Association between Screen Time and Children's Performance on a Developmental Screening Test," *JAMA Pediatrics* 173, no. 3 (2019): 244–250, https://jamanetwork.com/journals/jamapediatrics/fullarticle/2722666.

13. "Kids Competing with Mobile Phones for Parents' Attention," *AVG Now* (blog), June 24, 2015, https://web.archive.org/web/20230129084838/https://now.avg.com/digital-diaries-kids-competing-with-mobile-phones-for-parents-attention (site discontinued).

14. Claire M. Nightingale et al., "Screen Time Is Associated with Adiposity and Insulin Resistance in Children," *Archives of Disease in Childhood* 102, no. 7 (2017): 612–616, https://doi.org/10.1136/archdischild-2016-312016.

15. James Williams, *Stand out of Our Light: Freedom and Resistance in the Attention Economy* (Cambridge: Cambridge University Press, 2018), 7.

16. See, for example, Brian A. Primack et al., "Temporal Associations between Social Media Use and Depression," *American Journal of Preventative Medicine* 60, no. 2 (February 2021): 179–188, https://doi.org/10.1016/j.amepre.2020.09.014.

17. Jean M. Twenge, *iGen: Why Today's Super-Connected Youth Are Growing Up Less Rebellious, More Tolerant, Less Happy—and Completely Unprepared for Adulthood—and What That Means for the Rest of Us* (New York: Atria, 2017).

18. Jean M. Twenge, "Have Smartphones Destroyed a Generation?," *Atlantic*, September 2017, https://www.theatlantic.com/magazine/archive/2017/09/has -the-smartphone-destroyed-a-generation/534198.

19. Brendon Hyndman, "A New Study Sounds Like Good News about Screen Time and Kids' Health. So Does It Mean We Can All Stop Worrying?" *Conversation*, October 22, 2021, https://theconversation.com/a-new-study-sounds-like-good -news-about-screen-time-and-kids-health-so-does-it-mean-we-can-all-stop -worrying-170265.

20. Émile Durkheim, *The Elementary Forms of Religious Life*, trans. Carol Cosman, Oxford World's Classics (Oxford: Oxford University Press, 2008), 41.

21. Hilary Achauer, "What Is an 'Emotional Push-Up'? Exploring the Concept of Mental Health Gyms," *Washington Post*, October 19, 2021, https://www.washing tonpost.com/lifestyle/2021/10/19/mental-health-gym-emotional-fitness/.

22. Daniel Oppenheimer, "It's Easy for Lonely Rats to Get Addicted," *Futurity*, January 24, 2013, https://www.futurity.org/its-easy-for-lonely-rats-to-get-addicted.

23. "Chamath Palihapitiya, Founder and CEO Social Capital, on Money as an Instrument of Change," interview by Madeline Dangerfield-Cha, View from the Top, Stanford Graduate School of Business, YouTube video, 56:15, November 13, 2017, https://www.youtube.com/watch?v=PMotykw0SIk.

24. Trevor Haynes, "Dopamine, Smartphones and You: A Battle for Your Time," *Science in the News* (blog), Graduate School of Arts and Sciences, Harvard University, https://sitn.hms.harvard.edu/flash/2018/dopamine-smartphones-battle-time.

25. Jamie Waters, "Constant Craving: How Digital Media Turned Us All into Dopamine Addicts," *Observer*, August 22, 2021, https://www.theguardian.com /global/2021/aug/22/how-digital-media-turned-us-all-into-dopamine-addicts -and-what-we-can-do-to-break-the-cycle.

26. William Ross Perlman, "Rats Prefer Social Interaction to Heroin or Methamphetamine," *NIDA Notes*, August 23, 2019, https://archives.nida.nih.gov/news

-events/nida-notes/2019/08/rats-prefer-social-interaction-to-heroin-or-metham
phetamine.

27. Human Sustainability Inside Out, accessed September 9, 2023, https://hsio
.life.

28. Julie Jargon, "I Instagram Therefore I Am: Your Child's Brain on Social
Media," presentation, Harvard Alumni for Mental Health, June 10, 2022, You-
Tube video, 59:42, https://www.youtube.com/watch?v=ykt1EEbiJ3s.

29. Julie Jargon, "Doctors Have Been Seeing Teen Girls Develop Tics and
They Think TikTok Might Be Involved," interview by BuzzFeed Daily, *Buzz-
Feed*, October 27, 2021, https://www.buzzfeed.com/daily/tiktok-teen-girls-tics
-mental-health-social-media.

30. Tara Deliberto, "Why Are You Anxious about Gaining Weight?," *Psychology Today* (blog), October 23, 2019, https://www.psychologytoday.com/us/blog
/eating-disorder-recovery/201910/why-are-you-anxious-about-gaining-weight.

31. The Waldorf movement's history has been marred by racism and by a kind of
magical thinking that, even in 2020, was in some cases at odds with vaccination.

32. Matt Richtel, "A Silicon Valley School that Doesn't Compute," *New York
Times*, October 22, 2011, https://www.nytimes.com/2011/10/23/technology/at
-waldorf-school-in-silicon-valley-technology-can-wait.html.

33. "William Shunkamolah," Team Showcase, HEAL Initiative, University of
California–San Francisco, accessed September 9, 2023, https://healinitiative.org
/team-showcase/william-shunkamolah.

CHAPTER 5

1. Robert E. Lerner, "Apocalyptic Literature," in *Encyclopedia Britannica Online*, last
modified May 3, 2021, https://www.britannica.com/art/apocalyptic-literature.

2. Flavius Josephus, "From the Great Extremity to Which the Jews Were Reduced
to the Taking of Jerusalem by Titus," in *The Jewish War; or, the History of the Destruction of Jerusalem*, in *The Genuine Works of Flavius Josephus the Jewish Historian*, trans.
William Whiston (London, 1737), https://penelope.uchicago.edu/josephus/war
-6.html.

3. "About Macomb," Macomb Community College, accessed September 7, 2022,
https://www.macomb.edu/about-macomb/index.html.

4. See, for example, "Recognizing and Preventing the Strain of Hypervisibility," *Our Stories* (blog), *Bloomberg*, https://www.bloomberg.com/company/stories
/recognizing-and-preventing-the-strain-of-hypervisibility.

5. See Maureen T. Reddy, "Invisibility/Hypervisibility: The Paradox of Norma-
tive Whiteness," *Transformations: The Journal of Inclusive Scholarship and Pedagogy* 9,
no. 2 (Fall 1998): 55–64, https://www.jstor.org/stable/43587107.

6. Bill McGraw, "Riot or Rebellion? The Debate on What to Call Detroit '67," *Detroit Free Press*, July 4, 2017, https://www.freep.com/story/news/2017 /07/05/50-years-later-riot-rebellion/370968001.

7. Elizabeth Anne Martin, "Detroit and the Great Migration, 1916–1929" (honors essay, University of Michigan, 1992), https://web.archive.org/web/20140129 103021/http:/bentley.umich.edu/research/publications/migration/ch1.php.

8. Sandra Svoboda, "Detroit by the Numbers—the Truth about Poverty," Bridge Michigan, March 19, 2015, https://www.bridgemi.com/urban-affairs/detroit -numbers-truth-about-poverty.

9. Detroit Future City, "People," in *139 Square Miles* (Detroit, MI: Inland, 2017), https://detroitfuturecity.com/wp-content/uploads/2017/11/DFC_139-SQ-Mile _People.pdf.

10. See, for example, Chris Gilliard, "A Black Woman Invented Home Security. Why Did It Go So Wrong?," Ideas, *Wired*, November 14, 2021, https://www .wired.com/story/black-inventor-home-security-system-surveillance; Tawana Petty, "Defending Black Lives Means Banning Facial Recognition," Ideas, *Wired*, July 10, 2020, https://www.wired.com/story/defending-black-lives-means-banning -facial-recognition.

11. See Mark Binelli, "The Fire Last Time," *New Republic*, April 6, 2017, https:// newrepublic.com/article/141701/fire-last-time-detroit-stress-police-squad -terrorized-black-community.

12. Jamon Jordan, "Black Bottom: A Brief Lesson from Local Historian Jamon Jordan," produced by Clarence Tabb, Jr., *Detroit News*, March 23, 2021, video, 1:41, https://www.detroitnews.com/videos/news/local/detroit-city/2021/03/24 /i-375-black-bottom-paradise-valley/6969067002.

13. Ken Coleman, "The People and Places of Black Bottom, Detroit," *Humanities* 42, no. 4 (Fall 2021), https://www.neh.gov/article/people-and-places-black -bottom-detroit.

14. "How the Razing of Detroit's Black Bottom Neighborhood Shaped Michigan's History," Michigan Radio, *NPR*, February 11, 2019, https://www .michiganradio.org/arts-culture/2019-02-11/how-the-razing-of-detroits-black -bottom-neighborhood-shaped-michigans-history.

15. See Binelli, "Fire Last Time."

16. William K. Stevens, "Tactics of an Elite Police Unit Election Issue in Detroit," *New York Times*, June 11, 1973, https://www.nytimes.com/1973/06/11/archives /tactics-of-an-elite-police-unit-election-issue-in-detroit-friends.html; Doug Merriman, "A History of Violence: The Detroit Police Department, the African American Community and S.T.R.E.S.S. An Army of Occupation or an Army under Siege," *Doug Merriman* (blog), November 17, 2015, https://dougmerriman .org/2015/11/17/a-history-of-violence-the-detroit-police-department-the-african

-american-community-and-s-t-r-e-s-s-an-army-of-occupation-or-an-army-under
-siege/.

17. Dan Georgakas and Marvin Surkin, *Detroit: I Do Mind Dying: A Study in Urban Revolution*, updated ed., South End Press Classics (Boston, MA: South End, 1998).

18. The smaller Piquette Avenue in 1904, the larger Highland Avenue in 1910: Wikipedia, s.v. "Ford Piquette Avenue Plant," last modified July 6, 2023, https:// en.wikipedia.org/wiki/Ford_Piquette_Avenue_Plant; Wikipedia, s.v. "Highland Park Ford Plant," last modified April 3, 2023, https://en.wikipedia.org/wiki /Highland_Park_Ford_Plant.

19. See, for example, Johnathan A. Greenblatt, "Don't Whitewash Henry Ford's Anti-Semitism," letter to the editor, *New York Times*, February 5, 2019, https:// www.nytimes.com/2019/02/05/opinion/letters/henry-ford-anti-semitism.html, and Hasia Diner, "Ford's Anti-Semitism," interview by American Experience, PBS, accessed September 9, 2023, https://www.pbs.org/wgbh/americanexperience /features/henryford-antisemitism.

20. "Henry Ford and Anti-Semitism: A Complex Story," The Henry Ford, accessed September 9, 2023, https://www.thehenryford.org/collections-and -research/digital-resources/popular-topics/henry-ford-and-anti-semitism-a -complex-story; Nikita Stewart, "Planned Parenthood in N.Y. Disowns Margaret Sanger over Eugenics," *New York Times*, July 1, 2020, https://www.nytimes.com /2020/07/21/nyregion/planned-parenthood-margaret-sanger-eugenics.html.

21. Smruthi Nadig, "A Rough Year for Australia's Fortescue," *Mining Technology*, November 7, 2023, https://www.mining-technology.com/features/a-rough -year-for-australias-fortescue.

22. Domini Stuart, "Managing the Risks of Modern Slavery," Australian Institute of Company Directors, April 1, 2021, https://www.aicd.com.au/organisational -culture/business-ethics/issues/managing-the-risks-of-modern-slavery.html.

23. emleml, "The 200+ Best Apocalyptic Films Rated 6+ (Categorized!)," IMDb, updated March 11, 2022, https://www.imdb.com/list/ls072938013.

24. See, for example, Maria Farrell, "Do Tech Slogans Really 'Make the World a Better Place'?," *OneZero*, December 4, 2018, https://onezero.medium.com/do -tech-slogans-really-make-the-world-a-better-place-730836c2c3ec.

25. Justine Calma, "AI Suggested 40,000 New Possible Chemical Weapons in Just Six Hours," *Verge*, March 17, 2022, https://www.theverge.com/2022/3/17 /22983197/ai-new-possible-chemical-weapons-generative-models-vx.

26. Todd C. Frankel, "The Cobalt Pipeline: Tracing the Path from Deadly Hand-Dug Mines in Congo to Consumers' Phones and Laptops," *Washington Post*, September 30, 2016, https://www.washingtonpost.com/graphics/business/batteries /congo-cobalt-mining-for-lithium-ion-battery.

27. David Barboza, "In Chinese Factories, Lost Fingers and Low Pay," *New York Times*, January 5, 2008, https://www.nytimes.com/2008/01/05/business /worldbusiness/05sweatshop.html; "How Chinese Factory-Workers Express Their Views on Life," *Economist*, August 12, 2021, https://www.economist.com /china/2021/08/12/how-chinese-factory-workers-express-their-views-on -life; Chloe Taylor, "'Nightmare' Conditions at Chinese Factories Where Hasbro and Disney Toys Are Made," CNBC, December 7, 2018, https://www.cnbc .com/2018/12/07/nightmare-at-chinese-factories-making-hasbro-and-disney -toys.html.

28. Edmund Lee and John Koblin, "HBO Must Get Bigger and Broader, Says Its New Overseer," *New York Times*, July 8, 2018, https://www.nytimes.com/2018 /07/08/business/media/hbo-att-merger.html.

29. Chris Gilliard (@hypervisible), Twitter, July 9, 2018, 4:46 p.m., https:// twitter.com/hypervisible/status/1147469357792059392.

30. Chris Gilliard (@hypervisible), Twitter, July 6, 2019, 7:36 a.m., https:// twitter.com/hypervisible/status/1518570404360736768.

31. Chris Gilliard (@hypervisible), Twitter, April 25, 2022, 5:39 a.m., https:// web.archive.org/web/20220425123945/https://twitter.com/hypervisible/status /1518570404360736768.

32. Gavin Kenneally, "'We Created Robot Dogs to Patrol the U.S. Border,'" *Newsweek*, February 22, 2022, https://www.newsweek.com/robot-dogs-patrol-us -border-1681325.

33. Chris Gilliard (@hypervisible), Twitter, February 27, 2022, https://drive .google.com/file/d/19mzh93ykeH-GQdsTacTGMZ_AI1T7s1-9/view?usp=drive _link.

34. Chris Gilliard (@hypervisible), Twitter, September 6, 2021, 9:42 a.m., https:// web.archive.org/web/20210701000000*/https://twitter.com/hypervisible/status /1434874647447842816.

35. Chris Gilliard, in discussion with the author, June 8, 2021.

36. An example where problematic tech was actually inspired by a comic villain: Chris Gilliard (@hypervisible), "Reading this essay from a few years back by @cariatidaa and . . . find out that part of the inspiration of electronic monitoring system was . . . a comic where Kingpin . . ." Twitter, September 30, 2019, 3:51 p.m., https://web.archive.org/web/20210601015721/https://twitter.com /hypervisible/status/1178804534400950272.

37. Chris Gilliard (@hypervisible), "Someone please ask a tech CEO what they think the optimal number of cameras per household is," Twitter, September 19, 2021, 9:11 a.m., https://web.archive.org/web/20210909195637/https://twitter .com/hypervisible/status/1435999341903233024.

38. Chris Gilliard (@hypervisible), "Doctors are concerned that the "'app-ification" of mental health could sacrifice quality for the sake of cost and easy access,'" Twitter, April 28, 2022, 6:28 p.m., https://web.archive.org/web/20220429013028/https://twitter.com/hypervisible/status/1519850986348634112.

39. Chris Gilliard (@hypervisible), "It was hard to pick, but the worst thing in this article—about companies using AI to 'detect' your emotional state—is that Zoom is expected to integrate this . . ." Twitter, April 13, 2022, 8:16 p.m., https://web.archive.org/web/20220905114505/https://twitter.com/hypervisible/status/1514336857785524224.

40. Chris Gilliard (@hypervisible), "Every future imagined by a tech company is worse than the previous iteration," Twitter, April 28, 2022, 11:25 a.m., https://web.archive.org/web/20220428182617/https://twitter.com/hypervisible/status/1519744656782856193.

41. Chris Gilliard (@hypervisible), "Sigh," Twitter, June 3, 2022, 8:30 a.m., https://web.archive.org/web/20220603153319/https://twitter.com/hypervisible/status/1532746672878538754.

42. Chris Gilliard (@hypervisible), "Dept of Defense 'is now interacting with top AV startups to imagine the next era of potentially autonomous military vehicles,'" Twitter, April 15, 2022, 3:40 p.m., https://web.archive.org/web/20220415224020/https://twitter.com/hypervisible/status/1515097677322719240.

43. Chris Gilliard (@hypervisible), "Read it and weep," Twitter, April 4, 2022, 9:25 a.m., https://web.archive.org/web/20220404162637/https://twitter.com/hypervisible/status/1511017071345385472.

44. Shoshanna Zuboff, *The Age of Surveillance Capitalism: The Fight for a Human Future at the New Frontier of Power* (New York: PublicAffairs, 2019), 15.

45. Patrick Grother, Mei Ngan, and Kayee Hanaoka, *Face Recognition Vendor Test (FRVT) Part 3: Demographic Effects* (Washington, DC: National Institute of Standards and Technology, 2019), https://nvlpubs.nist.gov/nistpubs/ir/2019/NIST.IR.8280.pdf. Four datasets analyzed for this study included (1) domestic mug shots collected in the United States, (2) application photographs from a global population of applicants for immigration benefits, (3) visa photographs submitted in support of visa applicants, and (4) border crossing photographs of travelers entering the United States.

46. Sam Biddle, "Police Surveilled George Floyd Protests with Help from Twitter-Affiliated Startup Dataminr," *Intercept*, July 9, 2020, https://theintercept.com/2020/07/09/twitter-dataminr-police-spy-surveillance-black-lives-matter-protests; Steven Feldstein and David Wong, "New Technologies, New Problems—Troubling Surveillance Trends in America," *Just Security*, August 6, 2020, https://www.justsecurity.org/71837/new-technologies-new-problems-troubling-surveillance-trends-in-america.

47. Fabiana Sampaio, "80% das prisões errôneas por reconhecimento facial no RJ são de negros," *Radio Agência*, January 12, 2022, https://agenciabrasil.ebc.com .br/radioagencia-nacional/justica/audio/2022-01/80-das-prisoes-erroneas -por-reconhecimento-facial-no-rj-sao-de-negros. This article reports on "a survey by the public defender's office in Rio de Janeiro and the National College of General Public Defenders."

48. *The Age of Surveillance*, special report, *Politico Europe*, May 26, 2021, https:// www.politico.eu/special-report/the-age-of-surveillance; Nikos Smyrnaios, "In Europe, a Military-Industrial Complex Is Turning Smartphones into a Political Surveillance System," *Tech Policy Press*, September 7, 2022, https://techpolicy.press /in-europe-a-military-industrial-complex-is-turning-smartphones-into-a -political-surveillance-system/.

49. Vincent Manacourt and Mark Scott, "In Europe, a Coronavirus Boom for Foreign Surveillance Firms," *Politico Europe*, May 28, 2021, https://www.politico .eu/article/europe-surveillance-china-israel-united-states.

50. OVD-Info, *No to War*, updated April 14, 2022, https://en.ovdinfo.org/no -to-war-en; Anastasila Kruope, "Moscow's Use of Facial Recognition Technology Challenged," Human Rights Watch, July 8, 2020, https://www.hrw.org /news/2020/07/08/moscows-use-facial-recognition-technology-challenged; Anastasila Kruope, "The Russian Government's Advance on Biometric Data," Human Rights Watch, July 23, 2022, https://www.hrw.org/news/2022/07/23 /russian-governments-advance-biometric-data.

51. "New Evidence that Biometric Data Systems Imperil Afghans," Human Rights Watch, March 30, 2022, https://www.hrw.org/news/2022/03/30/new -evidence-biometric-data-systems-imperil-afghans; Paul Mozur, "In Hong Kong Protests, Faces Become Weapons," *New York Times*, July 26, 2019, https://www .nytimes.com/2019/07/26/technology/hong-kong-protests-facial-recognition -surveillance.html; Steven Feldstein, "Governments Are Using Spyware on Citizens. Can They Be Stopped?," Carnegie Endowment for International Peace, July 21, 2021, https://carnegieendowment.org/2021/07/21/governments-are-using -spyware-on-citizens.-can-they-be-stopped-pub-85019.

52. Dana Priest, Craig Timberg, and Souad Mekhennet, "Private Israeli Spyware Used to Hack Cellphones of Journalists, Activists Worldwide," *Washington Post*, July 18, 2021, https://www.washingtonpost.com/investigations/interactive/2021/nso -spyware-pegasus-cellphones.

53. TRAC Immigration, *Immigration Court Cases Now Involve More Long-Time Residents*, April 19, 2018, https://trac.syr.edu/immigration/reports/508.

54. *Annual Statistical Transparency Report: Regarding the Intelligence Community's Use of National Security Surveillance Authorities* (Washington, DC: Office of Civil Liberties, Privacy, and Transparency, Office of the Director of National Intelligence,

April 2022), https://www.intelligence.gov/assets/documents/702%20Documents
/statistical-transparency-report/2022_IC_Annual_Statistical_Transparency_Report
_cy2021.pdf.

55. Nick Bostrom, "The Vulnerable World Hypothesis," *Global Policy* 10, no. 4
(November 2019): 455–476, https://doi.org/10.1111/1758-5899.12718.

56. Joseph Cox, "Revealed: US Military Bought Mass Monitoring Tool That
Includes Internet Browsing, Email Data," *Motherboard*, September 12, 2022,
https://www.vice.com/en/article/y3pnkw/us-military-bought-mass-monitoring
-augury-team-cymru-browsing-email-data.

57. Cox, "Revealed: US Military Bought Mass Monitoring Tool."

58. Zuboff, *Age of Surveillance Capitalism*, 11–12.

59. Simone Browne, *Dark Matters: On the Surveillance of Blackness* (Durham, NC:
Duke University Press, 2015), 8.

60. Browne, *Dark Matters*, 11.

61. See, for example, Kade Crockford's work: "How Is Face Recognition Surveillance Technology Racist?," ACLU, June 16, 2020, https://www.aclu.org/news
/privacy-technology/how-is-face-recognition-surveillance-technology-racist.

62. Browne, *Dark Matters*, 12.

63. "Public Thinker: Virginia Eubanks on Digital Surveillance and People Power,"
interview by Jenn Stroud Rossman, *Public Books*, July 9, 2020, https://www
.publicbooks.org/public-thinker-virginia-eubanks-on-digital-surveillance-and
-people-power.

64. See Eubanks's books *Automating Inequality: How High-Tech Tools Profile, Police,
and Punish the Poor* (New York: St. Martin's, 2017) and *Digital Dead End: Fighting for
Social Justice in the Information Age* (Cambridge, MA: MIT Press, 2011).

65. See, for example, *The Feeling of Being Watched*, written and directed by Assia
Bandoui, featuring Gamal Abdel-Haiz and Christina Abraham (New York:
Women Make Movies, 2019).

66. Browne, *Dark Matters*, 10.

67. Zuboff, *Age of Surveillance Capitalism*, 189, 207, 221, 222, 226.

68. Chris Gilliard, in discussion with the author, January 17, 2024.

CHAPTER 6

1. James D. Tracy, "The Indulgences Controversy of Martin Luther," in *Encyclopedia Britannica Online*, last modified July 8, 2023, https://www.britannica.com
/biography/Martin-Luther/The-indulgences-controversy.

2. David B. Morris, "Luther the Outlaw," Martin Luther as Priest, Heretic and Outlaw, Library of Congress Research Guides, last updated December 6, 2023, https://guides.loc.gov/martin-luther-priest-heretic-outlaw/luther-the-outlaw.

3. Meredith Whittaker et al., "Organizing Tech" (panel discussion, AI Now 2019 Symposium, New York, NY, October 2, 2019), video, 31:47, October 16, 2019, https://www.youtube.com/watch?v=jLeOyIS1jwc.

4. Veena Dubal, "A Brief History of the Gig," *Logic*, no. 10 (May 2020), https://logicmag.io/security/a-brief-history-of-the-gig.

5. Dubal, "Brief History."

6. Dubal, "Brief History."

7. Dubal, "Brief History."

8. Veena Dubal, "Wage Slave or Entrepreneur?: Contesting the Dualism of Legal Worker Identities," *California Law Review*, no. 105 (2017): 101–159.

9. With certain exceptions. Rebecca Lake, "California Assembly Bill 5 (AB5): What's In It and What It Means," Investopedia, updated May 30, 2023, https://www.investopedia.com/california-assembly-bill-5-ab5-4773201.

10. *Dictionary.com*, s.v. "heresy," accessed July 5, 2023, https://www.dictionary.com/browse/heresy.

11. *Merriam-Webster*, s.v. "apostasy," accessed September 19, 2023, https://www.merriam-webster.com/dictionary/apostasy.

12. See, for example, Way of the Future and Anthony Levandowsky, as covered in section 1.

13. Harry S. Truman, quoted in "Truman Quotes," Truman Library Institute, accessed January 19, 2024, https://www.trumanlibraryinstitute.org/truman/truman-quotes.

14. Sal Bayat et al., "Letter in Support of Responsible Fintech Policy," updated June 1, 2022, https://concerned.tech.

15. Dubal, "Brief History."

16. See, for example, Afdhel Aziz, "The Power of Purpose: How Lyft Helps Improve Peoples Lives With 'The World's Best Transportation,'" *Forbes*, February 20, 2019, https://www.forbes.com/sites/afdhelaziz/2019/02/20/the-power-of-purpose-how-lyft-helps-improve-peoples-lives-with-the-worlds-best-transportation/?sh=5780e9fa1539.

17. *Global Gig Economy Industry Research Report, Competitive Landscape, Market Size, Regional Status and Prospect* (Baner, India: Market Growth Reports, 2022), https://www.marketgrowthreports.com/global-gig-economy-industry-research-report-competitive-landscape-market-21739387.

18. Kate O'Flaherty, "Apple Issues Stunning New Blow to Facebook as Google Joins the Battle," *Forbes*, February 19, 2022, https://www.forbes.com/sites/kateoflahertyuk/2022/02/19/apple-issues-stunning-new-blow-to-facebook-as-google-joins-the-battle.

19. Computing Task Force, *MIT Schwarzman College of Computing—Organizational Structure*, January 30, 2020, https://computing.mit.edu/wp-content/uploads/2022/10/SCC-Organizational-Structure.pdf.

20. Kate Kelly, "In Trump, Stephen Schwarzman Found a Chance to Burnish His Legacy," *New York Times*, January 19, 2021, https://www.nytimes.com/2021/01/19/business/schwarzman-blackstone-trump.html. It took Schwarzman until November 2022 to say he wouldn't back Trump for president again, and even then it was simply in the context of saying he would back another Republican for the nomination. Fredreka Schouten, "Donald Trump Faces Billionaires in Retreat and Tabloid Trolling a Day after Campaign Announcement," *CNN*, November 16, 2022, https://www.cnn.com/2022/11/16/politics/stephen-schwarzman-blackstone-trump-2024/index.html.

21. Matthew Martin and Bloomberg, "Wall Street CEOs Aren't Letting Biden's Oil Spat with Saudi Arabia Get in the Way of Their Trek to the Kingdom's 'Davos in the Desert,'" *Fortune*, October 24, 2022, https://fortune.com/2022/10/24/wall-street-ceos-biden-oil-spat-saudi-arabia-mbs-kingdom-davos-in-the-desert; David Gura, "Wall Street Eager to Strike Deals with Saudi Arabia, despite Political Concerns," NPR, October 25, 2022, https://www.npr.org/2022/10/25/1131396337/saudi-arabia-biden-wall-street-dimon-goldman-sachs-oil; Cristina Alesci, "Saudi Arabia Pledges $20 Billion to Blackstone for American Infrastructure," *CNN Business*, May 21, 2017, https://money.cnn.com/2017/05/21/news/companies/saudi-arabia-blackstone-deal/index.html; Patrick Range MacDonald, "Modern-Day Robber Baron: The Sins of Blackstone CEO Stephen Schwarzman," *Housing Is a Human Right* (blog), July 29, 2020, https://web.archive.org/web/20230430055818/https://www.housingisahumanright.org/modern-day-robber-baron-the-sins-of-blackstone-ceo-stephen-schwarzman; Patrick Range MacDonald, "Yes on Prop 21 Sues Big Real Estate to Stop Repeated Dirty Tricks," *Housing Is a Human Right* (blog), October 28, 2020, https://web.archive.org/web/20230523195516/https://www.housingisahumanright.org/yes-on-prop-21-sues-big-real-estate-to-stop-repeated-dirty-tricks.

22. "About Us," Center for Humane Technology, accessed January 19, 2024, https://www.humanetech.com/who-we-are.

23. Tristan Harris, "A Call to Minimize Distraction and Respect Users' Attention" (presentation, Google, February 2013), https://www.scribd.com/document/378841682/A-Call-to-Minimize-Distraction-Respect-Users-Attention-by-Tristan-Harris.

24. "Stop Killing Us" is a slogan associated with the Black Lives Matter movement, but it comes up at meetings of Mijente and allies. Sam Levin, "Revealed:

LAPD Officers Told to Collect Social Media Data on Every Civilian They Stop,"
Guardian (US edition), https://www.theguardian.com/us-news/2021/sep/08/revealed
-los-angeles-police-officers-gathering-social-media.

25. Cinthya Rodriguez et al., "Return of the Movement Jedi" (lecture, Instituto
de Formacion Politica, Mijente, livestreamed April 20, 2022), video, 1:02:36,
https://instituto.mijente.net/courses/tech-wars.

26. Cinthya Rodriguez et al., "Rogue One: The Blue Death Star" (lecture, Instituto de Formacion Politica, Mijente, livestreamed March 9, 2022), video, 1:10:59,
https://instituto.mijente.net/courses/tech-wars.

27. Cinthya Rodriguez et al., "The Global Menace" (lecture, Instituto de Formacion Politica, Mijente, livestreamed April 6, 2022), video, 59:08, https://instituto
.mijente.net/courses/tech-wars.

28. Carl T. Bergstrom and Jevin D. West, *Calling Bullshit: The Art of Skepticism in a
Data-Driven World* (New York: Random House, 2021), x.

29. Bergstrom and West, *Calling Bullshit*, x.

30. Ethan Zuckerman, *Mistrust: Why Losing Faith in Institutions Provides the Tools to
Transform Them* (New York: W. W. Norton, 2021), xvi.

31. The Daily Show (@TheDailyShow), "'The government is like a lawyer who
represents all of us . . . It's time for the advocate for all of us to actually fight on
our behalf.'—@AnandWrites on how to . . ." Twitter, October 7, 2019, 1:32 p.m.,
https://twitter.com/thedailyshow/status/1181260901103489025.

32. Debbie Lord, "Election 2022: If You Need a Ride to Vote, Lyft Is Offering
Discounts," Boston 25 News, November 4, 2022, https://www.boston25news
.com/news/trending/election-2022-if-you-need-ride-vote-lyft-is-offering-dis
counts/MQF23YACAJBJTK44B2HNSPCYK4.

33. Zuckerman, *Mistrust*, 64–66.

34. Dara Kerr, "'A Totally Different Ballgame': Inside Uber and Lyft's Fight
over Gig Worker Status," *CNET*, August 28, 2020, https://www.cnet.com/tech
/tech-industry/features/uber-lyfts-fight-over-gig-worker-status-as-campaign
-against-labor-activists-mounts.

35. Kerr, "'A Totally Different Ballgame.'"

36. See, for example, Allan M. Brandt, "Inventing Conflicts of Interest: A History of Tobacco Industry Tactics," *American Journal of Public Health* 102, no. 1 (January 2012): 63–71, https://www.ncbi.nlm.nih.gov/pmc/articles/PMC3490543.

37. Ben Smith, "Uber Executive Suggests Digging Up Dirt on Journalists," *Buzz-
Feed News*, November 17, 2014, https://www.buzzfeednews.com/article/bensmith
/uber-executive-suggests-digging-up-dirt-on-journalists.

38. "Memory of Dr. Sam Dubal '15 Honored through $1M Anthropology Fel-
lowship," Berkeley Letters and Science, updated September 7, 2021, https://
ls.berkeley.edu/news/dr-sam-dubal.

39. Ryan Byrne, "With Funding from Uber, Lyft, and DoorDash, Campaign behind California Proposition 22 Tops $180 Million," *Ballotpedia News*, September 9, 2020, https://news.ballotpedia.org/2020/09/09/with-funding-from-uber-lyft -and-doordash-campaign-behind-california-proposition-22-tops-180-million.

40. Initially, the coronavirus pandemic was disproportionately deadly for Black Americans, who were more likely to be exposed as frontline workers. It should be noted that, as vaccines became available, the virus became more likely to kill white Americans, perhaps because of greater tendencies among certain white communities to be "ideologically opposed to the vaccine." "COVID Death Rate Now Higher in Whites than in Blacks," In the News, Harvard T. H. Chan School of Public Health, October 21, 2022, https://www.hsph.harvard.edu/news/hsph-in -the-news/covid-death-rate-now-higher-in-whites-than-in-blacks.

41. Margot Roosevelt and Suhauna Hussain, "Prop. 22 Is Ruled Unconstitutional, a Blow to California Gig Economy Law," *Los Angeles Times*, August 20, 2021, https://www.latimes.com/business/story/2021-08-20/prop-22-unconstitutional. However, in March 2022, the state's appeals court reversed most of that ruling. Suhauna Hussain, "California Appeals Court Reverses Most of Ruling Deeming Prop. 22 Invalid," *Los Angeles Times*, March 13, 2023, https://www.latimes.com /business/story/2023-03-13/prop-22-upheld-california-appeals-court.

42. Yaël Eisenstat, the Facebook critic featured in chapter 8, has also noted in her Twitter bio that she has been called a Cassandra.

43. I was ordained by the International Institute for Secular Humanistic Judaism: IISHJ.org.

44. Veena Dubal, "Day 1: Power to the People—Not Just the President," *Fog City Journal*, January 21, 2009, http://www.fogcityjournal.com/wordpress/1003 /day-1-power-to-the-people-not-just-the-president.

45. Veena Dubal, "The New Racial Wage Code," preprint, submitted September 30, 2021, https://papers.ssrn.com/sol3/papers.cfm?abstract_id=3855094.

46. "Although available statistics are limited, Lyft estimates that 69% of their U.S. workforce identifies as racial minorities. In California, which is both the most diverse and most unequal state in the U.S., this percentage is likely much higher." Dubal, "New Racial Wage Code," 6.

47. Dubal, "New Racial Wage Code."

48. Dubal, personal communication with author, January 29, 2024.

CHAPTER 7

1. "TB in America: 1895–1954," American Experience, PBS, https://www.pbs .org/wgbh/americanexperience/features/plague-gallery.

2. Tom Ferrick, unpublished manuscript, 1997, https://drive.google.com/file/d /1TSN8PS9KgiJnCVPnRfpLhbYBg9_qXWLA/view?usp=sharing.

3. The original use of the word *humanism* was during the Renaissance: knowledge beyond the mind of god. There is also the use of the word *humanist* in academic contexts to refer to the opposite of a scientist.

4. The phrase is sometimes attributed to Sagan alone, but Druyan is better understood as a creator in her own right, who combined the mortal soul and depth of a poet with her partner's expansive intellect.

5. This quotation, from the ancient *Mettā Sutta*, was suggested for this purpose by Sebastian Rizzon, a Zen Master and artist from my city of Somerville, MA. Rizzon's translation.

6. *Yasna* 51.22, in *The Gāthās of Zarathustra*, trans. Stanley Insler, vol. 1 of *Textes et Mémoires*, Acta Iranica (Tehran: Bibliothèque Pahlavi, 1975). It's a hymn interpreted and translated by Stanley Insler from the ancient Old Avestan or Old Persian language; the line was suggested by my wonderful colleague Daryush Mehta, the Zoroastrian chaplain at Harvard and MIT.

7. "About Kate O'Neill," KO Insights, accessed January 19, 2024, https://www .koinsights.com/about-kate.

8. Kate O'Neill, *Tech Humanist: How You Can Make Technology Better for Business and Better for Humans* (self-pub., 2018), 5.

9. See, for example, Ryan Burge, "The 2022 Data on the Southern Baptist Convention is Out," *Graphs About Religion* (blog), May 10, 2023, https://www .graphsaboutreligion.com/p/the-2022-data-on-the-southern-baptist.

10. "Why AI Needs More Social Workers, with Columbia University's Desmond Patton," interview by Greg Epstein, *TechCrunch*, August 9, 2019, https:// techcrunch.com/2019/08/09/why-ai-needs-more-social-workers-with-columbia -universitys-desmond-patton.

11. Patton, "Why AI Needs More Social Workers."

12. Zack Whittaker, "Despite Controversies and Bans, Facial Recognition Startups Are Flush with VC Cash," *TechCrunch*, July 6, 2021, https://techcrunch.com /2021/07/26/facial-recognition-flush-with-cash; Ingrid Lunden, "AnyVision, the Controversial Facial Recognition Startup, Has Raised $235M Led by SoftBank and Eldridge," *TechCrunch*, July 7, 2021, https://techcrunch.com/2021/07/07/any vision-the-controversial-facial-recognition-startup-has-raised-235m-led-by-soft bank-and-eldridge.

13. Daniel Dennis Jones, "What Do We Owe to the Internet's 'First Responders?'" *Berkman Klein Center Collection*, April 17, 2019, https://medium.com/berk man-klein-center/what-do-we-owe-to-the-internets-first-responders-efc1889 cdbc3.

14. Or so I was told by Blake Lemoine via an unpublished Zoom interview on January 23, 2023.

15. With Angelina McMillan-Major. Emily M. Bender et al., "On the Dangers of Stochastic Parrots: Can Language Models Be Too Big?," in *FAccT '21: Proceedings of the 2021 ACM Conference on Fairness, Accountability, and Transparency* (New York: Association for Computing Machinery, 2021), 610–623.

16. Karen Hao, "We Read the Paper That Forced Timnit Gebru Out of Google. Here's What It Says," *MIT Technology Review*, December 4, 2020.

17. Distributed AI Research Institute (website), accessed May 23, 2023, https://dair-institute.org.

18. Timnit Gebru, in discussion with the author via Zoom, May 4, 2022.

19. Mobius (mobiusorg), LinkedIn, accessed January 12, 2024, https://www.linkedin.com/company/mobiusorg.

20. Jack Kornfield, *A Path with Heart: A Guide through the Perils and Promises of a Spiritual Life* (New York: Bantam, 1993); Sharon Salzberg, *Real Happiness: The Power of Meditation* (New York: Workman, 2011).

21. Guild of Future Architects (website), accessed May 23, 2023, https://futurearchitects.com.

22. Stanford Center on Philanthropy and Civil Society, "Digital Civil Society Lab Fellow's Work Nominated for New York Emmy Award," July 21, 2021, https://pacscenter.stanford.edu/news/digital-civil-society-lab-fellows-work-nominated-for-new-york-emmy-award; Pat Daly and Patricia Mahoney, "Eyes on You: Nuns Are Keeping Facial Recognition Companies in Check," interview, *Currents News*, February 27, 2020, YouTube video, 6:07, https://youtu.be/YSCnxH1JMVA.

23. Elizabeth Dwoskin, "Amazon Is Selling Facial Recognition to Law Enforcement—for a Fistful of Dollars," *Washington Post*, May 22, 2018.

24. A nonexhaustive list might include, in addition to Mozilla and Mozfest, Wikimedia and the Wikimania conferences, Processing Foundation, Personal Democracy Forum, RightsCon, the Stockholm Internet Forum, the Internet Governance Forum, and Ethan Zuckerman's work, formerly at the Center for Civic Media at MIT, now at the Institute for Digital Public Infrastructure at the University of Massachusetts Amherst.

25. In association with the "Please Don't Include Us" symposium organized by the wonderful Joan Donovan, formerly of Harvard's Shorenstein Center.

26. Payton Croskey, "The Augmented Undercommons and the Path to the Sun: An Exploration of Liberatory Technology and other Revolutionary Tools," April 27, 2022, video, 2:56, https://mediacentral.princeton.edu/media/The+Augmented+Undercommons+and+The+Path+to+The+SunA+An+Exploration+of

+Liberatory+Technology+and+other+Revolutionary+Tools%2C+Payton
+Croskey%2C+UG+%2723+%283963813%29/1_ciut22ep.

27. Such as when Dr. Joy Buolamwini donned a white mask, as shown in the Net-
flix/PBS documentary *Coded Bias*, to both reject and implicate the oppressiveness
of algorithmic facial recognition, or when Chris Gilliard, profiled earlier in this
book, obscures his own face in official photos for stories profiling his work.

28. Tong Ong, "Princeton Online Course Displays Bones of Black Children Killed
by Police," *College Post*, April 23, 2021, https://thecollegepost.com/princeton
-black-childrens-bones/.

29. Defend Democracy (website), https://defenddemocracy.eu.

30. Alice Stollmeyer, "The Vulnerable Cyborg," *Protego Press*, November 2, 2018,
https://www.protegopress.com/the-vulnerable-cyborg/.

31. Stollmeyer, "Vulnerable Cyborg."

32. Josh Constine, "Facebook Confirms It's Building Augmented Reality Head-
set," *TechCrunch*, October 24, 2018, https://techcrunch.com/2018/10/24/face
book-ar-headset; Lauren Golembiewski, "Are You Ready for Tech That Connects
to Your Brain?," *Harvard Business Review*, September 28, 2020, https://hbr.org
/2020/09/are-you-ready-for-tech-that-connects-to-your-brain; Timothy Revell,
"Mind-Reading Devices Can Now Access Your Thoughts and Dreams Using
AI," *New Scientist*, September 26, 2018, https://www.newscientist.com/article
/mg23931972-500-mind-reading-devices-can-now-access-your-thoughts-and
-dreams-using-ai, quoted in Stollmeyer, "Vulnerable Cyborg."

33. See, for example, "Brene Brown: How Vulnerability Can Make Our Lives
Better," interview by Dan Schwabel, *Forbes*, April 21, 2013, https://www.forbes
.com/sites/danschawbel/2013/04/21/brene-brown-how-vulnerability-can-make
-our-lives-better/?sh=57638b2b36c7.

34. Many thanks to Justin Hendrix, editor in chief of the excellent publication
Tech Policy Press, for curating a wonderful community of experts (and including me
in it) on the app Signal.

35. Stollmeyer, "Vulnerable Cyborg."

36. Kate O'Neill, *Surviving Death: What Loss Taught Me About Love, Joy, and Meaning* (self-pub., CreateSpace, 2015), chap. 7, Kindle.

37. O'Neill, *Surviving Death*, chap. 7.

38. Alexis Pauline Gumbs, "Evidence," in *Octavia's Brood: Science Fiction Stories from
Social Justice Movements*, ed. adrienne maree brown and Walidah Imarisha (Chico,
CA: AK Press, 2015), 34.

39. Gumbs, "Evidence," 39.

40. O'Neill, *Tech Humanist*, 4.

41. Timothy Jorgensen, *Spark: The Life of Electricity and the Electricity of Life* (Princeton, NJ: Princeton University Press, 2021); "Energy Cost Calculator," Electrical Calculators, Calculators, RapidTables, http://rapidtables.com/calc/electric/energy-cost-calculator.html.

CHAPTER 8

1. An earlier, abbreviated version of this chapter appeared in the fall 2023 "ethics issue" of *MIT Technology Review* magazine.

2. David Ryan Polgar et al., "Strengthening the Responsible Tech Ecosystem in 2023," December 6, 2022, video, 56:11, https://www.youtube.com/watch?v=YBi7zVjUAh8.

3. "About ATIH," All Tech Is Human, accessed January 12, 2024, https://alltechishuman.org/aboutatih.

4. United States Office of the Surgeon General et al., *Our Epidemic of Loneliness and Isolation: The U.S. Surgeon General's Advisory on the Healing Effects of Social Connection and Community* (Washington, DC, 2023), https://www.hhs.gov/sites/default/files/surgeon-general-social-connection-advisory.pdf. With thanks to my wonderful humanist chaplain colleague James Croft, with whom I originally researched these issues for what was to be a coauthored project that was never published.

5. Julianne Holt-Lundstad et al., "Loneliness and Social Isolation as Risk Factors for Mortality: A Meta-Analytic Review," *Perspectives on Psychological Science* 10, no. 2 (2015): 227–237, https://web.archive.org/web/20190325114311/http://www.ahsw.org.uk/userfiles/Research/Perspectives%20on%20Psychological%20Science-2015-Holt-Lunstad-227-37.pdf (site discontinued).

6. US Department of Health and Human Services, "Surgeon General Issues New Advisory About Effects Social Media Use Has on Youth Mental Health," press release, May 23, 2023, https://www.hhs.gov/about/news/2023/05/23/surgeon-general-issues-new-advisory-about-effects-social-media-use-has-youth-mental-health.html.

7. ATIH, "Responsible Tech Talent Pool," n.d., https://alltechishuman.org/responsible-tech-talent-pool-individuals.

8. Center for Humane Technology (website), https://www.humanetech.com.

9. Yaël Eisenstat, "I Worked on Political Ads at Facebook. They Profit by Manipulating Us," *Washington Post*, November 4, 2019, https://www.washingtonpost.com/outlook/2019/11/04/i-worked-political-ads-facebook-they-profit-by-manipulating-us.

10. Yaël Eisenstat and Nils Gilman, "The Myth of Tech Exceptionalism," *Noema*, February 10, 2022, https://www.noemamag.com/the-myth-of-tech-exceptionalism.

11. US Department of Health and Human Services, "Surgeon General Issues New Advisory."

12. Fairplay, accessed January 12, 2024, https://fairplayforkids.org.

13. Douglas Rushkoff, "Douglas Rushkoff: I Will Not Be Autotuned—Crashing Technosolutionism," filmed April 26, 2023 in New York, NY, YouTube video, 29:27, https://www.youtube.com/watch?v=RCVXaqSEqmI, 0:03–0:07.

14. Neil Postman, *Technopoly: The Surrender of Culture to Technology* (New York: Vintage Books, 1993).

15. Postman, *Technopoly*, 66.

16. Yaël Eisenstat, in discussion with the author via Zoom, January 16, 2024.

17. Renée Cummings, in discussion with the author via Zoom, January 22, 2024.

18. Chuck Schumer et al., "Statements from the Fifth Bipartisan Senate Forum on Artificial Intelligence," Washington, DC, November 9, 2023, https://www.schumer.senate.gov/newsroom/press-releases/statements-from-the-fifth-bipartisan-senate-forum-on-artificial-intelligence.

CONCLUSION

1. Lesley Hazleton, *Agnostic: A Spirited Manifesto* (New York: Riverhead, 2016), 21.

2. Phil Zuckerman, email message to author, January 20, 2024.

3. Robin Le Poidevin, "What Is Agnosticism?," in *Agnosticism: A Very Short Introduction* (Oxford: Oxford University Press, 2010), https://doi.org/10.1093/actrade/9780199575268.003.0002.

4. Hazleton, *Agnostic*, 5.

5. To return to Eusebius's description of Constantine's dilemma, as I explored at the beginning of the introduction to this book: Eusebius, *The Life of Constantine*, trans. Ernest Cushing Richardson, in *A Select Library of Nicene and Post-Nicene Fathers*, ed. Philip Schaff and Henry Wace, 2nd ser. (New York: Christian Literature, 1890), 1:489, https://archive.org/details/cu31924031002102.

6. Hazleton, *Agnostic*, 5–6.

7. Kimberly Marlowe Hartnett, "'Biography' of Jezebel Mixes Scripture, History and a Vivid Imagination," review of *Jezebel: The Untold Story of the Bible's Harlot Queen*, by Lesley Hazleton, *Seattle Times*, October 26, 2007, https://www.seattletimes.com/entertainment/books/biography-of-jezebel-mixes-scripture-history-and-a-vivid-imagination.

8. Gordon Pennycook, James Allan Cheyne, Nathaniel Barr, Derek J. Koehler, and Jonathan A. Fugelsang, "On the Reception and Detection of Pseudo-Profound

Bullshit," *Judgment and Decision Making* 10, no. 6 (2015): 549–563, https://psycnet .apa.org/record/2015-54494-003.

9. Antonio A. Arechar et al., "Understanding and Combatting Misinformation across 16 Countries on Six Continents," *Nature Human Behaviour* (2023), https:// www.nature.com/articles/s41562-023-01641-6.

10. Gordon Pennycook, Jabin Binnendyk, and David Rand, "Overconfidently Conspiratorial: Conspiracy Believers Are Dispositionally Overconfident and Massively Overestimate How Much Others Agree with Them," preprint, submitted December 6, 2022, https://psyarxiv.com/d5fz2.

11. David Rand and Gordon Pennycook, in discussion with the author via Zoom, May 30, 2023.

12. Madeline Ostrander, "How Do You Decide to Have a Baby When Climate Change Is Remaking Life on Earth?," *Nation*, March 24, 2016, https://www .thenation.com/article/archive/how-do-you-decide-to-have-a-baby-when-climate -change-is-remaking-life-on-earth.

13. Lesley Hazleton, "Jeff Bezos: How He Built a Billion Dollar Net Worth before His Company Even Turned a Profit," *Success*, July 1998, 58–60.

14. Lesley Hazleton, "On Reading the Koran," filmed at TEDxRainier, October 2010, in Seattle, WA, TED video, 9:17, https://www.ted.com/talks/lesley _hazleton_on_reading_the_koran.

INDEX

Note: Page numbers in italic type indicate illustrations.

AB 5. *See* Assembly Bill 5
Abortion, 191–192
Accumulation. *See* Increase and
  accumulation
Achievement. *See* Striving and
  achievement
Addiction, 151, 152–155, 162,
  164–168
African Americans
  in Detroit, 177–180
  enslavement of, 120
  and facial recognition technologies,
    178, 190–191
  hypervisibility of, 175–176
  intelligence of, 94–95
  religiosity of, 36–37
  religious discrimination against,
    125
  and surveillance technologies,
    193–194
  as tech leaders, 122
  US discrimination against, 126
AGI. *See* Artificial general intelligence
Agnatogenesis, 116
Agnosticism. *See also* Tech agnosticism
  atheism compared to, 98, 285
  characteristics of, 285
  criticisms of, 285–286
  and disinformation, 289
  in EA, 70
  embrace of unknowability in,
    286–288

humanism and, 17–18, 47, 79, 243–
  244, 285, 287
strong vs. weak, 286
AI. *See* Artificial intelligence
AI for the People, 256
AI Now Institute, 211, 220
Albright, Julie, 155, 157
Algorithms
  addictive behavior triggered by,
    154–155
  bias in, 131–134, 190
  health effects of, 156
  ridesharing, 212, 217–218
  social work and, 248
All Tech Is Human (ATIH), 269–284,
  271
  "Responsible Tech Guide," *275*
Alphabet, 3
Altman, Sam, 3
Altos Labs, 188, *189*
Altria, 137
Amazon, 3, 40, 61, 66, 127, 187, 256,
  281
American Academy of Child and
  Adolescent Psychiatry, 149
American Psychological Association,
  170
Ames, Morgan, 44–45
Anderson, Chris, 74
Andreessen, Marc, 1, 3, 84
Angelou, Maya, 231, 236
Anhalt, Emily, 159, 164, 170–171

Anxieties/fears. *See also* Uncertainty
  in contemporary society, 7, 25, 44,
    51
  eating as response to, 166–167,
    169–170
  about personal worth and happiness,
    21, 34, 36
  religion as response to, 40, 159–160
Apocalypse, 173–206
  ancient literature on, 173–175
  in popular culture, 183–186
  surveillance and, 175, 177–178,
    193–194, 199, 201, 204–205
  in tech, 185–186, 206
Apple, 131, 136, 220, 232
Aquinas, Thomas, 58
Arechar, Antonio, 289
Armstrong, Karen, 245
Artificial general intelligence (AGI), 89
Artificial intelligence (AI)
  aftermath of, 99, *100*
  alignment of, 89, 91
  beneficial AI, 99
  concept of, 16
  EA and, 75, 91, 112
  existential risk posed by, 91
  and ghost work, 249–250
  as god, 47–48, 99
  LessWrong and, 76
  politics and, 284
  skepticism about, 254
  tech's relationship to, 16
Artificial life, 13–14
Aslan, Mina, 280
Assassins, Order of, 81
Assembly Bill 5 (California), 212, 217,
    219, 229–231
Astronomical waste, 79, 88
AT&T, 186
Atheism
  agnosticism compared to, 98, 285
  the author and, 17
  and community, 272

criticisms of, 111
  in EA, 70
  Harvard Divinity School and, 31
  humanism and, 17–18, 47, 79, 118,
    186, 234–235, 243, 287
  in tech, 30, 49, 51, 105
ATIH. *See* All Tech Is Human
Atipica, 141
Augmented Undercommons, 258–260
Augury, 192
Augustine, Saint, 58
Authoritarianism/autocracy, 104–105,
    227, 283, 290
Ayatollah Khomeini of Iran, 210

Balance, 120, 144, 145, 164, 200
Bankers, 71
Bankman-Fried, Sam, 68–70, *69*, 90
Barrat, James, 89–90, 96
Bass, Diana Butler, 30–31
Beard, S. J., 87
Becker, Ernest, 51
Beers, Kendrea, 273
Bender, Emily, 253
Benjamin, Ruha, 223, 257–258, *258*,
    260, 266
Bensinger, Rob, 315n82
Bergstrom, Carl, 116, 226
Berkman Klein Center for Internet and
    Society, 220, 249
Betaworks, 282
Bezos, Jeff, 3, 40, 66, 187, 188, 294
Bhatlapenumarthy, Harsha, 277
Biden, Joe, 12, 69, 283
Binelli, Mark, 179
Binnendyk, Jabin, 289–290
Biometrics, 191, 193
Bitcoin, 84–85, 102, *103*
BITE Model of Authoritarian Control,
    104–105, *106*
Black Lives Matter, 190
Blade Runner (film), 200
Blair, Tony, 68, *69*

Book of Daniel, 173
Book of Revelation, 173–174, 184
Bostrom, Nick, 74–82, 88–89, 94–95,
    191–192
    "Are We Living in a Computer
        Simulation?," 74, 76
    "Astronomical Waste," 79
    "Existential Risk Prevention as
        Global Priority," 110
    "Existential Risks," 88
    "Letter from Utopia," 80–81
    Superintelligence, 89
    "Vulnerable World Hypothesis,
        The," 191–192
Bowie, David, "The Man Who Sold
    the World," 261
Brahmins, 120, 125
Branson, Richard, 29
Brennan, Jason, 94
Brooks, Cat, 223
Broussard, Meredith, Artificial
    Unintelligence, 130
Brown, Brené, 262
Browne, Simone, 193
Bruck, David, 133
Buber, Martin, 86, 245
Buddha, 67, 241
Buddhism
    and charity, 67
    conception of hell in, 87
    doctrine in, 59
    as a religion, 8
Building a Better Tech Future for
    Children, 278–282
Bullshit, 226–229, 289
Burek, Josh, 33
Buterin, Vitalik, 75
Butler, Octavia E., Parable of the Sower,
    184

Cahn, Albert Fox, 192
Callahan, David, 68
Calvin, John, 58

Calvinism, 33
Camp Glen Brook, 161, 168
Campus Crusade for Christ, 248
Cannon, Lincoln, 50
Capitalism. See also Surveillance
    capitalism
    gig economy and, 232
    inequalities in, 39–40
    Luddite response to, 225–226
    as religion, 19, 40, 45–46
    self-reinforcement of, 52–53
    tech as component of, 19
Carey, James, 98, 152
Carnegie, Andrew, The Gospel of
    Wealth, 45–46
Carson, Rachel, Silent Spring, 184
Carvaka school of skeptics, 67
Caste systems, 124–127. See also
    Hierarchies and castes
Castro, Fidel, 195
Catholic Church, 73, 107, 177, 195,
    209–210, 239–241, 243–244,
    276
Cellebrite, 191
Cellular regeneration, 188
Center for Humane Technology, 222,
    276
Centre for Effective Altruism, 69–71
Chalmers, David J., Reality+, 76–77
Chalom, Adam, 127
Chang, Shih-Fu, 247
Charismatic authority, 45
Charity/philanthropy, 66–73, 279
ChatGPT, 5, 132
Chew, Shou, 281
Chief Keef (Keith Farrelle Cozart),
    247
Chosenness, 91–97
Christianity
    and charity, 67
    conception of hell in, 87
    conception of religion promoted
        by, 8

Christianity (cont.)
  concept of doctrine originating in, 58
  Constantine and, 2, 4, *5*
  criticisms of, 79
  and the Crusades, 81
  declining membership of, 60
  impact of, 8
  as model for our experience of life, 8
  Pascal's wager and, 79–80
  violence conducted by, 81
  and white supremacy, 129
Citron, Danielle, 86
Civil War, 175
Clarke, Laurie, 85
Cleaners, The (documentary), 185
Climate change, 66
Clinton, Bill, 11, 68, *69*, 231
Cohanim, 127
Cohen, Shaye, 1
Colbert, Stephen, 29
Colonialism, 82–86
Comic-Con International (San Diego, 2016), *224*
Comics, 24, 186–190, 197–199
Community
  All Tech Is Human and, 269–284
  concern for, 170
  congregations as, 272
  happiness and fulfillment offered by, 161–162
  humanism in relation to, 37
  religion as means of experiencing, 9
  social media and, 158–159
  tech's avoidance of, 204–205
  tech's promise of, 3, 7, 17
Competition, 162, 167, 175. *See also* Striving and achievement
Competitive Enterprise Institute, 66
Computer manufacture, 131
Conceivable Future Project, 291

Congregations, 269–284
Conspiracy theories, 288, 290
Constantine the Great, 1–2, 4, *5*, 174, 287
Conway, Erik, *The Collapse of Western Civilization*, 184
Cosmopolitanism, 42–43
Costa, Maren, 66
Cotler, Irwin, 276–277
Couldry, Nick, *The Costs of Connection*, 83–85
COVID-19 pandemic, 147, 149, 229, 336n40
Cox, Harvey, "Religion and Technology: A New Phase," 16–17
Croskey, Payton, 258–261, *258*, 266
Crusades, 81
Cryptocurrency, 84, 105, 107, 214–215. *See also* Bitcoin
Cuba, 37, 195
Cults, 101–107, 136
Cummings, Renée, 284
Cushing, Richard, 240, 241
Cuvier, Georges, 87
Cybersecurity for Democracy, 283
Cyborgs, 57, 79, 86, 262

Dalits (untouchables), 120–121, 125
Data colonialism, 84
Dataminr, 190
Davis, Zachary, 81
Dawkins, Richard, 7, 79, 111, 288
DC Comics, 57, 190, 197
Dead Sea Scrolls, 173
Death
  anxiety over, 9, 49, 51, 160
  in contemporary society, 44, 203
  denial of, 51
  Freudian theory about, 23–24
  humanist approach to, 263–264
  loneliness and isolation leading to, 272

love as defeat of, 264
  tech's attempt to conquer, 12, 57, 80,
    88, 188, *189*
  Thanatos (death instinct), 23–24
Defend Democracy, 261
Dehumanization, 33, 84, 160, 211–
  212, 250
Deitsch, Chana, 273
Deliberto, Tara, 165–167, 169–170,
  171, 172
Delo, Ben, 75
Democracy, 62, 68, 124, 126, 147, 155,
  227–228, 261–262, 278, 283–284,
  290. *See also* Politics
Dennett, Daniel, 9
Descartes, René, *Meditations on First
  Philosophy*, 31
Detroit, Michigan, 24, 175–183, 205–
  206
Digital beings, 77–80
Disinformation/misinformation, 10,
  134, 155, 186, 192, 206, 237,
  276–277, 283, 288–290
Disney, 195, 197–198
Disruption, 84, 85, 169, 216, 227–228,
  233, 278
Distributed AI Research Institute, 254
Doctrine
  charity/philanthropy, 66–73
  chosenness, 91–97
  colonialism, 82–86
  cult, 101–107
  defining, 58
  fanaticism, 108–111
  God/gods, 97–101
  heaven, 80–82
  hell, 86–91
  moral values/ethics, 59–62
  prophecy, 63–66
Donig, Deb, 62
DoorDash, 231
Dopamine, 153, 162, 164
Dorsey, Jack, 3, 58

Druyan, Ann, 241, 337n4
Dubal, Sam, 230–231, 235–236
Dubal, Veena, 211–212, *213*, 215–216,
  218, 220, 226, 229–231, 234–237,
  256, 261
  "New Racial Wage Code, The,"
    236
Durkheim, Émile, 45, 149, 158
Dysgenics, 95

EA. *See* Effective altruism
Earn to give, 70–71
Eating and eating disorders, 164–167,
  169–170
Ecclesiastes (Koheleth), 98
Edwards, Jonathan, 33
Effective altruism (EA)
  and AI, 75, 112
  contributions of, to future risk and
    disaster, 185
  criticisms of, 78, 82
  declining reputation of, 112
  and existential risk, 3
  fanaticism displayed by, 108–111
  focus of, 90–91
  and philanthropy, 69–73
Ehrman, Bart, 2
80,000 Hours (website), 90–91
Einstein, Albert, 229
Eisenstat, Yaël, 277–278, 283–284
Electronic Frontier Foundation, 223
Element AI, 131
"Elephant in the Valley" study, 135
Ellington, Duke, 179
Elliott, Missy, 154, *154*, 170
Ellis, Albert, 234
El Oddi, Bahia, 165–166, 169, 172
EmTech Next, 63–64
Epstein, Greg M., *242*
  Good Without God, 17, 38, 60, 118,
    241
Erickson, Ingrid, 44–45
Eschatology, 17

Ethics/morality
    doctrine and, 59–62
    fanatical approaches to, 109
    humanism and, 17–18, 38, 118
    religion and, 60–61
    of ridesharing/gig economy, 212,
        215–219
    social media and, 61, 273
    in tech, 3, 219–221, 229, 257, 279
Eubanks, Virginia, 193–194
Eugenics, 75, 95, 182
Eusebius, 1–2
Existential risk, 3, 87–88, 91, *92*, 97–
        98, 110, 112
Ex-Moon Inc., 104
Experiential avoidance, 165, 167–168
Extinction. *See* Existential risk

Facebook, 3, 61, 131, 133, 220, 221,
        249, 277–278
Facial recognition technologies, 178,
        190–191, 193, 211, 247
Fairplay, 282
Fajardo, Glenn, 151
False consciousness, 124
Fanaticism, 108–111
Farha, Leilani, 221
Fears. *See* Anxieties/fears
Federal Bureau of Investigation (FBI),
        191, 194
Federal Communications Commission,
        187
Fernando, Randima, 222
Ferrick, Tom, 239–241, *242*, 244, 248,
        254, 255–256, 263
Flavius Josephus, 174
#fliplife, 232
Foerst, Anne, 22
    "Religion and Technology: A New
        Phase," 16–17
Forbes (magazine), 40
Ford, Harrison, 200
Ford, Henry, 181–182

Ford Foundation, 274
Ford Model T factory, Detroit, 180–
        181, *182*, *183*
Forrest, Andrew, 182–183
Fortescue, 182
Foundation of Humane Technology
        (course), 222
Founders Fund, 316n96
Fowler, Susan, 230
Francis, Pope, 245
Frankfurt, Harry, 226
Freedom Riders, 284
Freire, Paulo, 84
Freud, Sigmund, 23–24, 253
FTX, 71–72
FTX Foundation, 71, 94
Furman, Jason, 4, 6
Future Fund, 71–72
Future of Humanity Institute, 74

Gates, Bill, 169
Gebru, Timnit, 82, 91, 234, 253–254,
        257
Genius, 94–96
Geofencing, 192
Gershenfeld, Neil, *Fab: The Coming
        Revolution to Your Desktop*, 41
Ghost work, 249–250
Gibson, William, 206
Gig economy, 211–212, 215–219,
        230–233, 236. *See also* Ghost work
Gig Workers Rising, 218–219
Gilliard, Chris, 175–178, *176*, 180–182,
        186–190, *189*, 196–206, *201*, 256
Giridharadas, Anand, 39–41, 228
Glide Memorial Church, San
        Francisco, 29, 32
God/gods. *See also* Theology
    AI as, 47–48, 99
    attention-seeking by, 159
    doctrine and, 97–101
    Google as, 98–99
    jealousy as attribute of, 210

Goff, Trey, 85
Golden Rule, 60–61
Golin, Josh, 282
Gómez, Laura, 139–142, 144–146
Gonzalez, Jacinta, 223
Good vs. evil, 174–175
Google, 3, 47, 61, 98–99, 131–134,
    156, 194, 251, 253, 266
Google Gemini, 132
Google searches, 97–98, 132–133
Government surveillance, 190–192
Gramsci, Antonio, 146
Gray, Mary, 249–250, 253, 254, 257
    Ghost Work, 249–250
Greaves, Hilary, 77
    "Case for Strong Longtermism,
    The," 109–110
Growth. *See* Increase and accumulation
Guild of Future Architects, 256
Gumbs, Alexis Pauline, "Evidence,"
    266
Gupta, Meili, 64–65
Gwin, Cate, 282
Gypsies, 125

Ham (biblical figure), 125
Hamill, Mark, 219
Hạnh, Thích Nhất, 245
Happiness, 21, 34, 69, 138–139, 269
Harney, Stefano, 258–259
Harris, Sam, 7, 95, 111, 288
Harris, Tristan, 222, 276
Harry, Prince, 222
Harvard Alumni for Mental Health,
    165–166
Harvard Business School, 52
Harvard Divinity School, 17, 30–31,
    74, 220
Harvard University, 17–18, 33, 38, 40,
    118, 122, 135–136, 139, 199, 216,
    241, 248
    Berkman Klein Center for Internet
    and Society, 220, 249

Kennedy School of Government,
    25, 257
Hassan, Steve, 102, 104–107, 157–158
Haught, John F., 9, 304n25
Hazleton, Lesley, 285–288, *287*, 294–
    295
    Agnostic: A Spirited Manifesto, 285
HBO, 186
Health, mental and physical. *See
    also* Loneliness and isolation;
    Psychology of tech
    eating disorders, 165–167, 169–170
    of leaders in tech, 140–142
    loneliness's effect on, 272
    of lower-level tech workers, 185
    social media's effect on, 155, 157,
    162, 166, 272, 280
    tech's effect on, 155–159, 165–167
Heaven, 80–82
Hegel, G. W. F., 84
Heidegger, Martin, 10
Hell, 86–91
Helmreich, Stefan, 13–14
Hendrix, Justin, 281–282
Henry Ford Museum of American
    Innovation, 181–182
Heretics and apostates, 209–237
    defining, 213–214
    Luther, 209–211
    in tech, 214–237
Hernandez, Keith, 204
Herrnstein, Richard, *The Bell Curve*,
    95
Hewlett-Packard, 140
Hicks, Mar, 4, 6, 130
Hierarchies and castes, 117–146
    caste systems, 124–127
    denial of, 116, 119
    justifications of, 119, 121, 124, 125
    in religion, 119–122
    in Silicon Valley, 236
    in tech, 23, 119, 122–146, 236
    white male bias and, 127–134

Higher One, 117–118

Hill, Kashmir, 187

Hillis, Ken, 98

Hilton, Benjamin, 90

Hinduism
    caste system in, 120–121, 125
    and charity, 67
    conception of hell in, 87
    doctrine in, 59
    as a religion, 8
    violence conducted by, 81

Hitchens, Christopher, 7, 111

Hoelscher, David, 37–38

Homebrew Computer Club, 41

Homelessness, 32, 127, 232–233

Homer Simpson, *163*

House, Dave, 145

Humanism, 239–267. *See also* Secular
    humanism; Tech humanism
    agnosticism and, 17–18, 47, 79,
        243–244, 285, 287
    atheism and, 17–18, 47, 79, 118,
        186, 234–235, 243, 287
    characteristics of, 17–18, 37–38,
        241, 243
    manifestos of, 293
    meaning sought/found through, 18,
        37, 101, 263–264
    and morality, 17–18, 38, 118
    mysticism and, 252–253
    original meaning of, 337n3
    Renaissance, 101
    the sciences vs., 337n3
    and uncertainty, 292

Humanist chaplains, 17, 18, 21, 34, 38,
    118, 165, 213, 241, 255, 259, 264

Humanist Hub, 18

Humanity. *See also* Dehumanization;
    Existential risk; Human worth
    actions beneficial to, 51, 160–161
    fundamental questions of, 8, 11, 42
    human connection as a right for, 282
    meaning of, 241, 243, 265–266

Human worth. *See also* Humanity;
    Striving and achievement
    capacity for loving and being loved
        as source of, 22–23, 25, 52, 86,
        167, 199, 265
    feelings of lack of, 21–22, 52–54,
        139
    individualist emphasis in, 38
    measures of, 20–22, 162–163, 166–
        168
    tech as response to anxiety about, 21

Husain, Sarah, 273

Huxley, Aldous, *Brave New World*, 11

Huxley, Julian, 75

Huxley, Thomas Henry, 286

Hyman, Louis, 131

Hypervisibility, 175–176

IBM, 131

ICE. *See* US Immigration and Customs
    Enforcement

Increase and accumulation, 20, 43,
    205

Indigenous peoples, 83, 110, 120, 265,
    307n9

Individuality. *See* Self and individuality

Indulgences, 73, 209

Inequalities/inequities, 32–34, 36–40,
    42–43, 65

Inevitability, of tech leadership, 92–94

Ingersoll, Robert, 286

Instacart, 231

Intel, 144–145

Intelligence/IQ, 94–96

Isaac, Mike, 93

Islam and Muslims
    and charity, 67
    conception of hell in, 87
    doctrine in, 59
    heretics and apostates, 213–214
    and portrayals of Muhammad, 294
    violence conducted by, 81

Isolation. *See* Loneliness and isolation

Jackson, Sam, 134
JAMA Pediatrics, 149
January 6, 2021, insurrection, 129,
134, 158, 277
Jargon, Julie, 166
Jarrett, Kylie, 98
JavaScript, 130
Jews. *See* Judaism
Joan Ganz Cooney Center, 278
Jobs, Steve, 136, 169
John (author of Book of Revelation),
174
Johnson, Lyndon Baines, 24
Jordan, Jamon, 178
Judaism
apocalypticism in, 172
and Buberian thought, 86
and charity, 67
chosenness in, 92
conception of hell in, 87
defining, 24
and destruction of the Second
Temple, 174
doctrine in, 59
and family history (author's), 195–
196
hierarchies in, 120, 127
Zevi's messianic movement and, 55–
56, 107–108
Justitia, 200
JUUL, 136–137

Kalanick, Travis, 93–94, 230
Kamikaze pilots, 81–82
Kant, Immanuel, 12
Kardashian, Kim, 219
Keen, Andrew, 98–99
Kenneally, Gavin, 187–188
Kennedy, Elizabeth J., 95–96, 109
Kennedy, John F., 181, 240
Kerby, Richard, 123
Khalil, Sandra, 279, 282, 283
Khan, Lina, 220

Khosrowshahi, Dara, 93
King, Martin Luther, Jr., 10, 17, 135,
194, 240, 247
Klein, Naomi, 211
Kleiner Perkins, 134
Koch, David, 279
Koheleth (Ecclesiastes), 98
KO Insights, 243
Kornfield, Jack, 255
Kurzweil, Ray, 50, 57, 88, 89

Lacy, Sarah, 230, 234
LaMDA, 251–254
Lasater, Miles, 117–119, 122, 129, 144
Latinx peoples, 138, 140–141, 146,
222–223
Leaders/leadership
African American, 123
authoritarian, 227, 290
inevitability of, 92–94
mental and physical health of, 140–
142
mythologizing of, 42, 128–129
as prophets or saviors, 44–45, 51
values of, 137–139
white male bias in, 132, 137
women as, 64, 131
Leary, Timothy, 270
Lee, Bo Young, 135
Lemoine, Blake, 251–254, 257
Leo X, Pope, 209
LessWrong, 48–49, 74, 75–76
Levandowski, Anthony, 47–50
Leviim, 127
LGBTQ+ people, 127
Light Phone, 148, 154, 158, 172
Lil JoJo (Joseph Coleman), 247
Limitlessness. *See* Increase and
accumulation
Logic (magazine), 130, 225
Lokayata school of skeptics, 67
Loneliness and isolation, 162–163, *163,*
204, 218, 262, 272

Longtermism, 74, 82, 88, 94–95, 109–
    110, 185
Love
    death overcome by, 264
    human worth linked to, 22–23, 25,
        52, 86, 167, 199, 265
    meaning linked to, 266
    and meaning of human existence,
        241, 243, 265–266
    obstacles to, 265
    source of, 95
    technology in relation to, 265
Luddite Club, 225
Luddites, 99, 225–226
Luther, Martin, 58, 73, 209–211
Luttichuys, Isaack, portrait of Jacob
    Sasportas, *108*
Lyft, 212, 215–219, 228, 231–232, 236

MacAskill, William, 70–73, 77, 78, 90,
    108–110
    "Case for Strong Longtermism,
        The," 109–110
Macomb Community College,
    Michigan, 175
Mad Max: Fury Road (film), 202–203
Magic, 158
Maker movement, 41
Mandela, Nelson, 276
Manslaughter, 202–203
Manual labor, 203–204
Manusmriti, 125
Mao, Julie, 223
Markle, Meghan, 222
Martin, Trayvon, 133, 259
Marvel Comics, 24, 190, 197–198
Marx, Karl, 10
Masculinity, 202, 204. *See also* White
    male bias
Maslow, Abraham, 38
Massachusetts Institute of Technology
    (MIT), 17–18, 199, 216
    Artificial Intelligence Laboratory, 17

Fabrication (Fab) Lab, 41
Media Lab, 45
Stephen A. Schwarzman College of
    Computing, 220–221
Matrix, The (film trilogy), 17, 74
Maxentius, 1–2
Mayer, Oscar, 181
McCarthy, Caroline, 51
McKeown, Kathy, 247
McLeod, Pat, 248
Meaning. *See also* Happiness; Human
    worth; Theology
    humanism as means of seeking/
        finding, 18, 37, 101, 263–264
    love linked to, 241, 243, 266
    parenting and, 291–292
    religion and theology as means of
        seeking/finding, 9, 30–31, 42
    tech as means of finding, 42–44
Mehta, Abhishek, 84
Mejias, Ulises, 223
    Costs of Connection, The, 83–85
Men. *See* Masculinity
Meritocracy, 116, 119, 122–124, 130
Messianism, 55–58, 107–108. *See also*
    Saviors and salvation
Meta, 3, 134, 158–159, 220, 277,
    281
#MeToo movement, 135
Microsoft, 220, 249
Microsoft Encarta, 269
MicroStrategy, 101
Mijente, 222–223, 226, 234
Miller, Alice, *The Drama of the Gifted
    Child*, 20, 22
Milovidov, Elizabeth, 279
Minsky, Marvin, 17, 18
Mishnah, 76
Misinformation. *See* Disinformation/
    misinformation
Mistrust. *See* Trust/mistrust
MIT. *See* Massachusetts Institute of
    Technology

Mitchell, Margaret, 253–254
MIT Technology Review (journal), 62
Mobius, 255
Mohammed bin Salman, 221
Moon, Sun Myung, 104
Moore, Gordon, 145
Moore's Law, 145
Morality. *See* Ethics/morality
Moral licensing effect, 73
Moral Majority, 59–60
Moravec, Hans, 18
Mormon Transhumanism, 50
Morse, Samuel, 152
Moskovitz, Dustin, 75
Moten, Fred, 258–259
Muhammad (prophet), 294
Murray, Charles, *The Bell Curve*, 95
Murthy, Vivek H., 272
Musk, Elon, 3, 57–58, 61–62, 71, 75, 84, 175, 186, 187, 273
Muslims. *See* Islam and Muslims
Mysticism, 252–253

National Institute of Standards and Technology, 190
National Science Foundation, 272
Native Americans. *See* Indigenous peoples
Nazis, 182, 194, 196
Negative reinforcement, 167
Negroponte, Nicholas, 45
Newsom, Gavin, 219
New Testament, 173–174
New York Times (newspaper), 135, 225
Nichols, John, 179–180
Nine Dots Prize, 156
Nkonde, Mutale, 256–257
Noble, David, *The Religion of Technology*, 13
Noble, Safiya, *Algorithms of Oppression*, 132–134
Northrop Grumman, 256
Nuns, 256–257, 260

Nye, David, *American Technological Sublime*, 12–13, 43

Oates, Joyce Carol, 150, *150*
Obama, Barack, 41, 133, 236
Octavia's Brood (science fiction anthology), 266
Old Testament, 49, 159, 173
Oluo, Ijeoma, 117, 128–130, 142, *143*
    Mediocre: The Dangerous Legacy of White Male America, 129
    So You Want to Talk about Race, 128
Omidyar Network, 279–280
O'Neill, Kate, 243–244, *245*, 257, 263–264, 266
    Surviving Death, 263–264
One Laptop per Child project, 44–45
Operación Pedro Pan, 195
Ord, Toby, 91
Oreskes, Naomi, *The Collapse of Western Civilization*, 184
Orsi, Robert, 49
Orwell, George, *1984*, 11
Ostrander, Madeline, 291–292
Otto, 47
Özenç, Kürşat, 151

Palantir, 191, 223
Paley, Eric, 93
Palihapitiya, Chamath, 162, 168
PandoDaily (website), 230
Pao, Ellen, 134–138, 142, *143*, 144–146, 256
Parabiosis, 188
Parenting, 20, 22–23, 52, 152–153, 167–168, 177, 239, 291–292
Parks, Rosa, 181
Pascal's wager, 79–80, 97–98
Patton, Desmond, 246–249, 257
PayPal, 232
Pennacook people, 161
Pennycook, Gordon, 289–290

Perfectionism, 166

Petit, Michael, 98

Petty, Tawana, 178

Philanthropy. *See* Charity/
    philanthropy

Philip Morris, 137

Phillips Exeter Academy, 64

Pichai, Sundar, 169

Pinn, Anthony, 31, 59

Polgar, David Ryan, 269–270, 274–
    279, 283–284

Politics. *See also* Democracy
    AI and, 284
    cult behavior and, 104
    tech's effect on, 156–158, 283–284

Posner, Miriam, 130

Postapocalyptic art, 183–184

Postman, Neil, 11–12
    Amusing Ourselves to Death, 11
    Technopoly, 11–12, 283

Prayer, 11, 23, 115, 149, 151, 240, 264,
    305n28

Predestination, 33–34, 46

Print technologies, 210

Programming, 130–131

Project Green Light, 177–178, 205

Project Include, 135, 139

Pronomos Capital, 84

Prophets and prophecy, 45, 63–66, 77

Proposition 22 (California), 231, 232,
    235–236

Próspera, 85

Protestant work ethic, 20, 33–34

Protests, 228

Proyecto Solace (Proyecto Sol), 145–
    146

Psychology of tech, 159–166. *See also*
    Health, mental and physical

Puritans, 33–34, 41, 44, 46

Putin, Vladimir, 261

Putnam, Robert, 272

Putter, Bretton, 102

Python, 130

Quishpe, Salvador, 82

Race. *See also* African Americans;
    Indigenous peoples; Latinx
    peoples; White male bias; Women
    of color
    algorithm bias based on, 131–134,
        190
    distribution in tech employment,
        122
    gig economy and, 236
    hypervisibility based on, 175–176
    and surveillance technologies, 175,
        193–194

Racial justice, 175

Rand, Ayn, 130

Rand, David, 289–290

Raskin, Jamie, 104, 158

Rationality community, 74

Rationalization, 84, 121, 138

Reagan, Ronald, 60

Red State (news outlet), 231

Redwood City, California, 139–140,
    144–145

Religion
    analogues to, 10, 42
    capitalism as, 19, 40, 45–46
    colonialism buttressed by, 83–86
    defining, 7–10
    hierarchies in, 119–122
    meaning sought/found through, 9
    and moral values, 60–61
    as path to alternate reality, 9
    reassurance as promise of, 9
    science as, 13–14, 42, 165
    tech compared to, 4, 7, 9–14, 16–18,
        44–45, 47–50, 119, 136, 158–159
    who benefits from, 9

reMarkable, 147–148, 154, 158

Responsible tech, 220

Responsible Tech University Summit,
    273

Rich, Michael, 280–282

Richardson, Ernest Cushing, 2
Ridesharing, 211–212, 215–219, 230–233, 236
Riley, Jebb, portrait of Chris Gilliard, *201*
Rituals, 148–172
    addiction and, 151, 155
    of communication, 152
    importance of, 148–149
    religious, 23
    in tech, 23, 151–152, 155–164, 171
Rivero, Annette, 217–220, 226, 231–233
Robles-Anderson, Erica, 31
Robot dogs, 187–188
Robots, 131, 204, 250–251, 265, 266. *See also* Robot dogs
Roe v. Wade, 191
Rogers, Jean, 282
Roko's Basilisk, 48–49
Romani, 125
Roof, Dylann, 133–134
Rosner, Daniela, 44
Roth, Yoel, 61–62
Rushdie, Salman, 210
Rushkoff, Douglas, 282–283
Russell, Bertrand, 78

Sagan, Carl, 241, 260
Salesforce, 61, 220
Salvation. *See* Saviors and salvation
Salzberg, Sharon, 255
Sanger, Margaret, 182
San Jose, California, 127, 232–233
San Mateo County, California, 144–145
Santa Fe Institute, 13
Sarkissian, Alex, 273
Sasportas, Jacob, 107–108, *108*
Saviors and salvation, 33, 44–45, 56–58, 66, 76, 107, 112, 159, 211. *See also* Messianism
Saylor, Michael, 101–102, *103*, 107

Schmidt, Eric, 99
Schumer, Chuck, 284
Science
    fundamental human questions addressed through, 11, 13–14, 42
    as religion, 13–14, 42, 165
    teaching of, 304n25
Screen Time Action Network, 282
Seagate, 131
Second Temple, Jerusalem, 174
Secular humanism, 18, 24, 47, 127, 288
Seinfeld (television show), 204
Seinfeld, Jerry, 204
Self and individuality, 38, 170, 204
Sesame Workshop, 278–282
"Seven generations" principle, 110, 265
Sexism
    in religion, 121, 127
    in tech, 130–131, 134–135, 139–142
    at Uber, 230
Sharf, Robert, 8
Shelley, Mary, *Frankenstein*, 87
Shontell, Alyson, 48
Shoreline, Washington, 127
Shuar, 83
Shunkamolah, William, 170, *171*
Siegel, Dan, 255
Silicon Valley
    atheism in, 51
    contradictions of, 36
    gig workers' protest against, 216–219
    hierarchies in, 236
    personal computing culture of, 41
    privilege in, 65
    villainy of the leaders of, 186–188
    Waldorf education in, 169
Silicon Valley (television show), 188, 219
Simulation hypothesis, 74, 76–77
Sinclair, Upton, 38, 72
Singer, Adam, 49

Singer, Peter, 69–72
Singularity, 50, 57
Skepticism, 145, 215, 226, 229, 254, 260, 288–289
Slavery, 33–34, *35*, 182–183, 193, 307n9
Smartphones, 6, 32, 57, 148, 149, 151, 154, 161, 164, 280
SMART Recovery, 153
Smith, Jonathan Z., 121, 138
Smith, Joshua, 250–251, *251*, 257
Social media
    algorithms in, 154
    communal aspects of, 158–159
    disinformation spread on, 10, 11, 134, 192, 289
    ethics and, 61, 273
    health effects of, 155, 157, 162, 166, 272, 280
    obsession with/addiction to, 11, 19, 162
    societal influences of, 246–247
    surveillance uses of, 190–191
    tech cults and, 102
Social work, 246–249
Soni, Varun, 157
Souls, 3, 81, 87, 251, 253–254
Spiritual practitioners, 25, 245–246, 250, 255, 257, 264, 265, 294
Srinivasan, Balaji, 84
Stanford Design School, 137, 151
Stanford University, 36, 122
Stankey, John, 186–187
Starlin, Jim, 196–201
    Dreadstar series, 197–200
Star Wars (film series), 223
Stevens, Wallace, 51
Stewart, Patrick, 219
Stollmeyer, Alice, 261
STRESS (police unit), 179–180
Striving and achievement, 20, 22, 160, 163, 166–168, 199. *See also* Competition

Sublime, the, 12–13, 43
Suffering
    AI utopia as answer to, 80
    colonialist causes of, 83
    EA's goal of alleviating, 69
    eating disorders as response to, 167
    justifications of, 34, 132
    religion as response to, 159
    as reward for evil, 174–175
    systemic causes of, 201
    tech's ignorance and avoidance of, 186
Sufism, 169
Summers, Larry, 135–136
Supervillains, 186–190, 198, 202–203
Suri, Siddharth, *Ghost Work*, 249–250
Surveillance capitalism, 175, 190, 259
Surveillance technologies
    critiques of, 175, 193–194, 199, 201, 204–205, 259
    deportation using, 223
    in Detroit, 24, 177–178
    government use of, 190–192
    power gained through use of, 199, 204–205
    proposals for, 190
    racial component of, 175, 193–194
Swisher, Kara, 134

Tab, 16
Tallinn, Jaan, 75
Talmud, 76
Tang, Kaiwei, 148, 154, 164, 172
Tarfon, 145
Tarnoff, Ben, 225–226
Team Cymru, 192
Tech. *See also* Silicon Valley
    AI's relationship to, 16
    concept of, 14, *15*, 16
    control as goal and promise of, 1, 4, 6
    dangers of, 3, 43
    goals of, 16
    impact of, 4, 10, 16

merging/unification of experience
    with, 13, 50
as path to alternate reality, 9
portrayed as objective, 130, 133–134
promises of, 3, 44–45
Puritanism compared to, 41
as religion, 4, 7, 9–14, 16–18, 44–
    45, 47–50, 119, 136, 158–159
self-serving nature of, 7, 20, 43
ubiquity of, 6–7, 149
who benefits from, 7, 9, 65, 119,
    160–161, 185–186, 228, 254
Tech agnosticism. *See also* Agnosticism
action principles for, 164, 171, 260
ambivalence as characteristic of, 17,
    85, 120, 142, 145, 164, 221
manifesto of, 293–294
uncertainty as characteristic of, 99
TechCrunch (online newspaper), 39,
    216–217, 221, 225, 246
Tech ethics, 219–221, 229, 257, 279
Tech humanism
    characteristics of, 24–25, 244
    digital undercommons and, 257–260
    as reformist alternative to tech as
        religion, 244–267
    and social work, 246–249
    vulnerability embraced by, 261–266
Technochauvinism, 130
Technology
    history of, 14, 19–20
    purpose of, 19–20
    tech compared to, 14
Techno-optimism, 3
Tech Policy Press (online journal), 281
TECH WARS: A #NoTechForICE
    Saga (course), 222–223
Tegmark, Max, 95
    Life 3.0, 99–101
Teleology, 184
Temple-Raston, Dina, 134
Ten Commandments, 210
TESCREALism, 82

Thanatos (death instinct), 23–24
Thanos (comic book character), 24
Theology. *See also* God/gods
    defining, 30–31, 59
    meaning sought/found through,
        30–31, 42
    of tech, 42–44, 46–51
Theranos, 94
Thiel, Peter, 47, 76, 84, 188, 316n96
Thompson, Derek, 10
Threads (film), 184
Thrivous, 50
Thuggis (thugs), 81
TikTok, 159, 280, 281
TIME (magazine), 71
Time Warner, 186
Tomlin, Lily, 109
Torres, Émile, 77–78, 80, 87, 91, 94–95
Toxic masculinity, 202
Transhumanism, 74–75, 78, 82, 88. *See
    also* Mormon Transhumanism
Tresata, 84
Trudeau, Justin, 276
Truman, Harry, 214
Trump, Donald, 47, 60, 93, 104, 175,
    186, 191, 192, 221, 227, 231, 277,
    283–284
Trust/mistrust, 62, 157, 180, 204, 227–
    229, 246
Tufekci, Zeynep, 228
Turner, Fred, 41, 44
Tutu, Desmond, 245
Tweed, Rebekah, 274–275, 278
Twenge, Jean, 157
Twitter (X), 58, 61, 175, 186, 187, 190

Uber, 47, 93, 212, 215–219, 230–232,
    236
Uber Eats, 232
Uncertainty. *See also* Anxieties/fears
    in ancient times, 160
    in contemporary life, 25, 43–44,
        261, 290–292

Uncertainty (cont.)
in human condition, 63, 200, 244, 262
Undercommons, 258–260
Undue influence, 104–105, *106*
Unification Church, 102, 104–105
United Nations, 75
United States
caste system in, 126
government surveillance in, 191–192
inequalities/inequities in, 32–34, 36–40
Universal Declaration of Human Rights, 75
Unknowability, 286–288
Untouchables, 120–121, 125
Urban renewal, 179
US Department of Homeland Security, 257
US Federal Trade Commission, 220
US Immigration and Customs Enforcement (ICE), 191, 223
US Securities and Exchange Commission, 101
US Supreme Court, 191–192
Utilitarianism, 78
Utopias and utopianism, 45, 80, 117, 128–134

Van Noppen, Aden, 255, 257
Venture capitalists, 36–37, 53–54, 84–85, 117–118, 123, 137, 217
Ver, Roger, 84
Verge (website), 11
Vice (magazine), 192
Virgo Supercluster, 77
Virk, Rizwan, 76
Virtual reality, 76–77
Vorkink, Peter, 65
Vulnerability, 21, 22, 156, 166, 169, 184, 194, 261–263

Waldorf educational movement, 168–169

Waldorf School of the Peninsula, 169
Ward, Nathaniel, 33
Warner Media, 186–187
Warren, Wendy, 307n9
Washington, Dinah, 179
Watson, Sara, 220, 221, 233–234
WAYFARE (magazine), 81
Way of the Future, 47–48
Web 2.0, 17
Web 3.0, 214–215
Weber, Max, 45
Weigel, Moira, 130, 225
Weiss-Blatt, Nirit, 112
Wells, H. G., 98
West, Jevin, 226
White, E. B., 291
White, Molly, 214–215
White Christian nationalism, 60
White male bias
in tech, 94, 117, 122, 127–134
in venture capital, 122
White supremacy, 129
Whittaker, Meredith, 211–212, 236–237
Wikipedia, 269–270
Wilcox, Russ, 94
Wilkerson, Isabel, 126
Wilkinson, Hayden, 109
Williams, James, 156–158
Stand out of Our Light, 156
Wilson, Russell, 219
Wine, Sherwin, 47, 98, 99, 285
Wired (magazine), 131, 222
Women. *See also* Sexism; Women of color
intelligence of, 136
in leadership roles, 64, 131
and programming, 130–131
in tech, 140
Women of color
and computer manufacture, 131
in leadership roles, 64
as tech founders, 141

Woodward Avenue Dream Cruise, 181
Worth. *See* Human worth

Yang, Andrew, 65, 117
Yield farming, 72
YouTube, 139, 159, 249
Yudkowsky, Eliezer, 49, 88–91, 94,
    96, 110

Zealots, 81
Zevi, Sabbetai, 55–56, 107–108, 112–
    113
Zuboff, Shoshana, 190, 193, 194
Zuckerberg, Mark, 3, 138–139
Zuckerman, Ethan, 227, 228
Zuckerman, Phil, 286